The Chemistry and Biochemistry
of Nitrogen Fixation

The Chemistry and Biochemistry of Nitrogen Fixation

Edited by

J. R. Postgate

University of Sussex,
Falmer, Brighton BN1 9QJ,
Sussex.

PLENUM PRESS · London — New York · 1971

304153

Printed in Great Britain by
William Clowes & Sons Limited
London, Colchester and Beccles

Contributors

R. H. Burris   *Department of Biochemistry, University of Wisconsin, Madison 6, Wisconsin 53706, U.S.A.*

J. Chatt   *A.R.C. Unit of Nitrogen Fixation, University of Sussex, Falmer, Brighton BN1 9QJ, Sussex.*

H. J. Evans   *Department of Botany and Plant Pathology, Oregon State University, Corvallis, Oregon, U.S.A.*

G. J. Leigh   *A.R.C. Unit of Nitrogen Fixation, University of Sussex, Falmer, Brighton BN1 9QJ, Sussex.*

J. R. Postgate   *A.R.C. Unit of Nitrogen Fixation, University of Sussex, Falmer, Brighton BN1 9QJ, Sussex.*

R. L. Richards   *A.R.C. Unit of Nitrogen Fixation, University of Sussex, Falmer, Brighton BN1 9QJ, Sussex.*

S. A. Russell   *Department of Botany and Plant Pathology, Oregon State University, Corvallis, Oregon, U.S.A.*

W. S. Silver   *Department of Life Sciences, Indiana State University, Terre Haute, Indiana 47809, U.S.A.*

P. W. Wilson   *Department of Bacteriology, University of Wisconsin, Madison 6, Wisconsin 53706, U.S.A.*

M. G. Yates   *A.R.C. Unit of Nitrogen Fixation, University of Sussex, Falmer, Brighton BN1 9QJ, Sussex.*

CONTENTS

PREFACE

Understanding of biological nitrogen fixation has advanced with impressive rapidity during the last decade. As befits a developing area of Science, these advances have uncovered information and raised questions which will have, and indeed have had, repercussions in numerous other branches of science and its applications. This 'information explosion', to use one of to-day's cant idioms, was initiated by the discovery, by a group of scientists working in the Central Research laboratories of Dupont de Nemours, U.S.A., of a reproducibly active, cell-free enzyme preparation from a nitrogen-fixing bacterium. Full credit is due to them. But subsequent developments, albeit sometimes quite as impressive, have too often been marked by that familiar disorder of a developing field of research—the scramble to publish. It is a scramble which, at its best, may represent a laudable desire to inform colleagues of the latest developments; yet which too easily develops into an undignified rush for priority, wherewith to impress one's Board of Directors or Grant-giving Institution. This, in miniature, is the tragedy of scientific research to-day: desire for credit causes research to be published in little bulletins, notes and preliminary communications, so that only those intimately involved in the field really know what is happening (and even they may well not see the forest for the trees). Those outside the field, or working in peripheral areas, may glean something of what is going on from reviews and fragments presented at meetings, but the broad pattern of development is often elusive.

This book is an attempt to correct the situation within its own particular field. An explanation (some might even say an apology) is to-day necessary from everyone who presumes to add to the mounting deluge of scientific publications; for this book I offer the view that it is intended for the informed outsider. I asked contributors to stand back a little from the subject and to describe what the *real* advances of the last decade had been. What were they? What was their impact on our understanding of this particular subject? How did they influence our background knowledge of chemistry, biochemistry or physiology? The book was intended to be complementary to existing works on nitrogen fixation: an advanced but readily comprehensible survey of the last decade's innovations for students, teachers and research workers who would be reasonably well-informed about its background. Contributors, all recognized authorities in their particular fields, were asked specifically not to write reviews: there was no obligation to cite all known references bearing on a given aspect of the subject; historical continuity could be ignored and even the desire to be in all ways up-to-date could be resisted. Painstaking and exhaustive reviews of nitrogen fixation, in its biochemical, chemical and biological aspects, exist in plenty; contributors were asked to present their material as a survey which, while in no way 'talking down', would be useful and comprehensible to scientists whose training and interests would be broader and might range from purely chemical to wholly biological.

I thank my fellow contributors for interpreting my intentions so effectively. I am aware that some have not been able to resist adding their latest 'stop-press' item; that others, having been aware of an unseemly squabble for priority, may have felt it necessary to describe the history of a certain advance in a degree detail which may seem strange to those not involved; yet others have been overtaken by events (my own contribution contains information about heterocysts and nitrogen fixation by blue-green algae which became obsolete in press; fortunately Professor Burris's chapter came in later and amends it).* I have left these human touches intentionally, in the hope that the synoptic view taken by the book as a whole will compensate for the occasional idiosyncracy. (And what could be more tedious than contributions devoid of idiosyncracy?)

A final word about the first contributor. In 1940 Perry Wilson published 'The Biochemistry of Symbiotic Nitrogen Fixation', a seminal work which served as a springboard for a great deal of meaningful work in this field. The decade which the present volume covers ended with the thirtieth anniversary of Wilson's book and, at an early stage, we contributors had intended to dedicate our book to Perry Wilson as a kind of *festschrift*. Illness led to a re-shuffle of contributors and I found myself calling upon Perry, not to stand and be admired, but to do some more work.

So our book is graced by Perry's opening chapter. Our intention to honour him may have been thwarted but our esteem for his monumental contribution to the subject remains unaltered. We

* Compare pages 177 and 149

wished to do him formal honour; in the event we could not. But it is a great pleasure, and in some ways it feels more natural, to have him here with us.

John Postgate
University of Sussex
January, 1971

Acknowledgement. Four contributors (G. J. Leigh, J. Chatt, J. R. Postgate, M. G. Yates) are employees of the British Agricultural Research Council at its Unit of Nitrogen Fixation, University of Sussex. They acknowledge permission to contribute and emphasize that all opinions and interpretations expressed are their own and in no way commit the Council.

CHAPTER 1

The Background

PERRY W. WILSON

Department of Bacteriology
University of Wisconsin, Madison,
Wisconsin 53706, U.S.A.

1.1   INTRODUCTION

1.1.1   *The 1886-1906 period*

Beginning in the middle 1830's the roll-call of chemists interested in answering the important theoretical and practical question of whether green plants can use atmospheric N_2, includes such famous names as Boussingault and Ville in France, Lawes and Gilbert in England, Liebig in Germany and Atwater in the United States

(Wilson, 1957). When Hellriegel and Wilfarth in 1886-1887 reported their classical researches, interest of the chemist languished except for a few who undertook the rather thankless chore of confirming and extending the observations. The discoveries of Berthelot, Winogradsky and Beijerinck of biological nitrogen fixation by free-living soil bacteria and by Hiltner and others that even nonlegumes such as the alder in association with an appropriate endophyte can also use N_2 provided a backlog of basic science that could be exploited in agricultural practice. Such application was carried out appropriately in agricultural colleges and experimental stations; the research, dealing with laboratory, greenhouse and field experiments designed for solving very practical problems, was published as station bulletins and reports in agricultural journals. Studies concerned with more basic aspects usually appeared in an appropriate (i.e., nonmedical) journal of microbiology although many would have been acceptable in the chemical literature.

The indices of the basic chemical journals for the next forty years do contain references under the entry *nitrogen fixation,* but few of these refer to the biological process. For example, the total was four in the *Journal of the American Chemical Society* from 1886 to 1906, a period during which workers in the United States at many of the State agricultural experiment stations were actively engaged in research (Wilson, 1963). In the first of these, Coates and Dodson (1896) argued:

> 'As the cotton plant is highly nitrogenous in character, and as there seems to be no reason why the leguminosae should have a preemptive claim on the absorption of free nitrogen, it was decided to undertake certain experiments in the hope that something important would be discovered.'

Whether they regarded the negative results obtained as important is not stated.

Weber (1898) described an ingenious technique for demonstrating N_2-fixation by dwarf peas grown in an aerated water culture: another confirmation of Hellriegel and Wilfarth. Voorhees and J. G. Lipman (1905) contributed a lengthy paper dealing with nitrogen fixation in the soil, a subject typical of the material appearing in the agricultural bulletins, and Hopkins (1902), in a paper that was essentially a preprint for an Illinois agricultural bulletin, described nitrogen fixation by the alfalfa plant in the midwest prairies.

These are cited to suggest that the lack of publication in the professional journal did not reflect a snobbish editorial policy. There is additional evidence that the chemists were interested, philosophically at least, in the field even though they chose not to work there. The parochial coverage implicit in the title of *The Review of American Chemical Research* did not prevent this forerunner of *Chemical Abstracts* from publishing an abstract of Berthelot's paper in the *Bulletin Société Chimique de France* that told of the discovery of asymbiotic nitrogen fixation in the soil. Moreover, Friedburg (1890) furnished the journal a detailed translation of a contribution by A. Petermann describing nitrogen fixation in the lupine which originally appeared in the memoires of the Royal Academy of Belgium.

Finally, Wiley (1894) in his presidential address to the Society on the conservation and waste of plant foods reported with satisfaction that:

> 'Winogradsky and Warington have shown that an organism can grow in a sugar solution and excluding all nitrogenous matter save the free nitrogen of the atmosphere which it is capable of oxidizing and assimilating . . .'

He used this an an example in support of his optimistic prediction that:

> 'With the aid of scientific agriculture, with the help of the agricultural chemist, we may safely say that a thousand million people will not crowd our (i.e., USA) means of subsistence . . . the death of humanity is not to come from starvation but from freezing and many a geological epoch will come before this planet dies of cold.'

1.1.2 The 1906-1928 period

If the appearance of speciality journals heralds the birth of a new discipline, a fairly reliable estimate of the date can be set for biochemistry: 1906. Three journals appeared: *Biochemical Journal* (1906); *Biochemische Zeitschrift* (1906), and *Journal of Biological Chemistry* (1905-6). Their existence, however, did not appear to stimulate interest among the professionals in biochemical investigations of N_2-fixation. In the first 22 volumes of *Biochemical Journal,* only one paper was published directly related to the field. Florence Mockeridge (1915) investigated soil organic matter as a source of energy for the azotobacter. She reported fixation of 6 to

10 mg N_2/g substrate not only for glucose and sucrose but also for butyric acid, starch, dextrin, gum arabic and gum tragancanth. The most impressive result was obtained with ethylene glycol, but only one experiment was possible owing to the scarcity of this source of carbon. She stated that 20 or more days were required for the exhaustion of the one gram of substrate.

During this same period, *Biochemische Zeitschrift* published 202 (somewhat smaller) volumes, but papers on biological N_2-fixation were extremely rare. Kossowicz (1914) reported that, although he had claimed fixation by yeasts and fungi two years before, more careful experiments (chiefly filtering the air supply through acid and alkali) had failed to confirm the results. As a part of his extensive studies of the influence of radioactivity on plants, Stoklasa and his associates (1922, 1926, 1928) included trials on nitrogen fixation by *Azotobacter chroococcum*. Their chief interest was the effect of radiation; they alleged that N_2-fixation was stimulated by radium emanations and uranyl nitrate. Two other minor elements were also included: iodine stimulated, selenium inhibited. Of greater significance for the future developments in the field was a contribution from the Valio Laboratory in Finland by A. I. Virtanen (1928).

The Journal of Biological Chemistry had two entries: one by C. B. Lipman (1911) claiming fixation by yeast and other fungi, and finally, a paper from Wisconsin describing nitrogen fixation in fermenting manure (Tottingham, 1916).

1.2   PROPERTIES OF THE ENZYME SYSTEM IN *AZOTOBACTER*

1.2.1   *The method*

A paper by Meyerhof and Burk (1928) ushered in an era that saw a sharp revival of interest in the basic chemistry of biological N_2-fixation. Although their laboratories were often in agricultural colleges or experiment stations, the research workers attracted to the field, were trained in chemistry or biochemistry and bacteriology and published in the professional journals of these disciplines rather than station bulletins. Intact cells, either growing or 'resting', were the usual experimental material, but the goal always was verification and extension of the findings with cell-free preparations. This was accomplished with several of the auxiliary systems, e.g. the electron transport systems, including hydrogenase, but fixation of N_2 by such preparations remained elusive until 1960.

The contribution by Meyerhof and Burk was two-fold. The introduction of a new technique: the use of a microrespirometer for indirect measurement of N_2 fixed; and the development of experimental concepts that would allow critical examination of the properties of a specific enzyme system in growing cells. The instrument not only allowed experiments to be completed in hours that had required weeks with the traditional cultural methods but also provided a much more sensitive method for test of an hypothesis. It provided an essential technical back-up for the conceptual plan of attack on problems long neglected in this field.

Burk (1934) stated the basic assumptions as:

> 'Azotase is the enzyme system or complex in the aerobic organism Azotobacter that catalyzes the change of gaseous N_2 from a free to a fixed state at ordinary temperatures and pressures. Like the zymase complex, it consists of one or more specific enzymes, and of auxiliary substances of low molecular weight and relative greater stability. Nitrogenase is the specific enzyme within the azotase system that combines directly with N_2 with characteristic affinity.'

The task of the experimenter was to define the properties of the enzyme, nitrogenase, and to identify the auxiliary substances. To succeed, several essential criteria must be met in the experiments (Burk, 1937). The two most important were: (a) the use of appropriate velocity constants in measurement of the rate of reaction; (b) comparison of these constants when the cells are grown on both free and fixed forms of nitrogen. The appropriate velocity constant could be readily determined by observing the increase in rate of respiration with time during the period of exponential growth of the organism. During the 1930's the Fixed Nitrogen Laboratory group in the Bureau of Chemistry and Soils of the USDA (Burk, Allison, Lineweaver, Horner) performed the essential experiments; their significant findings concerned with the enzyme system are summarized here, but it is emphasized that equally important contributions were made dealing with other physiological aspects of both asymbiotic and symbiotic biological nitrogen fixation.

1.2.2 The pN_2 function

Experiments that measured the rate of N_2-fixation as a function of the pN_2 led to an estimate of 0.21 atm. for the Michaelis constant of fixation. From these experiments and others, Burk (1934) calculated the free energy of dissociation of N_2E, the heat of activation of N_2E association and other physical constants. However, in most of the experiments H_2 was used as the inert gas to replace the N_2. As will be discussed later, this led to a serious error and a controversy with the group at Wisconsin.

A byproduct of the research, however, has been of more significance than the precise evaluation of the K_{N_2}. Treatment of the extensive data gathered for its estimation resulted in an independent discovery of a method, earlier suggested by Barnett Woolf to Haldane (1957), that transforms the data so that a linear rather than a hyperbolic function results: the Lineweaver-Burk plot. This transformation, together with other methods suggested in their paper (Lineweaver and Burk, 1934) has played an important role in the development of enzyme kinetics. It is worth remembering that the extensive data gathered for estimating the K_{N_2} for nitrogen fixation provided the raw material for illustration of, together with the, often overlooked, appropriate statistical treatment, the usefulness of the method (Lineweaver, Burk and Deming, 1934).

1.2.3   The pO_2 function

The experiments described in the 1928 paper of Meyerhof and Burk dealt with the relationship between N_2-fixation in the azotobacter and the pO_2. In one experiment, the following values for μgN fixed/10 ml at the indicated pO_2 were obtained: 0.21 atm., 6; 0.13, 6; 0.07, 14; 0.04, 26; 0.013, 21. Although maximum fixation occurred at a pO_2 of 0.04 atm., the efficiency of fixation measured by (moles N_2 fixed)/(moles O_2 used) was greater at a pO_2 of about 0.01 atm. Burk (1930) confirmed these observations, but since the pO_2 functions with respect to respiration, growth and efficiency of growth were the same when the organism was grown on either free or fixed nitrogen, he concluded that 'they offer no indication of the chemical mechanism of nitrogen fixation'. Be that as it may, later investigators did establish definite possible roles for O_2. For example, the experiments of Parker and Scutt (1960) indicating competitive inhibition between O_2 and N_2, suggested that the two gases competed as terminal hydrogen (electron) acceptors. These questions are discussed more extensively in Chapter 5.

1.2.4   Role of minerals

Using the traditional stagnant culture method with long incubation times, Bortels (1930) reported that the growth of *Azotobacter chroococcum* was increased 2-3 fold through the addition of 0.0005% sodium molybdate. The microrespirometric technique was ideally suited for detailed study of this effect; Burk and his associates examined not only the effect of molybdenum but also a large number of other trace minerals using different strains and species of azotobacter (Burk, 1934, 1937). Their conclusions were that molybdenum (replaceable in part by vanadium) and calcium (replaceable by strontium) were specific requirements for N_2 fixation by the azotobacter (i.e. were auxiliary factors of azotase), but the requirement for iron was nonspecific.

However, in a 1941 review, his last contribution to the field, Burk stated:

> '... although all nitrogen-fixing organisms require molybdenum (or vanadium), iron and calcium (or strontium), in no case—regardless of earlier indications—can it not be regarded as probable that these elements are specifically required in fixation as distinguished from general nitrogen assimilation ... The only qualitative fixation specificity that can be regarded as definitely established at present is hydrogen inhibition ...' (Burk and Burris, 1941).

Other investigators have not accepted such a limited definition, and the cell-free work has established that both iron and molybdenum are constituents of the nitrogen-fixing enzyme system in essentially the sense that Burk originally defined their role. The term azotase, however, has been dropped in favour of nitrogenase for designation of the complete system. The specific role of calcium, if any, remains obscure (Jakobsons, Zell and Wilson, 1962).

1.3 THE PROPERTIES OF THE ENZYME SYSTEM IN LEGUMINOUS PLANTS

1.3.1 The method

In 1929 the departments of Agricultural Bacteriology and of Agricultural Chemistry at the University of Wisconsin received a grant from the Herman Frasch Foundation to investigate the biochemistry of symbiotic N_2-fixation. As greenhouse experiments involving the complicated two membered plant system obviously promised to be time-consuming as well as difficult to interpret, initial studies were directed toward attempts to obtain N_2-fixation by the bacteria when grown alone. When such efforts failed (Wilson, Hopkins and Fred, 1932), attention was turned to the intact plant system; a technique was developed that, it was hoped, would provide results similar to those found for the azotobacter using the microrespirometer. Briefly, clover plants were grown in a closed system on a sand substrate contained in 9 liter serum bottles. All plant nutrients were supplied except one-half of the cultures did not receive a source of combined nitrogen but instead were inoculated with an effective strain of *Rhizobium trifolii.* The closed system enabled reasonably close control of the gases in the atmosphere; details of the method have been described by Wilson (1936).

1.3.2   The pN_2 function

In experiments to estimate the K_{N_2} of the symbiotic system in red clover for comparison with that of the azotobacter, the pO_2 was held constant at 0.2 atm. and the pN_2 varied from 0.04 to 1.56 atm. in different trials. When necessary, H_2 replaced the N_2 removed. Under these conditions the N_2 fixed was a linear function of the pN_2 between 0.04 and 0.6 atm. When a cylinder of H_2 became empty early in one trial, the experiment was continued by growing those plants exposed to a pN_2 of less than 0.8 atm. under a partial vacuum. The result was strikingly different: a hyperbolic curve was obtained that indicated no detectable effect of the pN_2 above about 0.2 atm. This unexpected observation was verified and extended by using inert gases as helium and argon. After two years, 16 experiments had furnished 111 points for calculation of a K_{N_2} of 0.05 ± 0.005 atm. (Wilson, 1936, 1939).

1.3.3   The pO_2 function

Wilson and Fred (1937) reported the results from a large number of greenhouse experiments under different conditions of light, temperature, and the pO_2 in the atmosphere. Between a pO_2 of 0.1 and 0.4 atm., the dry weight formed and the assimilation of either free or combined nitrogen were unaffected. Above 0.4 atm. both were markedly decreased apparently from depletion of carbohydrate owing to increased respiration. Below 0.1 atm. a decrease was also observed which, in the absence of H_2, was independent of the source of nitrogen. In the presence of H_2, however, the decrease occurred at a higher pO_2 in the plants fixing N_2 than in those given NH_4NO_3. Nevertheless, the authors concluded, as had Burk with the azotobacter, that O_2 is not directly concerned with symbiotic N_2-fixation in red clover, only indirectly, for example, by its influence on the carbohydrate supply (Fred and Wilson, 1934).

The identification by Kubo (1939) of the characteristic red pigment in root nodules and the intensive studies by Virtanen and associates (Virtanen, 1947, 1948 and 1952) of its properties and possible function in N_2-fixation have led to a renewed interest in the

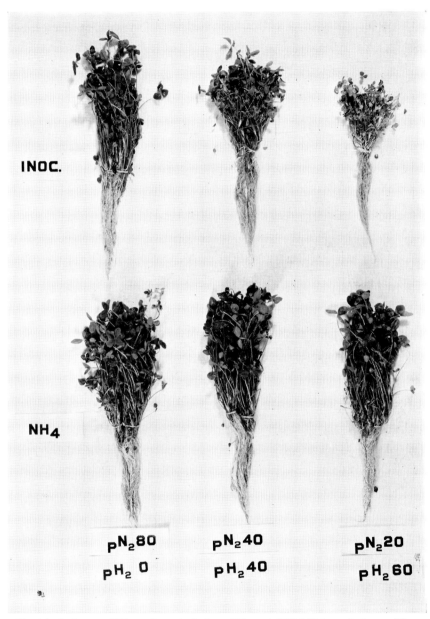

INOC.

NH₄

$_P$N₂80
$_P$H₂ 0

$_P$N₂40
$_P$H₂40

$_P$N₂20
$_P$H₂60

Fig. 1.1. Influence of H₂ on the assimilation of N₂ and NH₄⁺-N by red clover. (Figures 1.1 and 1.2 are from P. W. Wilson, *The Biochemistry of Symbiotic Nitrogen Fixation* (Madison: The University of Wisconsin Press; copyright 1940 by the Regents of the University of Wisconsin, pp. 198–206.) This monograph is out of print but is available in microfilm from University Microfilms, Ann Arbor, Michigan, USA).

(*To face page 8*)

Fig. 1.2. Inhibition by H_2 of N_2-fixation in red clover. The effect of a given concentration of the inhibitor depends on the concentration of the substrate, indicative of competitive inhibition.

role of O_2 in the symbiotic system. As will be discussed in Chapter 6, recent research favors the view that the hemoglobin functions to regulate the O_2 supply and not to participate directly in electron transport.

1.3.4 Effect of inhibitors

Detailed investigation of the effect of H_2 on N_2-fixation by red clover plants, discovered while attempting to determine the K_{N_2}, established that the inhibition was not only specific for assimilation of N_2 but also was competitive (Wilson, 1939). The advantage of the clover system over the bacterial was that the effects could readily be seen (Figs. 1.1, 1.2) although the quantitative aspects, of course, required chemical analysis. This relationship between H_2 and N_2 in the symbiotic system suggested that nitrogenase and hydrogenase might be someway related. A test of *Rhizobium trifolii* for hydrogenase, however, was negative. It was realized that this was not critical since cultures of the organism alone did not fix N_2. Using the reduction of methylene blue by H_2 for assay, Phelps and Wilson reported the presence of hydrogenase in the bacteroids from the pea plant, but later workers in the same laboratory using more sensitive methods were unable to detect hydrogenase in either bacteroids or intact nodules from pea, soybean, or cowpea plants (Wilson, Burris and Coffee, 1943).

In private communication, Burk stated that his group had not observed any effect of H_2 on azotobacter in their experiments. Re-examination of this bacterial system, however, demonstrated that H_2 was a specific, competitive inhibitor for N_2-fixation in the azotobacter as had been observed in the red clover symbiotic system (Wyss, Lind, Wilson, and Wilson, 1941). Moreover, cultures of *Azotobacter vinelandii* and *A. agile* were found to have a very active hydrogenase, which could be readily obtained as a cell-free preparation. As the specific activity, using either methylene blue or O_2 as the electron acceptor, was much higher in cells grown on N_2 than on NH_4NO_3 independent of which source of nitrogen had been used for previous culture, Lee, Wilson and Wilson (1942) proposed a definite relationship between the two enzyme systems. They noted, however, that the failure of N_2 to inhibit the activity of the hydrogenase was difficult to explain.

Although Burk originally was unconvinced, a series of joint experiments in his laboratory with Burris, confirmed that H_2 was a specific inhibitor (Burk and Burris, 1941). The K_{N_2} value of 0.21 atm. for the azotobacter system was discarded—0.02 atm. was later reported by Burris and collaborators (Wilson, 1958)—but the voluminous data obtained by Burk and his associates were not without value; they were used to calculate a highly accurate value for the K_{H_2}. The failure to detect a hydrogenase in the symbiotic system was for many years cited as a basic difference between the asymbiotic and symbiotic systems until Hoch, Schneider and Burris (1960) found an enzyme in soybean nodules that liberated H_2 and catalyzed the H_2-D_2 exchange reaction. Of major interest was the observation that N_2 inhibited the evolution of H_2 but stimulated the H_2-D_2 exchange.

1.3.5 CO and N_2O as specific inhibitors

Carbon monoxide, an isostere of N_2, was a logical inhibitor, but tests were not immediately made on this gas because of its well-established role as a respiratory inhibitor. When trials were undertaken, it was found that this activity did not interfere with the interpretation; CO inhibited nitrogen fixation in the red clover system at extremely low concentrations (Lind and Wilson, 1941). As little as 0.01% in the atmosphere brought about definite inhibition and 0.05 to 0.1% practically prevented all uptake of N_2. In this range growth on NH_4NO_3 was unaffected, but at 10-fold higher concentrations, a small but definite inhibition could be detected. Similarly, CO inhibited N_2-fixation in both the azotobacter and in the blue-green algae, *Nostoc*, although the concentrations necessary were higher than in the plant system. Surprisingly, the competition was essentially non-competitive in the plant and bacterial systems, but in the alga some competitive inhibition was observed (Burris and Wilson, 1946).

Burk and Burris (1941) suggested that the H_2 inhibition may arise from physical competition for sites on the enzyme surface. If so, then the ratio of the dissociation constants should approximate that of the van der Waals constants *a*. This was true of the H_2 inhibition, both being about 5.5, but tests using other gases with appropriate van der Waals constants offered no evidence in support of the suggestion. However, the experiments did establish another specific competitive inhibitor, N_2O (Molnar, Burris and Wilson, 1948; Repaske and Wilson, 1952). Wilson (1939) lists 14 compounds that were tested for inhibition of the symbiotic clover system. Most were toxic and, except for CO, were equally toxic for plants using either free or combined nitrogen thus preventing any conclusion regarding their specificity; the list includes both cyanide and azide.

1.4   THE CHEMICAL PATHWAY

1.4.1   Early conjectures

Winogradsky in 1894 proposed that nitrogen fixation by *Clostridium pasteurianum* was the reduction of N_2 to NH_3 by 'nascent' H_2 formed in the butyric acid fermentation. Although this mechanism seemed reasonable for an anaerobe, it was not until 1925 when Kostytschev suggested the same for the azotobacter based on finding small amounts of ammonia in stagnant cultures that the view received experimental consideration. In 1930, Winogradsky published results of experiments on the formation of NH_3 by cultures of azotobacter grown on silica gel plates under alkaline conditions. Burk and Horner (1936) provided a critical survey of the previous investigations and reported extensive investigations of their own that led them to the conclusion that:

> '. . . *extracellular ammonia observed so far, in our own and in previous investigations,* is derived in all probability entirely from decomposition of normal cell nitrogen . . . The occurrence of any ammonia as an essential intermediate product in the fixation of nitrogen, *although possible, still remains to be demonstrated.*' [their italics]

1.4.2   Hydroxylamine as the key intermediate

Blom in 1936 had proposed that N_2-fixation in the azotobacter proceeded via hydroxylamine based on detecting trace amounts of this compound in cultures. Burk (1937) summarily rejected this view not only because the observation had not been confirmed, but also because none of seven criteria that he had proposed for proof of specificity had been met.

At this time, however, the Finnish school under the leadership of Professor A. I. Virtanen was providing much more impressive support for this view based on a study of excretion of nitrogen compounds by the pea plant. Although J. G. Lipman had studied this phenomenon in 1912 (Wilson, 1955), his interest was strictly agronomic: in its role in mixed cropping. Virtanen (1938) and his associates developed an ingenious technique for growing the pea plant under sterile conditions that allowed them to identify the excreted products. A series of experiments indicated that these consisted almost entirely of aspartic acid, β-alanine together with 1 to 2% of an oxime identified as oximinosuccinic acid. On the basis of this finding, he proposed that N_2 was fixed in the pea plant by formation of hydroxylamine which combined with oxalacetic acid from breakdown of carbohydrate to form the oxime which was then reduced to aspartic acid. The root nodule bacteria produced the β-alanine by decarboxylation of the aspartic acid.

Although these data were the most impressive of any offered to that time in support of any chemical pathway, Wilson (1939) criticized their specificity for hydroxylamine arguing they could be used equally in favor of the ammonia hypothesis. An exchange of views between the two protagonists might have taken place at the meeting of the Third Commission of the International Society of Soil Science held in New Brunswick, New Jersey in late August, 1939, but as neither attended, the reading of their position papers (Virtanen, 1939; Wilson and Wyss, 1939) *in absentia* hardly constituted a confrontation. However, Virtanen's report did clarify why he regarded the excreted aspartic acid as uniquely implying NH_2OH. He stated:

> 'That glutamic acid is not formed seems to be ascribable to the fact that the reaction velocity of hydroxylamine with oxaloacetic acid is many times greater than with ketoglutaric acid. Should nitrogen fixation occur through the ammonia stage, the formation of glutamic acid together with aspartic acid would be very likely.'

After the end of World War II, Virtanen (1947) reported in a review that in new experiments both aspartic and glutamic acids had been found in the excreted products. Quoting his 1939 statement, he suggested that the oxime might have arisen from a side reaction in the formation of ammonia from N_2 via NH_2OH, a view that was in essential agreement with a suggestion by Burris and Wilson (1945).

1.4.3   Ammonia as the key intermediate

Using $^{15}N_2$ as a tracer, Burris (1942) furnished the first clear-cut evidence that NH_3 might be the key intermediate in biological N_2-fixation, defined as the compound that terminates the inorganic phase of fixation and which then is converted into some organic compound and eventually is incorporated into cell protein. Cells of *Azotobacter vinelandii* strain O, growing in air, were placed in a closed system and exposed to an atmosphere of 60% $^{15}N_2$ (35 atom % excess)—40% O_2 for 90 min. After harvesting, the cells were hydrolyzed with H_2SO_4 and the hydrolysate subjected to various laborious chemical manipulations that eventually provided three pure amino acids—arginine, histidine, and glutamic acid—together with an aspartic acid fraction, an 'amide' fraction, as well as such ill-defined mixtures as 'H$_2$O soluble, MeOH insoluble Cu salts'. Each compound or fraction was analyzed for its ^{15}N content; the highest label was in the glutamic acid, followed by aspartic acid.

Only half of the paper is devoted to experimental details and a single table of data; the remaining half summarizes the argument that the results indicate ammonia as the key intermediate. The support at this point was shaky, but four years later another prop was added from more extensive experiments that demonstrated that the distribution of the label among the fractions was similar in cells supplied ^{15}NH$_4$$^+$ and those fixing ^{15}N$_2$. Obviously, the methods then available for the research were not likely to produce a mass of convincing evidence in the immediate future, but the development of column chromatography for separation of amino acids and the availability of commercial mass spectrometers (the home-made one used in these early studies was usually being torn down for repair) overcame these limitations. During the next decade, research that has been discussed in detail in several reviews (Burris, 1956; Wilson and Burris, 1947, 1953; Wilson 1958) provided the following evidence in support of the ammonia hypothesis:

1. Representatives of all major groups of nitrogen-fixing organisms were surveyed for the pathway of fixation using ^{15}N$_2$ as a tracer. In all, the highest level of the label was found in glutamic acid followed usually, but not always, by aspartic acid. No other amino acid was found to have a level that would implicate it in the direct pathway of fixation.

2. Of equal, or perhaps greater, significance was that the distribution of ^{15}N in nitrogen-fixing agents is quite similar whether supplied ^{15}N$_2$ or ^{15}NH$_4$$^+$.

3. Ammonia immediately supplants N$_2$ when supplied to a culture actively fixing the free form (an application of the concept later described as simultaneous adaptation or sequential induction).

4. If the supply of dicarboxylic acids is low, cultures of *Clostridium pasteurianum* W5 fixing ^{15}N$_2$ excrete recently fixed nitrogen compounds including ammonia. Ammonia itself has the highest ^{15}N concentration followed by glutamine amide-N and asparagine amide-N. A similar result was obtained with *A. vinelandii*.

The simplest conclusion from these results appeared to be that N$_2$ goes to NH$_3$ *via* an unknown pathway; the NH$_3$ is converted to glutamic acid by reductive amination in those agents possessing glutamic dehydrogenase, and the other amino acids are formed by transamination reactions; in agents with an active aspartase, aspartic acid may be formed directly.

1.4.4   Search for other intermediates

With the demonstration of the role of ammonia furnished by the direct isolation of the compound in cultures of *Clostridium* and *Azotobacter*, interest shifted to possible compounds between N_2 and NH_3. Even if NH_2OH was not the key compound, it remained a possible precursor. Efforts to demonstrate such a role for it and for several other possibilities, including hyponitrite, nitramide and hydrazine, met with little success. It was generally concluded that if one of the postulated compounds was an intermediate it was in a form, e.g. bound to an enzyme, so that it could not be detected with the techniques available, which were primarily ^{15}N experiments with the free compound added as carrier.

An article of faith was that the preparation of a reliable cell-free enzyme system would solve this problem. In the 1950's, several groups of research workers were actively pursuing this goal, encouraged by the isolation of other enzyme systems from both the *Azotobacter* and *Clostridium*. Most of the experiments with the azotobacter, usually carried out in the presence of O_2, had a built-in factor of failure. With the anaerobe, *C. pasteurianum* W5 it is likely that fixation was occasionally obtained (Burris, 1966). If so, until 1960 this was probably a matter of the chance implied in the saying: 'A person may find a pearl in an oyster twice, but he has to open many oysters'.

Although for obvious reasons the symbiotic system received less attention in the cell-free research, what might be regarded as the first step was the development of a technique for obtaining fixation by excised nodules from the soybean plant. Re-examination of the properties of the enzymes in these gave results in reasonable agreement with those reported for the intact red clover plant including inhibition by H_2 (Burris, 1956).

1.4.5   Agents of fixation

Although the research worker of the 1886-1928 period would probably find many of the discussions in this volume bewildering, one aspect of the field would still be familiar: efforts to demonstrate N_2-fixation in various biological agents. The basic question first raised by Boussingault and others more than a century ago still remains of interest. Soon after the discoveries of Berthelot, Beijerinck and Winogradsky, claims of fixation by numerous species of bacteria, yeasts and molds appeared. (Lipman, 1911). By 1930, the repeated failure to verify such claims resulted in the conservative belief that nitrogen fixation was restricted to a limited number of bacterial genera, probably only *Azotobacter* and *Clostridium,* blue-green algae (*Nostoc* and *Anabaena*), legumes in association with the root nodule bacteria, and to some nonlegumes in association with unknown endophytes.

The possibility that hydrogenase could be used as a marker for N_2-fixation, and the development of the sensitive $^{15}N_2$ technique for detection led to an examination of typical organisms that possessed the H_2 enzyme including *Escherichia coli, Scenedesmus* adapted to H_2, *Hydrogenomonas* and others. It may be that the uniform negative results (Wilson, 1958) were caused by usually carrying out the experiments aerobically owing to the mistaken belief that *Azotobacter* was much superior to *Clostridium* for fixation of N_2. From experiments designed to investigate photo-evolution of H_2 by *Rhodospirillum rubrum,* Kamen and Gest (1949) observed inhibition by N_2, a finding that led to the demonstration of N_2-fixation by this photosynthetic bacterium. Examination of other photosynthetic bacteria established that members of the three families could use N_2, but only anaerobically. The work with this group led to a re-investigation of the relationship between photo-synthesis and N_2-fixation, first studied by Wilson and Fred in 1934 (Wilson, 1958).

Success with the photosynthetic bacteria was followed by demon-strating fixation in several other organisms which contain a hydro-genase; these include: *Klebsiella pneumoniae* (*Aerobacter aerogenes*), *Bacillus polymyxa, B. macerans, Methanobacterium omelianskii* and, more recently, *Mycobacterium flavum* (Biggins and Postgate, 1969) and several species of *Desulfovibrio* (see Chapter 5). The recent publication on nitrogen fixation in the gut of animals including man (Bergersen and Hipsley, 1970) recalls the early studies of Tóth (1948, 1949) on nitrogen-fixing bacteria in the rumen of cows, sheep and goats and in the caecum of horses. The responsible organisms in the recent work appear to be not only known N_2-fixers as *Klebsiella* but also, possibly, *Escherichia coli* and *Enterobacter cloacae.*

The period has not been free of erroneous claims involving species of *Pseudomonas, Azotomonas,* and *Achromobacter* and, of course, yeasts and molds (Millbank, 1969). Procter and Wilson (1959) claimed fixation by *Achromobacter,* but later trials established that the only strain that fixed consistently, one isolated by V. Jensen, was an anaerogenic strain of *Klebsiella pneumoniae* (Mahl, Wilson, Fife and Ewing, 1965). Perhaps wishful thinking accounts for the not infrequent reports of fixation by plants independent of bacteria. These include that of Lipman and Taylor (1924) on alleged fixation by wheat and barley, a strange feature of which was that the best fixation was often, although not always, achieved by the plants given more than enough nitrogen as $NO_3{}^-$. Ruben, Hassid and Kamen (1940) published a latter-day confirmation of this using barley and radioactive $^{13}N_2$ (half-life, 10.5 min), but repetition of the experiment using $^{15}N_2$ was negative (Burris, 1941). A variation was Vita's claim of fixation by germinating pea seeds supplied low level of alkaloids (Wilson, 1939).

These questionable results can usually, though not always, be explained by some technical deficiency (Hill and Postgate, 1969; Wilson 1939, 1952, 1958). Certainly the analyses of the wheat and barley cultures supplied nitrate were open to question since losses were also observed. Vita's results likewise appear to have arisen from misleading Kjeldahl analysis (Wilson, 1939). It was thought that use of $^{15}N_2$ would correct the deficiencies (Wilson, 1958), but what appear to be spurious claims vouched for by use of this tracer still are published. Bremner (1965) has emphasized the several pitfalls in its use, but little evidence is supplied in the suspected reports that his warnings have been heeded. Although the recently developed C_2H_2 technique may prove to be more reliable with fewer false claims, after examining many papers on this subject during the past forty years, the author of this chapter does not share such optimism.

REFERENCES

F. J. BERGERSEN and E. H. HIPSLEY, *J. Gen. Microbiol.*, **60**, 61 (1970).

D. R. BIGGINS and J. R. POSTGATE, *J. Gen. Microbiol.*, **56**, 181 (1969).

H. BORTELS, *Arch. Mikrobiol.*, **1**, 333 (1930).

J. M. BREMNER, Methods of Soil Analysis. Agronomy Monograph 9 (C. A. Black, ed.) American Society of Agronomy, Madison, Wisconsin (1965).

D. BURK, *J. Phys. Chem.*, **34**, 1195 (1930).

D. BURK, *Ergeb. Enzymforsch.*, **3**, 23 (1934).

D. BURK, *Biochimica*, **2**, 312 (1937).

D. BURK and R. H. BURRIS, *Ann. Rev. Biochem.*, **10**, 587 (1941).

D. BURK and C. K. HORNER, *Soil Sci.*, **41**, 81 (1936).

R. H. BURRIS, *Science*, **94**, 238 (1941).

R. H. BURRIS, *J. Biol. Chem.*, **143**, 509 (1942).

R. H. BURRIS, A Symposium on Inorganic Nitrogen Metabolism. (W. D. McElroy and B. Glass, eds.), p. 316. The Johns Hopkins Press, Baltimore (1956).

R. H. BURRIS, *Ann. Rev. Plant Physiol.*, **17**, 155 (1966).

R. H. BURRIS and P. W. WILSON, *Ann. Rev. Biochem.*, **14**, 685 (1945).

R. H. BURRIS and P. W. WILSON, *Bot. Gaz.*, **108**, 254 (1946).

C. E. COATES and W. R. DODSON, *J. Amer. Chem. Soc.*, **18**, 25 (1896).

E. B. FRED and P. W. WILSON, *Proc. Natl. Acad. Sci. U.S.*, **20**, 403 (1934).

L. H. FRIEDBURG, *J. Amer. Chem. Soc.*, **12**, 145 (1890).

J. B. S. HALDANE, *Nature*, **179**, 832 (1957).

S. HILL and J. R. POSTGATE, *J. Gen. Microbiol.*, **58**, 277 (1969).

G. E. HOCH, K. C. SCHNEIDER and R. H. BURRIS, *Biochem. Biophys. Acta*, **37**, 273 (1960).

C. G. HOPKINS, *J. Amer. Chem. Soc.*, **24**, 1155 (1902).

A. JAKOBSONS, E. A. ZELL and P. W. WILSON, *Arch. Mikrobiol.*, **41**, 1 (1962).

M. D. KAMEN and H. GEST, *Science*, **109**, 560 (1949).

A. KOSSOWICZ, *Biochem. Z.*, **64**, 82 (1914).

H. KUBO, *Acta Phytochim.*, **11**, 195 (1939).

S. B. LEE, J. B. WILSON and P. W. WILSON, *J. Biol. Chem.* **144**, 273 (1942).

C. J. LIND and P. W. WILSON, *J. Amer. Chem. Soc.*, **63**, 3511 (1941).

H. LINEWEAVER and D. BURK, *J. Amer. Chem. Soc.*, **56**, 658 (1934).

H. LINEWEAVER, D. BURK and W. E. DEMING, *J. Amer. Chem. Soc.*, **56**, 225 (1934).

C. B. LIPMAN, *J. Biol. Chem.*, **10**, 169 (1911).

C. B. LIPMAN and J. K. TAYLOR, *J. Franklin Inst.*, **198**, 475 (1924).

M. C. MAHL, P. W. WILSON, M. A. FIFE and W. H. EWING, *J. Bact.*, **89**, 1482 (1965).

O. MEYERHOF and D. BURK, *Z. phys. Chem.*, **139A**, 117 (1928).

J. W. MILLBANK, *Arch. Mikrobiol.*, **68**, 32 (1969).

F. A. MOCKERIDGE, *Biochem. J.* **9**, 272 (1915).

D. MOLNAR, R. H. BURRIS and P. W. WILSON, *J. Amer. Chem. Soc.*, **70**, 1713 (1948).

C. A. PARKER and P. B. SCUTT, *Biochim. Biophys. Acta*, **38**, 230 (1960).

M. H. PROCTOR and P. W. WILSON, *Arch. Mikrobiol.*, **32**, 254 (1959).

R. REPASKE and P. W. WILSON, *J. Am. Chem. Soc.*, **74**, 3101 (1952).

S. RUBEN, W. Z. HASSID and M. D. KAMEN, *Science*, **91**, 578 (1940).

J. STOKLASA, *Biochem Z.*, **130**, 604 (1922); **176**, 38 (1926).

J. STOKLASA and J. PĔNKAVA, *Biochem. Z.*, **194**, 15 (1928).

L. TÓTH, *Experientia*, **4**, 395 (1948); **5**, 474 (1949).

W. E. TOTTINGHAM, *J. Biol. Chem.*, **24**, 221 (1916).

A. I. VIRTANEN, *Biochem. Z.*, **193**, 300 (1928).

A. I. VIRTANEN, Cattle Fodder and Human Nutrition, p. 26. Cambridge University Press, London (1938).

A. I. VIRTANEN, *Third Comm. Intern. Soc. Soil Sci., Trans.*, **A**, 4 (1939).

A. I. VIRTANEN, *Ann. Rev. Microbiol.*, **2**, 485 (1948).

A. I. VIRTANEN, *Biol. Rev.*, **22**, 239 (1947).

A. I. VIRTANEN, *Ann. Acad. Sci. Fenn. Ser. A II Chemica*, **43**, 3 (1952).

E. B. VORHEES and J. G. LIPMAN, *J. Am. Chem. Soc.*, **27**, 556 (1905).

H. A. WEBER, *J. Amer. Chem. Soc.*, **20**, 9 (1898).

H. H. WILEY, *J. Amer. Chem. Soc.*, **16**, 1 (1894).

P. W. WILSON, *J. Amer. Chem. Soc.*, **58**, 1256 (1936).

P. W. WILSON, *Ergeb. Enzymforsch.*, **8**, 13 (1939).

P. W. WILSON, Perspective and Horizons in Microbiology. (S. A. Waksman, ed.), p. 110. Rutgers University Press, New Brunswick, New Jersey (1955).

P. W. WILSON, *Bacteriol. Rev.*, 21, 215 (1957).

P. W. WILSON, Encyclopedia of Plant Physiology. (W. Ruhland, ed.), Vol. VIII, p. 9. Springer-Verlag, Berlin, Göttingen and Heidelberg (1958).

P. W. WILSON, *Bacteriol. Rev.*, 27, 405 (1963).

P. W. WILSON, R. H. BURRIS and W. B. COFFEE, *J. Biol. Chem.*, 147, 475 (1943).

P. W. WILSON and R. H. BURRIS, *Bacteriol. Rev.*, 11, 41 (1947).

P. W. WILSON and R. H. BURRIS, *Ann. Rev. Microbiol.*, 7, 415 (1953).

P. W. WILSON and E. B. FRED, *Proc. Natl. Acad. Sci. U.S.*, 23, 503 (1937).

P. W. WILSON, E. W. HOPKINS and E. B. FRED, *Arch. Mikrobiol.*, 3, 322 (1932).

P. W. WILSON and O. WYSS, *Third Comm. Intern. Soc. Soil Sci., Trans.*, B, 13 (1939).

O. WYSS, C. J. LIND, J. B. WILSON and P. W. WILSON, *Biochem. J.*, 35, 845 (1941).

CHAPTER 2

Abiological Fixation of Molecular Nitrogen

G. J. LEIGH

A.R.C. Unit of Nitrogen Fixation, University of Sussex,
Falmer, BN1 9QJ, Sussex, England

2.1 INTRODUCTION

The most recent interest in the reactions of molecular nitrogen has been stimulated by the conviction that it should be possible to reproduce in the laboratory the reactions whereby micro-organisms can convert molecular nitrogen to ammonia at normal ambient temperatures and partial pressures, and probably in an aqueous environment. Chemists have also been intrigued for many years by the possibility of making complexes of dinitrogen (Orgel, 1966) analogous to those of dioxygen (Bayer and Schretzmann, 1967) or carbon monoxide (Abel, 1963), which have been known for a comparatively long time. Although progress in both these areas has been rapid of late, this is only a continuation of studies of the fixation of molecular nitrogen which have proceeded for more than a century, prompted, in part, by the very early recognition of the apparent ability of some plants to fix dinitrogen (Davy, 1836) and, in part, by the need for nitrogenous fertilizers to augment and replace supplies from natural sources.

2.2   THE PROPERTIES OF DINITROGEN

Dinitrogen is the most inert of the common diatomic molecules. The reason for this, and for the consequent difference from other similar diatomic molecules, lies in its electronic structure. The energy levels of dinitrogen and some other diatomic molecules are shown in Fig. 2.1. They were determined principally by photo-electron spectroscopy (Al-Joboury et $al.$, 1965). Dioxygen is the classical example for demonstrating the formation of molecular orbitals, in orderly sequence of energy, from atomic orbitals. This sequence is: $\sigma_g 1s$, $\sigma_u 1s$, $\sigma_g 2s$, $\sigma_u 2s$, $\sigma_g 2p$, $\pi_u 2p$, $\pi_g 2p$, $\sigma_u 2p$. Simple theory thus explains why dioxygen with sixteen electrons is paramagnetic (Coulson, 1961). Dinitrogen, with two electrons less, should, at first sight, be rather similar, and several writers have been tempted to make the simple extrapolation from dioxygen to dinitrogen. This is unfortunate. Although nitric oxide and acetylene (which is not strictly comparable, being a four-atom molecule) do follow the dioxygen pattern, dinitrogen and carbon monoxide do not, the $\sigma_g 2p$ and $\pi_u 2p$ levels of dinitrogen and the equivalent orbitals of carbon

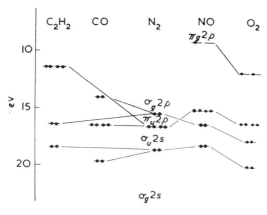

Fig. 2.1. Energy level correlation diagram of acetylene, carbon monoxide, dinitrogen, nitric oxide and dioxygen using adiabatic ionization potentials (reprinted with permission from M. I. AL-JOBOURY, D. P. MAY and D. W. TURNER, $J.$ $Chem.$ $Soc.$, 616 (1965)).

monoxide being inverted. The π-electrons are thus not highest in energy. In ethylene, the π-electrons are the highest in energy, having an ionization potential similar to that of the lone pair in ammonia. Ethylene can use these π-electrons to form donor-acceptor complexes with metals, the ethylene molecule being bonded 'sideways on'. Dinitrogen might also be expected to form donor-acceptor complexes, but the potential lone-pair electrons constitute the σ-bond (i.e. they are in $\sigma_g 2p$) and they are localized principally between the nitrogen atoms. Nitrogen should not form complexes of this type very easily and the interaction should be end-on, not side-ways.

The reason for this inversion of the $\sigma_g 2p$ and $\pi_u 2p$ levels may be simply understood as follows. Energy levels of similar symmetry and energy tend to interact, pushing each other apart. This is seen at its simplest when two atomic orbitals on separate atoms, but with the same symmetry with respect to the molecular axis, overlap to form new molecular orbitals. One of the new orbitals is of higher energy than either of the two from which it is formed (the antibonding orbital) and the other is of lower energy (the bonding orbital). If the energies of the constituent atomic orbitals are very different, then the interaction is weak. Ultraviolet spectroscopy allows the rough estimation of the separation of $2s$ and $2p$ *atomic* orbitals (see Fig. 2.2) (Jaffé, 1956). In the oxygen atom the $2s$ and $2p$ orbitals are so

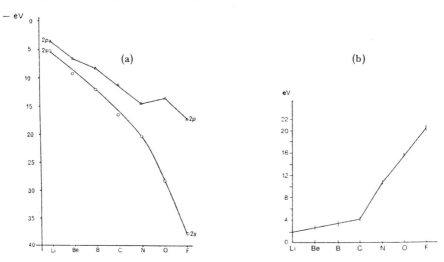

Fig. 2.2. (a) Energy levels of s and p orbitals and (b) Separation of s and p orbitals (reprinted with permission from *Recent Aspects of Nitrogen Metabolism in Plants*, E. J. HEWITT and C. V. CUTTING ed., 1968 [Academic Press, London and New York] p. 5).

far apart that $\sigma_g 2s$ and $\sigma_g 2p$ molecular orbitals of the dioxygen
molecule interact relatively weakly although they are of similar
symmetry. On the other hand, the $2s$ and $2p$ orbitals in the nitrogen
atom are much closer, and consequently so would be $\sigma_g 2s$ and $\sigma_g 2p$
in the dinitrogen molecule if they did not interact. The strong
interaction resulting from the small energy separation pushes $\sigma_g 2p$
and $\sigma_g 2s$ apart, so that $\sigma_g 2p$ is found above $\pi_u 2p$. Further to the left
of nitrogen in this Period of the Periodic Table the separation of $2s$
and $2p$ is even less and the consequent interaction of the σ_g orbitals
in the diatomic molecules even greater. The B_2 molecule has only
two bonding electrons (apart from those in closed shells) and due to
the inversion of levels it is held together by a single π-bond (Bender
and Davidson, 1967).

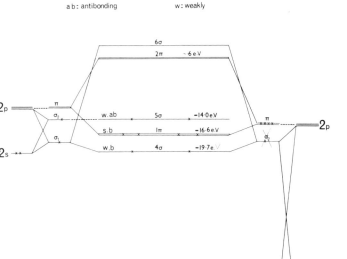

Fig. 2.3. (a) Correlation diagram for carbon monoxide.

Carbon monoxide and dinitrogen emerge as the most similar of the diatomic molecules under consideration. Correlation diagrams are shown in Fig. 2.3 (Mulliken, 1958). These are very much alike, although the resultant electron distributions must be different in the two molecules because carbon and oxygen have different radii, electronegativities etc., whereas dinitrogen is homonuclear. Calculations show a considerably higher lone pair electron density on carbon than on oxygen and carbon monoxide is consequently a better donor molecule than dinitrogen. The energy levels themselves are also not precisely similar. Thus $5\sigma(CO)$ derived principally from an sp hybrid of carbon (it contains the 'lone-pair' electrons) is weakly antibonding whereas the corresponding N_2 orbital ($3\sigma_g$) is weakly bonding and hence N_2^+ has a bond length of 1.117 Å, just

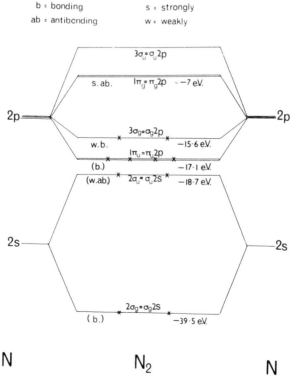

ENERGY LEVELS IN THE NITROGEN MOLECULE

Fig. 2.3. (b) Correlation diagram for dinitrogen (reprinted with permission from *Recent Aspects of Nitrogen Metabolism in Plants*, E. J. HEWITT and C. V. CUTTING, ed., 1968 [Academic Press, London and New York] p. 6).

greater than that of N_2 (1.098 Å) whereas CO^+ has a length (1.115 Å) just shorter than that of the parent molecule (1.128 Å) (The Chemical Society, 1958). Table 2.1 shows comparative physical

TABLE 2.1. Comparative physical properties of dinitrogen, dioxygen, acetylene, carbon monoxide and nitric oxide.

	N≡N	O=O	HC≡CH	CO	NO
Bond length (Å)	1.098	1.207	1.208	1.128	1.150
Ionization energy (eV)	15.6	12.3	11.4	14.0	9.3
Dissociation energy (kcal)	225	118	200	235	150
Stretching frequency (cm^{-1})	2331	1555	1974	2143	1876

properties of dinitrogen and some analogues. It is clear that the nitrogen-nitrogen triple bond is exceedingly strong (225 kcal $mole^{-1}$), and only the bond in carbon monoxide requires more energy to bring about homolytic dissociation (235 kcal $mole^{-1}$). A reaction requiring the homolytic fission of the dinitrogen is thus likely to require a considerable activation energy, and will probably occur only at elevated temperatures. The ionization energy of dinitrogen is also exceptionally high (15.58 eV, 367 kcal $mole^{-1}$), approximately as great as that of argon. This would suggest that the activation energy for simple oxidation is likely to be prohibitively high. The bond length is very short, and reflects again the strength of the binding. Finally, the nitrogen-nitrogen stretching frequency observed in the Raman spectrum, 2331 cm^{-1}, is higher than that of any other simple diatomic molecule.

Dinitrogen is also unusual in that so much energy is required to convert the nitrogen-nitrogen triple bond to a nitrogen-nitrogen double bond. Thus the bond energy of a nitrogen-nitrogen double bond is about 100 kcal $mole^{-1}$. For comparison, the carbon-oxygen double bond energy is about 170 kcal $mole^{-1}$. Other relevant bond energies are shown in Table 2.2. Most systems where multiple bonds are possible apparently fit the pattern of C–C and C–O. Only N–N and N–O have such large contributions of the third bond to he total triple bond energy. The reason for this is not known, although it has been attributed to he absence of lone pair-lone pair repulsions in N_2 and NO^+. These repulsions are much more significant in molecules containing N=N and N–N, and are also apparently responsible for

TABLE 2.2.   Comparison of single and multiple bond energies.

	C–C	C–N	C–O	N–N	N–O	O–O
Triple bond	194	210	235	225	252	
Ratio, triple:double	*1.32*	*1.43*	*1.31*	*2.25*	*~3.0*	
Double bond	147	147	170	100	~80	96
Ratio, double:single	*1.77*	*2.10*	*2.01*	*2.63*	*~2.0*	*2.91*
Single bond	83	70	84	38	~40	33

the very low bond energies of the N—N, O—O and F—F single bonds (Pauling, 1960; *idem*, 1962).

There is considerable uncertainty about the electron affinity of dinitrogen. Assuming a similarity between the pair NO^+ and NO and the pair N_2 and N_2^- (N_2 and NO^+ are isoelectronic, as are N_2^- and NO), a value of the electron affinity of the dinitrogen molecule of 84 kcal $mole^{-1}$ has been calculated (Borodko and Shilov, 1969). Despite the relatively high value of Pauling electronegativity assigned to the nitrogen atom (3.0; c.f. F = 4.0, Cl = 3.0), direct measurement of the electron scattering by dinitrogen suggests an electron affinity of not less than 35 kcal $mole^{-1}$ (Golden, 1966). Evidently, only very strong reducing agents will be able to produce N_2^-.

The reactions of dinitrogen will be classified somewhat arbitrarily to simplify discussion. Some reactions do not fit easily into any classification and the division does not necessarily imply any qualitative similarities. Fixation of dinitrogen is understood to mean any reaction of dinitrogen, and not solely conversion to ammonia. However, the formation of transition-metal dinitrogen complexes directly from dinitrogen will not be dealt with here because it is discussed in Chapter 3.

2.3   MODES OF REACTION OF DINITROGEN

2.3.1   *Addition reactions*
In organic chemistry, multiple bonded systems, as in acetylene, undergo characteristic addition reactions with, for example, alkyl halides. No comparable reaction has been observed for dinitrogen. Dr. A. Diamantis, in unpublished work in the A.R.C. Unit of Nitrogen Fixation, sought a possible reaction of dinitrogen with di-imide. Di-imide is a short-lived species which rapidly decomposes to give dinitrogen and hydrazine. It also reduces many unsaturated systems as in azo compounds and ethylene (Hunig *et al.*, 1965), and even in dioxygen (van Tamelen *et al.*, 1961)

$$C_2H_4 + N_2H_2 \rightarrow C_2H_6 + N_2$$

Diamantis exposed di-imide, generated from potassium azodicarboxylate, to $(^{15}N)_2$ and examined the hydrazine produced for its content of ^{15}N. No enrichment above natural abundance was observed, so that the hydrogen transfer reaction below does not occur.

$$^{14}N_2H_2 + (^{15}N)_2 \rightarrow (^{14}N)_2 + {}^{15}N_2H_2$$

The formation of di-imide from dinitrogen is endothermic (49 kcal mole^{-1}), which is almost the inverse of the hydrogenation of acetylene to ethylene (42 kcal mole^{-1}, exothermic). The formation of aqueous hydrazine would be endothermic to the extent of only 8 kcal mole^{-1}. However, the direct formation of hydrazine from dinitrogen in the biological system is unlikely unless the reduction proceeds in symmetrical, two-electron steps.

2.3.2   Reaction with electrophiles

Dinitrogen does not react with usual Lewis acids such as BF_3 to form stable complexes. An interaction between BF_3 and N_2 has been inferred from line-broadening in the infrared spectrum of gaseous mixtures of the two (Borodko and Krylova, 1967). This is interpreted as being due to charge-transfer in the sense $N\equiv\overset{\delta+}{N} \rightarrow \overset{\delta-}{B}F_3$ but it is very far from complex formation.

However, dinitrogen does react with the very reactive electrophile, carbene, $:CH_2$ (Borodko *et al.*, 1966; Shilov *et al.*, 1968). Diazomethane, CH_2N_2, decomposes rapidly upon exposure to ultraviolet irradiation. When a mixture of $(^{15}N)_2$ and $CH_2{}^{14}N_2$ was irradiated and the reaction stopped before the diazomethane had been completely decomposed, then the remaining diazomethane was found to contain ^{15}N. This can only mean that the decomposition is, in fact, reversible

$$CH_2 + N_2 \rightleftharpoons CH_2N_2$$

Reaction has also been observed with $(CF_3)HC:$ and $(CN)HC:$. The activation energy for the reaction of carbene with dinitrogen is very small (10 kcal mole^{-1}).

The very reactive phenylsulphenium ion, PhS^+, generated from phenylsulphenium chloride and silver perchlorate also binds dinitrogen (Owsley and Helmkamp, 1967). The product is believed to be $PhSN_2{}^+$, but it can be recognized only by colour tests typical of diazonium salts; for example, coupling reactions. Only trace amounts of dinitrogen are absorbed and no compound was isolated. The pressure of dinitrogen in these experiments was only 1 atm., much lower than that used in some of the work described below.

The ion $N_2{}^+$ may itself be considered as a product of an electrophilic attack. It is present in dinitrogen gas which has been exposed to ultraviolet irradiation and its physical properties have been studied intensively. However, no compounds containing this ion have ever been isolated.

The reactions of dinitrogen with various low-oxidation-state transition metal compounds and with polyhydrides (e.g. $[OsCl_2(THF)(PMe_2Ph)_2]$ and $[CoH_3(PEtPh_2)_3]$) may be considered to be reactions with electrophiles, similar to the interaction with BF_3. The products are dinitrogen complexes, and whether or not their formation involves overall charge transfer from dinitrogen to the metal is not yet clear. Suffice it to say that 'end-on' complex formation to give products such as $[OsCl_2(N_2)(PMe_2Ph)_3]$ (Chatt et al., 1969a) and $[CoH(N_2)(PEtPh_2)_3]$ (Sacco and Rossi, 1967) has now been widely observed. The complexes themselves are discussed in detail in Chapter 3.

Dinitrogen also reacts with singlet (i.e. excited state) dioxygen (Anbar, 1966). The ground state is, of course, a triplet. The reaction of hydrogen peroxide with chlorine or hypochlorous acid produces singlet dioxygen ($^1\Sigma$), and in the presence of about 100 atm. of dinitrogen, nitrite and nitrate are eventually produced. This has been suggested as a plausible model for the route of biological fixation. Singlet *atomic* oxygen also reacts with dinitrogen (Katakis and Taube, 1962; Demore and Davidson, 1959; Demore and Raper, 1963), forming N_2O, and in solid dinitrogen the yield of N_2O may reach 35%.

The conditions for these last two reactions are rather extreme. The oxygen reactants are apparently in electronic states far from normal-temperature ground states. Excited dinitrogen molecules, on the other hand, have been long recognized. They are formed in ultraviolet irradiated dinitrogen gas in the upper atmosphere, for example, and they are relatively long-lived. Chemical studies suggest that excited dinitrogen molecules are still exceedingly stable and unreactive, and their principal function in reacting mixtures would appear to be that of an energy-transfer agent (Fotin, 1967).

Dinitrogen reacts with alumina at $500°$, to form nitric oxide. This is a facile oxidation which at first sight shows commercial promise, but it has not yet been exploited (Parkyns and Patterson, 1965).

Finally, the results of certain chemisorption studies should be mentioned. Dinitrogen is absorbed by many metal surfaces (see below). On tungsten and tantalum (10^{-5} mm Hg and at $-183°$) dinitrogen is negatively charged (Sachtler and Dorgelo, 1957; Mignolet, 1950b). However, on nickel dinitrogen is positive (Gundry et al., 1962; Mignolet, 1950a).

The interpretation of these observations is that dinitrogen adsorption on nickel is probably molecular and N_2^+, or something like it, may be formed, but that on tungsten and tantalum it is likely to be dissociative, forming nitrogen atoms. This type of interaction will be discussed in a later section.

2.3.3   Reactions with nucleophiles

Any reaction of dinitrogen leading to the eventual production of ammonia must involve reduction at some stage and may, therefore, have involved nucleophilic attack on N_2. However, this is rarely a useful way of looking at the systems. The same is true of the conversion of dinitrogen to nitrides. For convenience, however, I consider here solely reactions in which a nucleophilic reagent attacks dinitrogen, or where electrons are transferred to dinitrogen, without breaking the nitrogen-nitrogen triple bond.

The ion N_2^- is isoelectronic with NO and O_2^+. However, whereas derivatives of the latter two are well known (e.g. NOCl, $O_2[PtF_6]$) no derivatives of N_2^- have been isolated. Nevertheless, it has been identified in certain crystalline azides (KN_3, NaN_3) which have been damaged by γ-irradiation. The metastable ion is probably stabilized by the crystal lattice. Polynuclear anions such as N_4^- have also been postulated (Gelerinter and Silsbee, 1966; Sander, 1962; Shuskus et al., 1960; Wylie et al., 1962; Brezina and Gelerinter, 1968). N_2^- should, of course, be paramagnetic, and it is the e.s.r. spectrum which has enabled it to be identified.

Otherwise N_2^- has only been identified in the mass spectrometer (Golden, 1966). There is, however, persuasive evidence that N_2^- is an intermediate in the nitridation of metal surfaces. A kinetic investigation of the nitridation of calcium (Roberts and Tompkins, 1959) suggests that the initial step is transfer of an electron from the metal surface to dinitrogen to give N_2^-. Adsorbed dinitrogen can take up electrons at $450°$ from negatively charged silver, copper, cobalt, and also, apparently, from iron (Krasilshchikov et al., 1963; Krasilshchikov and Antonova, 1965). The products are presumably also N_2^- ions. The loss of electrons to give N_2^+ was much less ready under these conditions.

N_2^- ions, or possibly N_2^{m-}, where m is a small integer, also occur quite frequently in azide systems. Thus, the thermal decomposition of barium azide, $Ba(N_3)_2$, is claimed to proceed via an unstable nitride Ba_2N_2, formed by the reaction of barium metal with electronically excited dinitrogen. Ca_2N_2 is apparently more stable (Torkar and Spath, 1967). Calcium, strontium and barium all form pernitrides, M_3N_4, and these, upon solvolysis in hydroxylic solvents, yield about 20% of their nitrogen as hydrazine and the rest as ammonia. It is tempting to ascribe this to the presence of anions of the type N_2^{m-} (Linke and Linkmann, 1969).

Lithium is the only alkali metal to form a nitride directly from N_2. It can do so under very mild conditions, at about 30° or below. Although Li_3N is formally an ionic compound, and an X-ray investigation has shown that the crystal contains isolated lithium and nitrogen atoms (Bishop *et al.*, 1966), Chatt and Leigh (unpublished) showed that the dense, red solid is paramagnetic and that its hydrolysis yields ammonia, hydrogen, nitrogen and a trace of hydrazine. Once again, diatomic nitrogen anions appear to be implicated.

A much more significant development, which may involve genuine nucleophilic attack upon dinitrogen to form well defined compounds, has been reported by Ellerman *et al.* (1969). The rather unexpected reaction is as follows. The tetraphosphine below,

$$\begin{array}{cc} Ph_2PCH_2 & CH_2PPh_2 \\ & \diagdown C \diagup \\ Ph_2PCH_2 & CH_2PPh_2 \end{array}$$

hereafter referred to as tetraphos, reacts with sodium in liquid ammonia to form a tetrasodium salt, with cleavage of one phenyl group from each phosphorus. The sodium salt of tetraphos reacts with carbon disulphide at 60° to form a product formulated as:

$$C\left[\begin{array}{c} S^- \\ CH_2-P=C=S, Na^+ \\ Ph \end{array}\right]_4$$

This, in the solid state or in solution, is said to absorb two moles of dinitrogen 'greedily' at room temperature to yield a complex of which one of the canonical forms is A.

There is one disturbing aspect to this report. Salts such as sodium diphenylphosphide react with carbon disulphide at −50° to give the sodium salt of a thiocarboxylic acid, $NaSSCPPh_2$, in which both sulphur atoms are bonded to carbon. Indeed, carbon dioxide, the oxygen analogue of carbon disulphide, reacts with the sodium salt from tetraphos to yield $C(CH_2PPhCO_2Na)_4$, which is inert to dinitrogen. The oxygen atoms are, of course, bound to carbon. It is not clear why temperature has such a significant effect upon the carbon disulphide reaction. Chatt and Leigh (unpublished) found that triphos, $HC(CH_2PPh_2)_3$, and diphos, $Ph_2PCH_2CH_2PPh_2$, do

$$\left[\begin{array}{c} \text{Ph} \\ \backslash \quad \text{S}^- \ \text{S}^- \\ \quad \diagup \ \diagup \\ \text{CH}_2-\text{P}=\text{C} \\ \diagup \qquad\qquad \backslash \\ \text{C} \qquad\qquad\qquad \text{N} \\ \backslash \qquad\qquad\qquad \| \\ \qquad\qquad\qquad \text{N} \\ \qquad\qquad \diagup \\ \text{CH}_2-\text{P}-\text{C} \\ \diagup\!\diagup \quad | \quad \backslash \\ \text{S} \quad \text{Ph} \quad\ \text{S} \end{array} \right]_2 \quad 4\text{Na}^+$$

A

not give dinitrogen-fixing derivatives. It is not clear why this should be so, and further information on this system is eagerly awaited.

Nucleophilic attack on dinitrogen to give simple products in which the nitrogen-nitrogen bond is retained are thus not very common. On the other hand, nitridation reactions which may be a consequence of nucleophilic attack are much more usual, and may be involved in the commercial fixation of dinitrogen. These are discussed below (Section 2.3.6.).

2.3.4   Reactions of 'active nitrogen'

Dinitrogen which has been passed through an electrical discharge is much more reactive than ordinary dinitrogen. This is because it contains molecules and atoms in excited states which are sometimes long-lived. Considerable amounts of energy may be involved. Thus the excitation of dinitrogen to the first vibrational level requires 0.43 eV, which would require a temperature of several thousand degrees for thermal excitation. Electronic excitation requires 6.17 eV. Nevertheless, several species of excited dinitrogen molecules and nitrogen atoms have been identified in 'active nitrogen' (Fotin, 1967).

The reactions of nitrogen atoms with organic molecules do not seem very relevant to the problem of biological fixation. All seem to give rise to hydrogen cyanide; many lead to nitrogen-containing polymers, and some also yield ammonia. Chlorinated compounds may yield ClCN. In all cases, less ammonia than hydrogen cyanide is formed. Few reactions have indicated that less highly degraded derivatives of the original organic compound might be obtainable. No general reaction mechanism has been postulated, although the processes may involve initial attack of a nitrogen atom on a carbon atom of the organic molecule to form a transition complex, which then decomposes to hydrogen cyanide and organic radicals (Fotin, 1967).

Excited dinitrogen molecules in 'active nitrogen' also react with molecules such as methane, ethane and ethylene to give products very similar to those produced by nitrogen atoms. In these cases the dinitrogen is believed to hand on energy to the hydrocarbon producing an excited molecule which dissociates into radicals which then react with atomic nitrogen in the 'active nitrogen' giving hydrogen cyanide, ammonia and other products (Fotin, 1967; March and Schiff, 1967). It is thus evident that even excited dinitrogen molecules are not very reactive.

It seems unlikely that 'active nitrogen' can give any important information regarding fixation processes under normal laboratory or industrial conditions. This also applies to the conversion of dinitrogen to nitric oxide by atmospheric electrical discharges, relevant though this may be in the terrestrial nitrogen cycle.

2.3.5 Dissociative reactions of dinitrogen: (a) Industrial N_2 fixation

As early as 1836, Humphrey Davy wrote in his 'Elements of Agricultural Chemistry' that '... when glutinous and albuminous substances exist in plants, the azote they contain may be suspected to be derived from the atmosphere'. Of course, it is not now generally believed that plants themselves can fix dinitrogen, but the fact that dinitrogen fixation occurs under chemically very mild conditions was early recognized. Davy even synthesized nitric oxide by passing air over an electrically heated wire. Nevertheless, the first commercially successful process for fixing dinitrogen by this route, the Birkland-Eyde or arc process, went into operation as late as 1903 (Ernst, 1928; Mellor, 1964; Mellor, 1967).

This process involves essentially the oxidation of dinitrogen by dioxygen.

$$N_2 + O_2 \rightleftharpoons 2NO$$

The process is endothermic (about 43 kcal mole^{-1}) and requires energy. In practice, this was supplied by passing dinitrogen and dioxygen through an electric arc furnace parts of which reached temperatures of 3300°. The process was never very efficient in energy terms, requiring about 60 000 kWh for every ton of dinitrogen fixed. The theoretical amount of heat energy required by the above equation is only 1 630 kWh for every ton of dinitrogen fixed. The nitric oxide was not produced in the full equilibrium amount (about 6% by volume) and, in any case, a considerable amount decomposed as the gases were cooled on leaving the furnace. Oxidation of the nitric oxide with dioxygen at about 50° produced nitrogen dioxide, and this was absorbed in water or alkali to give nitrite and nitrate.

The arc process consumed a large amount of energy, and it was only commercially successful in Norway where hydroelectric power is cheap. The process presumably involves dissociation of the reactants, and combination of atoms of oxygen and nitrogen.

Another general route for fixing dinitrogen which achieved some commercial success involves the formation of cyanides. This process had its origin in an observation of Scheele that a mixture of potassium carbonate and carbon reacts with dinitrogen at high temperature to form potassium cyanide. No commercial process was based directly upon this reaction, although a synthesis for sodium cyanide was developed.

$$Na_2CO_3 + 4C + N_2 \rightarrow 2NaCN + 3CO$$

The variant which was exploited commerically involved calcium cyanamide. This necessitated three stages, represented by the following equations:

$$CaCO_3 \xrightarrow{400°} CaO + CO_2$$

$$CaO + 3C \xrightarrow{2200°} CaC_2 + CO$$

$$CaC_2 + N_2 \xrightarrow{1000°} CaCN_2 + C$$

The first two steps are endothermic, and the second, in particular, requires a large amount of energy. The reaction of pure dinitrogen (produced by fractional distillation of liquid air) with calcium carbide is exothermic and is self-sustaining. The process overall required about one-fifth of the energy consumed in the arc process. Nevertheless, it was still much less efficient than the direct synthesis of ammonia from dinitrogen and dihydrogen, which is the only significant process in use today.

Dinitrogen and dihydrogen were observed in 1865 to form ammonia when heated to 1300°. The reaction was not commercially useful until the correct catalyst was discovered by Haber, and its first exploitation, rapidly copied throughout the world, was in Germany. The reaction

$$N_2 + 3H_2 \rightleftharpoons 2NH_3$$

is exothermic (12 kcal mole^{-1} of ammonia) and is carried out at elevated temperatures (say in the region of 500°) to ensure that it is sufficiently rapid. The equilibrium yield of ammonia at this temperature is only about 35% compared with 100% at 0°. Pressures of several hundred atmospheres are employed. A catalyst based on iron (initially in the form of iron oxide) is used, and it normally contains promoters such as potassium, aluminium and calcium. The synthesis gas is often produced today by the gasification of oil by air under pressure.

The mechanism of this synthetic reaction has never been entirely clarified. The rate-determining step under most conditions would appear to be the chemisorption of dinitrogen by the iron catalyst, and the Temkin-Pyzhev theory of ammonia synthesis which has wide currency at present takes this as a basic assumption, the hydrogenation of the adsorbed nitrogen being fast (Bond, 1962). Iron apparently adsorbs dinitrogen physically and reversibly at low temperatures, and chemisorbs it molecularly at about room temperature. Breakage of the nitrogen-nitrogen bond occurs above room temperature, as on a Haber synthesis catalyst, in which the promoter metal atoms in the iron catalyst may help the dissociation (Trapnell, 1953). Whereas dissociation of dinitrogen upon chemisorption has been assumed by many workers, secondary ion ion-emission studies on pure iron (admittedly at low pressures, 10^{-4} mm Hg, but in a temperature range 20-800°) showed no indication of dissociation of dinitrogen and little evidence for the presence of a surface species other than FeN$_2$ (Schvachko et al., 1966). The infrared spectrum of mixtures of N$_2$ and H$_2$ adsorbed on an iron catalyst at about 1 atm. pressure and at 410° is very similar to that of hydrazine. This indicates that intermediates such as M–N$_2$H, M–N$_2$H$_2$ and M–N$_2$H$_3$ might be involved, although mononitrogen species, M=NH, M–NH$_2$ and M–NH$_3$ cannot be excluded (Brill et al., 1969). On the other hand, similar studies using an iron-silica catalyst were taken to show positively the presence of mononitrogen species (Nakata and Matsushita, 1968). Field emission studies on iron single crystals have apparently failed to detect species containing one nitrogen atom, but showed large numbers of hydrogenated dinitrogen species (Schmidt, 1968). Isotope scrambling reactions (e.g. $^{14}N_2 + {}^{15}N_2 \rightleftharpoons 2^{14}N–{}^{15}N$) do not occur on doubly promoted iron at 410° and they are slow even at 750° (Joris and Taylor, 1939; Kummer and Emmett, 1951). There is no doubt that both atomic and molecular chemisorption of dinitrogen on iron can occur, but it is not yet clear what actually happens on a Haber process catalyst.

The basic picture of dinitrogen chemisorption being rate-determining must also be modified because it is now clear that under some conditions there is a hydrogen isotope effect on the rate of ammonia synthesis (Bond, 1962; Logan and Philp, 1968). The rate-determining step in ammonia synthesis is evidently hydro-genation of adsorbed nitrogen under certain conditions (Tamaru, 1965). Ponec and Knor (1968) have shown that, at 0°, the prior chemisorption of dinitrogen on a clean iron surface does not affect subsequent chemisorption of dihydrogen. In contrast, when dihydro-gen is pre-adsorbed at 0°, no dinitrogen can be subsequently chemisorbed. One of the basic assumptions of the currently accepted theory is that the adsorption of dinitrogen is not influenced by the pressure of either dihydrogen or ammonia (Bond, 1962). This has led Carra and Ugo (1969) to suggest that the increase in the rate of dinitrogen chemisorption in the presence of dihydrogen found under catalytic conditions (i.e. elevated temperatures and pressures) is probably due to a specific combination of dinitrogen and dihydrogen chemisorption. Carra and Ugo (1969) postulate a non-dissociative interaction of dinitrogen with a chemisorbed hydride giving species of the type:

$$M-N=N-H$$

These are subsequently reduced to ammonia. In support of this suggestion, a complex

has recently been characterized by X-ray crystallography, although it decomposes at room temperature to give a platinum(II) hydride plus dinitrogen (Dobinson *et al.*, 1968).

There is now substantial evidence that it may be possible to develop Haber catalysts which work at much lower temperatures. Thus a 1:1 mixture of dinitrogen and dihydrogen with a total pressure of 15 cm Hg forms ammonia slowly at 50° over sodium- or potassium-graphite donor-acceptor complexes. Dinitrogen and dihydrogen are chemisorbed by a sodium-iron phthalocyanine donor-acceptor complex. A 1:3 molar mixture (total pressure 40 cm Hg) forms 0.26 ml ammonia when circulated for 20 hours over this catalyst at 110°. At 260° in the same conditions, 4.60 ml ammonia are formed. The mechanisms of these catalyses have not been explained, and the reactions are still very slow but they hold much promise for the future (Sudo *et al.*, 1969).

2.3.6   *Dissociative reactions of dinitrogen: (b) The formation of nitrides, of ammonia and amines*

Dinitrogen reacts with many metals, usually at elevated temperatures, to yield nitrides (e.g. TiN, Cr_3N (Juza, 1966)). These nitrides can often be synthesized from the metal and ammonia also. With transition elements the nitrides are often very hard refractory materials, known as interstitial nitrides. They are in many ways analogous to carbides, and would conceivably have a similar solvolysis chemistry were the N—N single bond stronger. Thus the solvolysis of carbides can give rise to a variety of hydrocarbons, Mn_3C yielding hydrogen, methane and ethane, and Fe_3C mixtures, including unsaturated hydrocarbons. Similarly, the hydrolysis of lithium nitride (Chatt and Leigh, unpublished), and of calcium, strontium and barium pernitrides all give hydrazine as well as ammonia. Sometimes as much as 20% of the nitrogen is converted to hydrazine (Torkar and Spath, 1967; Linke and Linkmann, 1969). Because nitrides can be made directly from dinitrogen, and because nitridation represents the reduction of the dinitrogen molecule by a metal, then these observations of hydrolysis products may give some indication of the route of the reduction reaction.

The reduction of the dinitrogen by aluminium to give aluminium nitride was once suggested as a method for commercial dinitrogen fixation (Serpek process) (Ernst, 1928). This involved heating bauxite, coke and dinitrogen in an electric furnace at 1800°.

$$Al_2O_3 + 3C + N_2 \rightarrow 2AlN + 3CO$$

Subsequently, the aluminium nitride was hydrolyzed to ammonia, regenerating the alumina. The chief objections to this process were the excessive consumption of electrical energy and the difficulty of finding furnace materials capable of withstanding the reaction conditions. Consequently, it was never exploited on a large scale.

Nitridation of metals does not necessarily require such extreme conditions. Chatt and Leigh (unpublished) found that nitridation of lithium can occur in refluxing pentane (35°) when an alkyl halide is present to expose new surface to dinitrogen attack by forming the alkyl-lithium compound. Thus reports of dinitrogen fixation under mild conditions must always be considered seriously. Many claims of the reduction of dinitrogen to ammonia in aqueous conditions are probably incorrect, and there are other reports which have yet to be confirmed. On the other hand, the apparently common observation that Ziegler-Natta catalysts (these are non-aqueous systems, often, a mixture of a titanium halide and an aluminium alkyl) can absorb dinitrogen from the gas phase has been amply confirmed in the last few years (Volpin and Shur, 1966a); these are also reductive systems, converting dinitrogen to nitrides.

One of the most intriguing reports of abiological fixation in aqueous systems concerns the formation of 'jeewanu'—particles with a definite microstructure, and containing amino acids, among other constituents (Bahadur and Rangnayaki, 1963). The particles are claimed to 'duplicate' (i.e. reproduce) and to exhibit 'peculiar movement which is not Brownian motion'. The 'jeewanu' (from the Sanskrit, and meaning 'particles of life') can be produced by sunlight irradiation (in pyrex flasks), during eight hours, of mixtures of metal oxides and salts and citric acid buffered to pH 6. Both iron and molybdenum are involved. To our knowledge, no laboratory has succeeded in exactly reproducing these observations, but the significance of the formation of amino acids in a system where there is no source of fixed nitrogen cannot be over-estimated. Our own attempts to produce 'jeewanu', admittedly carried out in a somewhat sceptical frame of mind, have met with no success.

Dhar (1959) claimed that metal oxides, in the presence of an energy source (e.g. glucose), are capable of producing ammonia from dinitrogen when aqueous mixtures are allowed to stand for about six months. Zinc oxide, aluminium oxide, ferric oxide, nickel oxide, cobaltous oxide, cupric oxide and manganese dioxide are all equally effective catalysts. Irradiation of the pyrex containers with daylight approximately doubled the yield of ammonia. These systems are apparently sterile but it is very easy to make mistakes, and very difficult to avoid contamination, when experiments may last for several months. These results have been reported from India. The

East indeed appears mysterious, and one would dearly like to see such experiments confirmed in the West.

There are two less dramatic accounts of N_2-fixation in aqueous solutions. Haight and Scott (1964) reported that solutions of molybdate and tungstate in concentrated hydrochloric acid, when reduced with zinc or stannous chloride, can convert dinitrogen to ammonia. The conversion was found to be inhibited by catechol, which might well be expected to complex with transition element ions and affect a catalytic reaction. Blanks were also run. A dinitrogen gas stream was passed through the solutions for about twenty-four hours in these experiments. Similarly, Yatsimirskii and Pavlova (1966) claimed that a large number of transition-metal compounds (titanium(II) chloride, vanadyl(IV) sulphate, niobium pentachloride, tantalum pentachloride, potassium dichromate, sodium molybdate, sodium tungstate and potassium perrhenate) are able to convert dinitrogen to ammonia after solutions in 3N hydrochloric acid of 4N sulphuric acid have been reduced by zinc or zinc amalgam.

Hardy and Knight (1968) and Shilov and Shilova (1969) have cast doubt upon both these reports. Ammonia produced in these reductions is not enriched with ^{15}N when $(^{15}N)_2$ is used as starting material, and it seems likely that the positive reports are due to the scavenging of nitrogen oxides from the dinitrogen gas stream (Shilov and Shilova, 1969).

Parker (1955) has made another claim of this type. Iron wool has been used to create anaerobic aqueous environments for bacterial cultures, because fresh iron surfaces are very rapidly oxidized by atmospheric oxygen. Under certain conditions, not completely specified, ammonia may be produced at the same time. This system has not been reinvestigated.

Apart from the claims discussed above, the only other report of fixation in an aqueous environment is of the reaction of the aquopenta-ammineruthenium(II) cation with dinitrogen (Harrison and Taube, 1967). The product is a dinitrogen complex, and it is discussed in Chapter 3.

$$[Ru(NH_3)_5(H_2O)]^{2+} + N_2 \rightleftharpoons [Ru(NH_3)_5(N_2)]^{2+} + H_2O$$

In contrast to the current situation regarding fixation in aqueous solutions, there is a large and growing literature describing fixation in non-aqueous systems. These generally involve nitrides. The major contribution to the work on non-aqueous systems has been made by Russian workers, particularly Volpin and Shur. They have shown that a large number of transition metal compounds, generally in an ethereal solvent and in the absence of dioxygen, will fix dinitrogen after treatment with a reducing agent. The transition-metal compounds successfully employed have included titanium tetrachloride, dichlorobis(π-cyclopentadienyl)titanium(IV), titanium tetraethoxide, vanadyl(IV) bis(acetylacetonate), chromium tris(acetylacetonate), chromium trichloride, molybdenum pentachloride, tungsten pentachloride, manganese tris(acetylacetonate), ferric chloride, trichlorobis(triphenylphosphine)iron(III), cobaltous chloride, palladous chloride and cuprous chloride. The reducing agents used include Grignard reagents, alkyl-lithium compounds, alkali metals, aluminium alkyls and aluminium metal. Yields of 1 mole ammonia per atom of metal have been claimed for some titanium systems (Volpin and Shur, 1966a). In addition, it has been shown that solutions of bis(π-cyclopentadienyl)dimethyltitanium, $[(\pi\text{-}C_5H_5)_2$ $Ti(CH_3)_2]$, and bis(π-cyclopentadienyl)diphenyltitanium, $[(\pi\text{-}C_5H_5)_2Ti(C_6H_5)_2]$, in pentane at 90-100°, as well as of titanocene, $[(\pi\text{-}C_5H_5)_2Ti]$, react with dinitrogen at a pressure of about one hundred atmospheres, to give ammonia after hydrolysis (Volpin et al., 1966). The fixation reactions all apparently take place at one atmosphere pressure of dinitrogen, but elevated pressures are claimed to give increased yields of ammonia after hydrolysis.

American workers have come much later onto the scene, and have added niobium and tantalum to the list of transition elements which fix dinitrogen (Gray and Brubaker, 1969). Titanium has also been investigated, but generally alkoxides have been used, with an alkali metal or sodium naphthalene as the reducing agent (van Tamelen et al., 1968; van Tamelen et al., 1967). By careful solvolysis with a limited amount of solvolytic agent, and then addition of more reducing agent, the systems have been made cyclic. This may be an appropriate model for the biological system which undoubtedly involves a complex cycle of reactions. A simple catalytic system in the chemical sense (i.e. reducing agent + protonating agent + catalyst, capable of producing ammonia without some regenerating device to restore it) seems much too simple a model.

An electrolytic reduction of dinitrogen has also been claimed (van Tamelen and Akermark, 1968; van Tamelen and Seeley, 1969). The fixing agent was titanium tetra-isopropoxide plus aluminium trichloride in 1,2-di(methoxy)ethane. However, the voltage and current (90 volts, 50 milliamps) are such that a reduction of the titanium and/or aluminium compounds is also a possibility. This would result in a fixing system analogous to those produced by chemical reducing agents.

Russian work has shown that very simple chemical reducing agents can be effective. Thus a mixture of magnesium and magnesium(II) iodide, claimed to be essentially magnesium(I) iodide, is quite efficient (Volpin *et al.*, 1965a), and a mixture of aluminium, aluminium tribromide and titanium tetrachloride shows true catalytic fixing activity (Volpin *et al.*, 1968a; Volpin *et al.*, 1968b). This latter mixture produces what is best regarded as a catalytic nitridation of aluminium. For example, titanium tetrachloride, aluminium and aluminium tribromide in the proportions 1:600:1000, yield, after exposure to dinitrogen and subsequent hydrolysis, 286 moles of ammonia per mole of titanium compound. The reaction is carried out at 130°and under 100 atm. of dinitrogen. The intermediates involved in the reaction are complex, and probably contain titanium and aluminium linked through halogen bridges. Thus, the compound $[C_6H_6 \cdot TiX_2 \cdot 2AlX_3]$ (X = halogen) under dinitrogen, when added to a mixture of aluminium and aluminium tribromide, can convert 80% of the aluminium to the nitride AlN (Volpin *et al.*, 1968a; Volpin *et al.*, 1968b).

The most intriguing aspect of these fixation reactions is that under certain conditions amines may be formed via nitriding reactions. Thus one mole of $[(\pi\text{-}C_5H_5)_2TiCl_2]$ or $[(\pi\text{-}C_5H_5)_2Ti(C_6H_5)_2]$ and five moles of phenyl lithium in ether under dinitrogen (one atm.) yield 0.17 moles of ammonia and 0.03 moles of aniline, plus traces of *o*-aminodiphenyl, after hydrolysis. The use of *o*-tolyl lithium leads to all three isomeric toluidines. Alkyl lithium compounds produce only traces of amines, and a benzyne intermediate produced by breakdown of titanium-aryl bonds has been suggested as the active fixing species. Direct evidence of benzyne formation has recently been obtained from the study of the decomposition of $[(\pi\text{-}C_5H_5)_2Ti(C_6H_5)_2]$ in hexane solution under argon (Latyaeva *et al.*, 1968). An alternative possibility is that the reaction involves insertion of dinitrogen into the titanium-aryl bond (Volpin and Shur, 1966b; Volpin *et al.*, 1968c).

Volpin, Belii and Katkov (1969) have claimed that lithium naphthalene (a charge-transfer compound, consisting of lithium cations and naphthalene anions, which accommodate electrons in anti-bonding orbitals) reacts with dinitrogen, giving, after hydrolysis, α-naphthylamine and a lesser amount of β-naphthylamine. The yield of amine is increased ten-fold when titanium tetrachloride is added to the lithium naphthalene. The mechanism of these amine-producing reactions is, at present, obscure.

2.3.7 Dissociative reactions of dinitrogen: (c) Mechanism of nitriding reactions

It would appear that titanium compounds give the most efficient nitrogen-fixing systems, although a stoichiometric formation of ammonia has been achieved using sodium naphthalene and salts of transition elements such as vanadium and chromium (Henrici-Olivé and Olivé, 1967). These systems have some properties in common with the biological nitrogen-fixing systems. Thus carbon monoxide inhibits fixation by chromium trichloride + ethylmagnesium bromide (Volpin and Shur, 1966a). Dihydrogen also inhibits some fixation reactions, although titanium-containing systems all give higher yields of ammonia when dihydrogen is mixed with the dinitrogen (Volpin et al., 1966; Volpin et al., 1965b). Diphenylacetylene and 1-hexene are also inhibitors of fixation for the chromium system. The fixation process is relatively slow and reaction times as long as thirty hours have been used (Volpin and Shur, 1966a).

The reaction mixtures are evidently complex, and the fixation process proceeds in at least two steps. The titanium systems have been most investigated and will be discussed in detail. The reaction of a titanium halide with an alkyl Grignard reagent is likely to form unstable titanium alkyls, which would decompose to give alkanes and alkenes, via direct homlytic fission of the titanium-carbon bonds and/or by olefin elimination forming an equally unstable metal hydride. The titanium-containing species will thus eventually be in an oxidation state well below plus four. Nechiporenko et al., (1965) and Maskill and Pratt (1966) observed that hydrocarbons are produced when the reactants are mixed. If the reaction of ethylmagnesium bromide and $[(C_2H_5)_2TiCl_2]$ is carried out between $-70°$ and $-100°$ under argon, and the reaction mixture is then exposed to dinitrogen, a change in the ultraviolet spectrum of the mixture is observed. A blue nitrogen-containing complex with two titanium atoms per molecule of dinitrogen is formed. It decomposes at $-60°$, and dinitrogen is released (Shilov et al., 1969). This reaction is unlikely to involve cleavage of the dinitrogen, and presumably represents the first step in the fixation process.

An apparently similar dinitrogen adduct has been prepared from a characterized titanium(II) compound. It contains one dinitrogen molecule and two atoms of titanium and has a band in its infrared spectrum assigned to $\nu(N{\equiv}N)$ at 1960 cm^{-1}. It can be isolated from the prolonged reaction of an ethereal solution of titanocene with dinitrogen under pressure (van Tamelen et al., 1969a).

If the ethylmagnesium bromide + $[(\pi\text{-}C_5H_5)_2\,TiCl_2]$ reaction is carried out at $20°$, ethane and ethylene are evolved rapidly, and a fixing species is generated. The fixation is most rapid in fresh mixtures and is irreversible. Ammonia is not formed in this fixation process, and only a trace of the hydrogen on the final ammonia is derived from the solvent ether. Ammonia is first generated upon addition of water or sulphuric acid. About 20% of the fixed nitrogen is converted to hydrazine. The solvolytic reagents also produce dihydrogen. Deuterium oxide produces HD. If the reaction mixture is not exposed to dinitrogen, the yield of dihydrogen is greater. Infrared spectroscopy suggests the cyclopentadienyl groups are cleaved from the titanium. Evidently metal hydrides are formed in the reaction mixture, as expected, and, under these conditions, the fixation appears a typical nitridation (Nechiporenko *et al.*, 1965).

The identity of the fixing species has never been properly settled. The reaction mixture is paramagnetic, and on the basis of electron spin resonance spectroscopy Brintzinger (1966a,b,c) suggested that the fixing species is a paramagnetic titanium(III) hydride, formulated as

$$\left[\begin{array}{c} (\pi\text{-}C_5H_5) \\ (\pi\text{-}C_5H_5) \end{array} Ti \begin{array}{c} H \\ H \end{array} Ti \begin{array}{c} (\pi\text{-}C_5H_5) \\ (\pi\text{-}C_5H_5) \end{array} \right]$$

A most elegant theory was devised for the fixation process, and suggested as a model for the biological system. This involved addition of a dinitrogen molecule across the hydride bridges forming a species

$$\left[(\pi\text{-}C_5H_5)_2\,Ti \begin{array}{c} H \\ N \\ N \\ H \end{array} Ti(\pi\text{-}C_5H_5)_2 \right]$$

Subsequent addition of acid to this species should yield ammonia and a titanium(IV) complex, which can be reduced to the titanium hydride again to recommence the cycle. However, the hydride has now been obtained in a reasonable degree of purity, and it is diamagnetic. It does apparently react with dinitrogen to give a product which decomposes to ammonia upon hydrolysis, but there is now no real evidence to support the insertion mechanism (Bercaw and Brintzinger, 1969).

Henrici-Olivé and Olivé (1968) have reassigned the e.s.r. signal from the bridged hydride to

$$[(\pi\text{-}C_5H_5)_2TiH_2]^- \quad \text{or} \quad [(\pi\text{-}C_5H_5)_2Ti\diagdown\overset{H}{\underset{H}{\diagup}}Ti(\pi\text{-}C_5H_5)_2]^{n-}.$$

and Brintzinger (1967) believes that the signal originally assigned to the dinitrogen-insertion product arises from $[(\pi\text{-}C_5H_5)_2Ti(C_2H_5)_2]^-$. These results emphasize that the fixing mixtures are complex. It will also be difficult to investigate the properties of the components because their isolation from the reaction mixture, or their independent synthesis, is likely to be difficult. For example, whereas trimethylamine extracts BH_3 residues from $[(\pi\text{-}C_5H_5)_2M(BH_4)_2]$ (M = Zr or Hf) to give $[(\pi\text{-}C_5H_5)_2MH_2]$, there is no reaction of trimethylamine with $[(\pi\text{-}C_5H_5)_2Ti(BH_4)_2]$ (James et al., 1966).

Maskill and Pratt (1968) made an attempt to investigate the kinetics of dinitrogen uptake. The interpretation of the measurements is somewhat difficult, because, of the reactants, only dinitrogen is well-defined. They found that in the fixing system generated by $[(\pi\text{-}C_5H_5)_2TiCl_2]$ and ethylmagnesium bromide, the rate of fixation reaches a maximum at p_{N_2} of about 0.5 atm. Higher pressures of dinitrogen have no effect on the rate. This behaviour roughly parallels the natural system, and conflicts with earlier reports (see above). The maximum amount of fixation is about one nitrogen atom per atom of titanium, as found by others. The rate of fixation is proportional to the square of the total titanium concentration, but it was not shown whether only one, or more than one, fixing species is involved. The observation is certainly in accord with a two-metal fixing site, and species containing a titanium(III) dihydride or titanium(II) hydride anion were suggested as being involved.

A thorough reinvestigation of these systems by Bayer and Schurig (1969) has aimed at identifying the oxidation state of the titanium in the fixing species. It was shown that fixation reaches a maximum when 4 to 5 moles of reducing agent have been added per mole of original titanium compound (the kinetic measurements suggested 10). In any case, never more than one mole of ammonia per mole titanium compound is produced. This suggests that the dinitrogen-fixing species contains titanium(0). Other experiments suggest that the formal oxidation state is less than plus two. The e.s.r. signals are apparently due to species present in only small concentrations, and both the fixing species and the product of fixation are diamagnetic.

The fixing species can react with dihydrogen as well as dinitrogen. Titanocene is suggested as an intermediate in the initial reduction, but it is dismissed as a fixing species because, containing titanium(II), it cannot reduce dinitrogen all the way to nitride. Only a hydridic titanium(II) species could provide three electrons per atom of nitrogen, and no dihydrogen is evolved during fixation.

This interpretation is open to criticism. In particular, it may not be the case that dinitrogen is reduced all the way to N^{3-}. The discovery of hydrazine as one of the solvolysis products by van Tamelen et al. (1969b) suggests the diatomic intermediates in various states of reduction may be involved. Hydrides may react with dinitrogen by an insertion mechanism, and not release dihydrogen. The weight of evidence now favours a binuclear species as the major component of the fixing system, probably containing titanium(0) or titanium(II), and possibly hydridic.

Yamamoto et al. (1969) have actually isolated nitride-containing materials from some systems. These are amorphous, yet homogeneous and reproducible. Thus magnesium reduces the complex $[TiCl_3(tetrahydrafuran)_3]$ to a titanium(II) species which takes up dinitrogen to give a product formulated as $[TiNMgCl_2(tetrahydrofuran)]$. A similar vanadium-containing material has been prepared. The catalytic nitridation of aluminium (Volpin et al., 1968a; Volpin et al., 1968b) involves species such as $[(TiCl_2 \cdot 2AlCl_3)_3N \cdot C_6H_6]$ and $[N(TiCl_2)_3]$. The significance of these materials in the dinitrogen-fixing process is not clear.

2.3.8   The interactions of the elements with dinitrogen

Titanium gives rise to the most efficient nitrogen-fixing systems. Some insight into the likely efficiency of low-oxidation-state derivatives in fixing dinitrogen can be gained from a general discussion of the nitrides of the elements. The Periodic Table in Table 2.3 shows the heats of formation of the binary nitrides of he elements per atom of combined nitrogen. Where, as with many transition elements, more than one nitride is recognized, then an average value is selected. In fact, the heat of formation does not vary much among the nitrides of a given element when reckoned on the basis of a single atom of combined nitrogen. Thus the heat of formation of γ–CrN is 29 kcal mole^{-1} and of Cr_2N is 31 kcal; the heat of formation of ζ–$Mn_{2.5}N$ is 24 kcal and of ϵ–Mn_4N is 30 kcal. There are no values available for those elements where no figures are supplied in Table 2.3. This may mean that no binary nitride is known for the element (for example, platinum), or else that the quantity has

TABLE 2.3.   Reactivity of the elements with dinitrogen.

H*
−11
450

	1 A	2 A	3 A	4 A	5 A	6 A	7 A	8			1 B	2 B	3 B	4 B	5 B	6 B	7 B
2	Li −47 / 0	Be −68 / 900	B −32 / 1300											C +37 / 3000	N	O +22 / 1500	F −26
3	Na −36	Mg −58 / 660	Al −57 / 820											Si −41 / 1250	P§ −17 / 850	S +32	Cl +55
4	K +20	Ca −52 / −70	Sc −68	Ti −80 / 800	V −52 / 600	Cr −30 / 500	Mn −27 / 390	Fe −4 / 500	Co† 0 / 500	Ni‡ 0 / 500	Cu +78	Zn −3	Ga −25	Ge −4 / 850	As +34	Se +42	Br +80
5	Rb +43	Sr −46 / 380	Y −72	Zr −82 / 700	Nb −59 / 600	Mo† −17 / 500	Tc	Ru	Rh	Pd	Ag +68	Cd +19	In −5	Sn	Sb	Te	I +65
6	Cs +75	Ba −45 / 260	La −71 / 320	Hf −78	Ta −59 / 600	W −17 / 600	Re −17	Os	Ir	Pt	Au	Hg +2	Tl +20	Pb	Bi	Po	At
7	Fr	Ra	Ac	Th −68	U −77												

* Over an iron catalyst.
† At 50 atm. N₂ pressure.
‡ There metals react with N₂ at 500° only if they are impure.
§ At 1000 atm. N₂ pressure.

not been determined or estimated. The table also provides estimates of the temperature at which the reaction of the given element and dinitrogen becomes significant (Juza, 1966; Gmelin, 1934). The diagonal line means that there is no direct reaction between dinitrogen and that element, but this does not necessarily mean that no nitride can be synthesized. For example, nickel only reacts with dinitrogen to form a nitride when doped with lithium; sulphur nitrides may be synthesized, generally via the reaction sulphur or sulphur halides with ammonia or ammonium halides; and cadmium nitride may be synthesized by the pyrolysis of $Cd(NH_2)_2$. There are no data available for elements shown together with horizontal lines.

The following points arise from Table 2.3. Of the alkali metals, only lithium is able to reduce dinitrogen. There is evidently no correlation with aqueous solution electrode potentials (Li^+/Li, -3.02; Na^+/Na, -2.71; K^+/K, -2.92; Rb^+/Rb, -2.99; Ca^+/Cs, -3.02 volts) and the stability of the product must be the most important factor. In general, in the Transition Series, there is a maximum exothermicity in Group IV. With few exceptions, the elements to the right of the iron triad do not react with dinitrogen below 400-500°. The break in reactivity coincides with the change-over from exothermic nitrides to endothermic nitrides. There appears to be an asymmetric exothermic 'island' in the B sub-groups but by Group VIIB the nitrides are generally very endothermic and often explosive (e.g. nitrogen tri-iodide). The reasons for this pattern of behaviour are not known, but it can be used, nevertheless, to rationalize many observations.

The heat of formation of titanium nitride, TiN, is greater than or equal to that of any other nitride. It is tempting to correlate this with the observed efficiency of titanium compounds in ammonia formation via nitridation. Those fixing systems may involve reactions which are essentially nitridations, but in type they are more akin to a surface reaction, or a metal corrosion reaction. It is therefore necessary to demonstrate a connection between a surface reaction with dinitrogen and bulk nitridation. This kind of connection has been demonstrated for reactions of gases such as dioxygen with metals (Bond, 1962; Sachtler and Van Reijen, 1962).

The normal process of a gas-solid reaction may be conveniently divided into three stages. These are not to be considered as distinct, or even, necessarily, to proceed consecutively. It is assumed that the reaction, a chemisorption, is governed not by the bulk properties of the solid (e.g. Fermi levels, d-band holes, etc.) but by the chemical properties of the individual atoms in the surface and of their neighbours. The three stages are described as follows:

(1) chemi-adsorption (the surface atoms are still on their lattice positions, and they interact individually with gas atoms, or molecules)

(2) corrosive chemisorption (regrouping of surface atoms and of adsorbate atoms, exposing new surface atoms to attack; this type of re-arrangement of metal atoms has been observed even at liquid-nitrogen temperature)

(3) ligand chemisorption (metal atoms or ions achieving co-ordination saturation by further chemisorption).

These stages have been observed directly using the field-ion microscope. It is possible to test this model quantitatively. The heat of adsorption of a gas at a metal surface gives some measure of the intensity of the gas-metal interaction. It is not possible to define a unique heat of adsorption because it varies with the particular crystal face which is being covered when metal single crystals are investigated, and also with the degree of coverage and the method of determination. In an attempt to counter some of these difficulties, only initial heats of adsorption are used in the subsequent discussion. Consider a case where the gas adsorbed is diatomic and dissociates on the surface. Then one can define a property

$$Z_A = \tfrac{1}{2}(Q_A - D_{X-X}),$$

where Q_A is the heat of adsorption per mole of adsorbate and D_{X-X} is its dissociation energy, which represents the intensity of interaction between the metal and the adsorbate molecule. Similarly, one can define another property

$$Z_F = \tfrac{1}{2}(Q_F - D_{X-X}),$$

where Q_F is the heat of formation of the bulk compound per mole of diatomic adsorbate. Z_F represents the intensity of interaction between the diatomic molecule and relatively widely separated metal atoms. These properties have been used to discuss the behaviour of dioxygen in some detail (Sachtler and Van Reijen, 1962). For dinitrogen, a plot of Z_A against Z_F (or of Q_A against Q_F) is essentially linear, within the limitations of the few data at present available (Fig. 2.4). In other words, the bulk heat of formation of a nitride would appear to be a measure of the intensity of dinitrogen-metal interaction, the more exothermic the nitridation, the stronger the interaction. This then, accords with intuitive feelings about the efficiency of titanium compounds in dinitrogen-fixing systems.

At first sight, too, it would apparently explain the efficiency of iron as a catalyst in the Haber synthesis. Iron interacts with dinitrogen just strongly enough to bring about the dissociation of the dinitrogen, and the resultant nitride is relatively unstable and can be hydrogenated easily. Unfortunately, the observed course of the

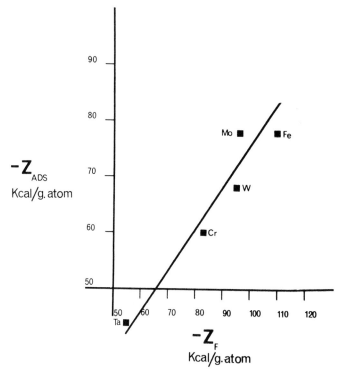

Fig. 2.4. The relationship between the heat of adsorption of dinitrogen on a metal and the bulk heat of formation of its nitride.

Haber synthesis discounts such a facile rationalization, and dinitrogen may actually insert into an iron-hydrogen bond (see above).

Even assuming that insertion into a metal-hydrogen bond is the key step in the Haber synthesis, it is also true that not all metals which activate dihydrogen are Haber catalysts. Platinum and palladium are cases in point. The consideration of nitridation is thus still worthwhile. A complete theory of chemisorption would consider such factors as the work function of the metal, the structure of the adsorbing crystal face, the dissociation energy and electronic configuration of the adsorbate, the electronic structure of the metal (e.g. the number of electron holes in the d-band, or in the d-orbitals) and even steric effects. Unfortunately, there is no such theory available and one must rest content with generalizations of questionable significance. For example, metals with a low work function (W, 4.53; Ta, 4.13; Mo, 4.24; Ti, 4.16; Zr, 3.93; Fe, 4.63; Ca, 3.20; Ba, 2.52 eV) chemisorb dinitrogen. Those which have a higher work function (Ni, 4.91; Pd, 4.98; Rh, 4.65; Pt, 5.36) do not do so under comparable conditions (Trapnell, 1953). However, silver has a work function which is lower than that of tungsten, yet it still does not chemisorb dinitrogen. These generalizations also depend on how the work function is defined and measured, and this introduces uncertainties. Calcium forms a nitride rapidly under relatively mild conditions ($-78°$), so d-orbital vacancies are not a necessity. Both potassium and calcium can form body-centred cubic lattices. Potassium has a much lower work function, yet Ca_3N_2 forms and K_3N does not. This suggests that certain geometrical criteria have also to be met.

The discussion has shown how the tendency to form a nitride from dinitrogen falls on moving from left to right along a Transition Series. Nitridation is apparently paralleled by dissociative chemisorption. There is, however, no real break in behaviour at or about the iron group. The elements to the right of the iron group in the transition series show predominantly molecular adsorption, which may be either physical or chemical. It is generally not easy, nor particularly rewarding, to attempt to distinguish between the two. The chemisorbed molecules may be detected by a variety of techniques, but, most interestingly, by infrared spectroscopy.

The investigations have posed some unexpected difficulties. Thus the crystallites of metal must have a radius between 15 and 70 Å for any considerable adsorption to occur in the cases of palladium and platinum. Nickel is similar, and this condition initially led to considerable contradictions in the literature (van Hardeveld and van Montfort, 1966). The nitrogen-nitrogen stretching frequencies reported are: cobalt 2190 cm^{-1} (Borodko *et al.*, 1966), iridium 2185 cm^{-1} (Ravi *et al.*, 1968), nickel 2202 cm^{-1} (Eischens and Jacknow, 1965; Chernikov *et al.*, 1968), palladium 2260 cm^{-1} (van Hardeveld and van Montfort, 1966), platinum 2230 cm^{-1} (*idem, ibid.*). That this is indeed a nitrogen-nitrogen stretching frequency arising from a dinitrogen co-ordinated end-on to the metal surface was confirmed using $^{14}N-^{15}N$, $^{14}N-^{14}N$ and $^{15}N-^{15}N$. The $^{14}N-^{15}N$ species gives rise to two infrared bands, which can only be consistent with end-on rather than sideways co-ordination (Chernikov *et al.*, 1968). Sideways co-ordination might, of course, also occur at the same time, but

the nitrogen-nitrogen stretching frequency of a species such as $\underset{N}{\overset{N}{|||}} \rightarrow M$
would not be infrared active. The lowering of $\nu(N{\equiv}N)$ from the value found in gaseous dinitrogen (2331 cm^{-1}) parallels exactly the behaviour of co-ordinated dinitrogen in molecular dinitrogen complexes (see Chapter 3). The values of $\nu(N{\equiv}N)$ on nickel, palladium and platinum may indicate a greater tendency for π-bonding from nickel to dinitrogen than from platinum and palladium. Significantly for our understanding of the Haber synthesis, $\nu(N{\equiv}N)$ on nickel is moved some 60 cm^{-1} to higher frequencies when dihydrogen is added to the system (Eischens and Jacknow, 1965; Chernikov et al., 1968). On iridium, with $\nu(N{\equiv}N)$ lower than on nickel, dinitrogen was immediately displaced by dihydrogen, and could not be re-absorbed on top of it (Ravi et al., 1968).

These results make it clear that nitridation is a consequence of a strong metal-dinitrogen interaction. Weaker interaction (and possibly less donation of electrons from the metal to dinitrogen) can lead to molecular chemisorption (or complex formation). It is likely that most transition elements can be made to form dinitrogen complexes. A strong interaction, such as would cause nitridation on a metal surface, would be forced to stop at the complex stage on an isolated atom where there are no neighbours available to take up the second nitrogen atom of the dinitrogen molecule. The interaction would also be modified if the formal oxidation state of the metal atom were raised above zero (its value in the metal). On the other hand, inability to form nitrides is obviously no barrier to dinitrogen-complex formation (see Chapter 3). With these considerations, we now move on to discuss a mechanism for biological N_2-fixation.

2.4   A POSSIBLE MECHANISM FOR BIOLOGICAL N_2-FIXATION

Nitrides are normally formed by systems which are strong reducing agents. These systems are always very sensitive to water and dioxygen as well as to dinitrogen, and this makes it highly unlikely that the biological system involves direct nitridation of a metal if it operates under aerobic and aqueous conditions. There is strong evidence, however, that even in aerobic microbes nitrogenase functions in an intracellular environment which is anaerobic (see Chapters 4 and 5). The aqueous nature of the molecular environment has been questioned (see Chapter 4), but this may not be a serious theoretical problem because the ion $[Ru(NH_3)_5(H_2O)]^{2+}$ prefers dinitrogen to water as a ligand, even in aqueous solution. In addition, dioxygen is not able to prevent completely dinitrogen from forming some dinitrogen complex when air is passed into an aqueous solution containing $[Ru(NH_3)_5(H_2O)]^{2+}$ (Allen and Bottomley, 1968). It is reasonable to assume, therefore, that the primary step in a biological fixation involves dinitrogen-complex formation.

If, once the complex had been formed, the protein co-ordinating the metal were to change its conformation, so that the metal at the active centre becomes much more like an isolated metal ion with a low co-ordination number and a low oxidation state, then the metal would tend to nitride, but, because the dinitrogen is already co-ordinated, access to the metal atom of water and dioxygen, which would oxidize the site preferentially, is prevented. The metal would reduce the dinitrogen, and simple hydrolysis would lead to ammonia.

An attempt has been made to devise a cyclic system based upon this model. Thus, rhenium forms a series of molecular nitrido-complexes $[ReNCl_2(PR_3)_n]$, where n = 2 or 3, and PR_3 = tertiary organic phosphine (Chatt *et al.*, 1969b). A series of dinitrogen complexes $[ReCl(N_2)(PR_3)_4]$ is also known. These can be synthesized directly from molecular N_2 (Chatt *et al.*, 1969c). One can envisage a cycle

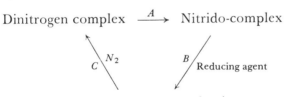

Dinitrogen complex $\xrightarrow{A}$ Nitrido-complex

Low-oxidation-state rhenium
compound + NH_3

Of these steps, both B and C have been realized. Step A has not. One possible way of achieving this (not yet tested) would be to remove one of the four phosphines from the dinitrogen complex using another acceptor, this being equivalent to the postulated change in conformation of the protein. The inner nitrogen atom would form a nitride, but one is then left with the problem of the uncoordinated nitrogen atom of the co-ordinated dinitrogen molecule. This might pick up hydrogen from water to form ammonia or it could be accepted by another metal atom. In any case, the necessity for the interaction of another entity with the dinitrogen is clear.

As is made clear in later chapters, both iron and molybdenum are believed to be involved in the biological system. Iron has not a wide-enough range of easily accessible oxidation states to be involved in nitrido-complex formation. The change from a co-ordinated dinitrogen molecule to a co-ordinated nitrido-anion requires an oxidation-state change of three units. Molybdenum is quite capable of accommodating such changes. I therefore suggest, as have others, that the dinitrogen is co-ordinated between iron and molybdenum. Whether iron, or molybdenum, or perhaps both, are involved in the initial capture of dinitrogen is not clear. At present iron is preferred, because its congener, ruthenium can complex dinitrogen in aqueous conditions. Reduction is envisaged as involving nitridation of the molybdenum, and also protonation of the nitrogen atom attached to the iron. Ammonia is not a good ligand for iron, and subsequent hydrolysis would remove both it and the nitride ligand.

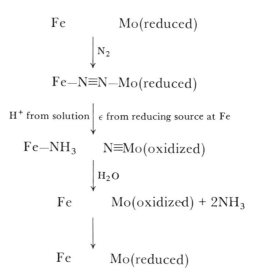

The precise sequence of events cannot be defined. This system has the advantage that it provides a specific function for both molybdenum and iron, that it suggests an asymmetric reduction of dinitrogen, which has always seemed more likely than pairs of electrons and protons, or pairs of hydride ions, jumping together on to adjacent nitrogen atoms, and that it avoids unlikely intermediates, such as di-imide or hydrazine. Most important, a test of this kind of system with appropriate model substances should become possible in the near future.

REFERENCES

E. W. ABEL, *Quart. Rev. (London)*, **17**, 133 (1963)
M. A. I. AL-JOBOURY, D. P. MAY and D. W. TURNER, *J. Chem. Soc.*, 616 (1965).
A. D. ALLEN and F. BOTTOMLEY, *Can. J. Chem.*, **46**, 469 (1968).
M. ANBAR, *J. Am. Chem. Soc.*, **88**, 5924 (1966).
K. BAHADUR and S. RANGNAYAKI, *Vijnana Parishad Anasandahan Patrika*, **6**, 63 (1963).
E. BAYER and P. SCHRETZMANN, *Structure and Bonding*, **2**, 181 (1967).
E. BAYER and V. SCHURIG, *Chem. Ber.*, **102**, 3378 (1969).
C. F. BENDER and E. R. DAVIDSON, *J. Chem. Phys.*, **46**, 3313 (1967).
J. E. BERCAW and H. BRINTZINGER, *J. Am. Chem. Soc.*, **91**, 7301 (1969).
S. G. BISHOP, P. J. RING and P. J. BRAY, *J. Chem. Phys.*, **45**, 1523 (1966).
G. C. BOND, Catalysis in Metals, Butterworth, London (1962).
Yu. G. BORODKO, A. E. SHILOV and A. A. SHTEINMAN, *Dokl. Akad. Nauk, S.S.S.R.*, **168**, 581 (1966).
Yu. G. BORODKO and A. E. KRYLOVA, *Zh. Strukt. Khim*, **8**, 246 (1967).
Yu. G. BORODKO and A. E. SHILOV, *Russ. Chem. Rev.*, **35**, 355 (1969).
G. W. BREZINA and E. GELERINTER, *J. Chem. Phys.*, **41**, 3293 (1968).
R. BRILL, P. JIRU and G. SCHULZ, *Z. Physik. Chem. (Leipzig)*, **64**, 215 (1969).
H. BRINTZINGER, *J. Am. Chem. Soc.*, **88**, 4305 (1966a).
H. BRINTZINGER, *J. Am. Chem. Soc.*, **88**, 4307 (1966b).
H. BRINTZINGER, *Biochemistry*, **5**, 3947 (1966c).
H. BRINTZINGER, *J. Am. Chem. Soc.*, **89**, 6871 (1967).
S. CARRA and R. UGO, *J. Catalysis*, **15**, 435 (1969).
J. CHATT, G. J. LEIGH and R. L. RICHARDS, *Chem. Commun.*, 515 (1969a).
J. CHATT, C. D. FALK, G. J. LEIGH and R. J. PASKE, *J. Chem. Soc. (A)*, 2288 (1969b).
J. CHATT, J. R. DILWORTH and G. J. LEIGH, *Chem. Commun.*, 687 (1969c).
Chemical Society Special Publication 11, *Interatomic Distances*, The Chemical Society, London (1958).
S. S. CHERNIKOV, S. G. KUZMIN and Yu. G. BORODKO, *Zhur. Fiz. Khim.*, **42**, 2038 (1968).
C. A. COULSON, Valence, 2nd ed. Oxford University Press (1961).
H. DAVY, Elements of Agricultural Chemistry, 5th ed. Longmans, London (1836).
W. B. DEMORE and N. DAVIDSON, *J. Am. Chem. Soc.*, **81**, 5869 (1959).
W. B. DEMORE and O. F. RAPER, *Can. J. Chem.*, **41**, 808 (1963).
N. R. DHAR, *J. Chim. Phys.*, **55**, 980 (1959).
G. C. DOBINSON, R. MASON, G. B. ROBERTSON, R. UGO, F. CONTI, R. MORELLI, S. CENINI and F. BONATI, *Chem. Commun.*, 1539 (1968).
R. P. EISCHENS and J. JACKNOW, *Proc. Intern. Congress Catalysis Amsterdam 1964*, 627 (1965).
J. ELLERMAN, F. POERSCH, R. KUNSTMANN and R. KRAMOLOWSKY, *Angew. Chem.*, **81**, 183 (1969).
F. A. ERNST, Fixation of Atmospheric Nitrogen, van Nostrand Company, New York (1928).
O. FOTIN, *Russ. Chem. Rev.*, **36**, 725 (1967).

E. GELERINTER and H. SILSBEE, *J. Chem. Phys.*, **45**, 1703 (1966).

Gmelins *Handbuch der Anorganischen Chemie*, 8th ed., System Number 4, Nitrogen, Part 1. Verlag Chemie, Weinheim/Bergstr (1934).

D. E. GOLDEN, *Phys. Rev. Letters*, **17**, 847 (1966).

D. R. GRAY and C. H. BRUBAKER, *Chem. Commun.*, 1239 (1969).

P. M. GUNDRY, J. HABER and F. C. TOMPKINS, *J. Catalysis*, **1**, 363 (1962).

G. P. HAIGHT and R. SCOTT, *J. Am. Chem. Soc.*, **86**, 743 (1964).

R. W. F. HARDY and E. KNIGHT, Progress in Phytochemistry. (L. Rheinhold, ed.), p. 387. John Wiley and Son, New York (1968).

D. E. HARRISON and H. TAUBE, *J. Am. Chem. Soc.*, **89**, 5706 (1967).

G. HENRICI-OLIVÉ and S. OLIVÉ, *Angew. Chem.*, **79**, 897, 898 (1967).

G. HENRICI-OLIVÉ and S. OLIVÉ, *Angew. Chem.*, **80**, 398 (1968).

S. HUNIG, H. R. MÜLLER and W. THIER, *Angew. Chem.*, **77**, 368 (1965).

H. H. JAFFE, *J. Chem. Educ.*, **33**, 25 (1956).

B. D. JAMES, R. K. NANHA and M. G. H. WALLBRIDGE, *Chem. Commun.*, **897** (1966).

G. G. JORIS and H. S. TAYLOR, *J. Chem. Phys.*, **7**, 893 (1939).

R. JUZA, *Advances in Inorganic and Radiochemistry*, **9**, 95 (1966).

D. KATAKIS and H. TAUBE, *J. Chem. Phys.*, **36**, 416 (1962).

A. I. KRASILSHCHIKOV, L. G. ANTONOVA, Z. M. BIRYUKOVA, I. M. KARATAEVA and T. T. FILCHENKOVA, *Russ. J. Phys. Chem.*, **37**, 102 (1963).

A. I. KRASILSHCHIKOV and L. G. ANTONOVA, *Russ. J. Phys. Chem.*, **39**, 895 (1965).

J. T. KUMMER and P. H. EMMETT, *J. Chem. Phys.*, **19**, 289 (1951).

V. N. LATYAEVA, L. I. VISHINSKAYA, V. B. SHUR, L. A. FEDEROV and M. E. VOLPIN, *Dokl. Akad. Nauk S.S.S.R.*, **179**, 875 (1968).

K. H. LINKE and H. LINKMANN, *Z. Anorg. Allgem. Chem.*, **366**, 89 (1969).

R. S. LOGAN and J. PHILP, *J. Catalysis*, **10**, 1 (1968).

R. E. MARCH and H. I. SCHIFF, *Can. J. Chem.*, **16**, 1891 (1967).

R. MASKILL and J. M. PRATT, *J. Chem. Soc.* (A), 1914 (1968).

J. W. MELLOR, Comprehensive Treatise on Inorganic and Theoretical Chemistry, Vol. 8, Supplement 1, Part 1, 1964, and Part 2, 1967. Longmans Green and Co., London.

J. C. P. MIGNOLET, *Discussions Faraday Soc.*, **8**, 105 (1950a).

J. C. P. MIGNOLET, *Discussions Faraday Soc.*, **8**, 326 (1950b).

R. S. MULLIKEN, *Can. J. Chem.*, **36**, 10 (1958).

T. NAKATA and S. MATSUSHITA, *J. Phys. Chem.*, **72**, 458 (1968).

G. N. NECHIPORENKO, G. M. TABRINA, A. K. SHILOVA and A. E. SHILOV, *Dokl. Akad. Nauk S.S.S.R.*, **164**, 977 (1965).

L. E. ORGEL, Ligand Field Theory—An Introduction to Transition Metal Chemistry, 2nd ed., p. 141. Methuen, London (1966).

D. C. OWSLEY and G. K. HELMKAMP, *J. Am. Chem. Soc.*, **89**, 4558 (1967).

C. A. PARKER, *Aust. J. Exp. Biol. Med. Sci.*, **33**, 33 (1955).

N. D. PARKYNS and B. C. PATTERSON, *Chem. Commun.*, 530 (1965).

L. PAULING, The Nature of the Chemical Bond, 3rd ed. Cornell University Press (1960).

L. PAULING, *Tetrahedron*, **17**, 229 (1962).

V. PONEC and Z. KNOR, *J. Catalysis*, **10**, 73 (1968).

A. RAVI, D. A. KING and N. SHEPPARD, *Trans. Faraday Soc.*, **64**, 3358 (1968).

M. W. ROBERTS and F. C. TOMPKINS, *Proc. Roy. Soc. (London) A*, **251**, 369 (1959).

A. SACCO and M. ROSSI, *Chem. Commun.*, 419 (1967).

W. M. H. SACHTLER and G. J. H. DORGELD, *J. Chim. Phys.*, **54**, 27 (1957).

W. M. H. SACHTLER and L. L. van REIJEN, *J. Res. Institute Catalysis*, 87 (1962).

W. SANDER, *Z. Physik*, **169**, 353 (1962).

W. A. SCHMIDT, *Angew. Chem.*, **80**, 151 (1968).

V. I. SCHVACHKO, YA. M. FOGEL and V. YA. KOLOT, *Kinetics Catalysis (U.S.S.R.)*, **7**, 734 (1966).

A. E. SHILOV, A. A. SHTEINMAN and M. B. TYABIN, *Tetrahedron Letters*, 4177 (1968).

A. E. SHILOV and A. K. SHILOVA, *Kinetika i Kataliz*, **9**, 1163 (1969).

A. E. SHILOV, A. K. SHILOVA, and E. F. KVASHINA, *Kinetika i Kataliz*, **10**, 1402 (1969).

A. J. SHUSKUS, G. C. YOUNG, O. R. GILLIAM and P. W. LEVY, *J. Chem. Phys.*, **33**, 622 (1960).

M. SUDO, M. ICHIKAWA, M. SUMO, T. ONISHI and K. TAMARU, *J. Phys. Chem.*, **73**, 1174 (1969).

K. TAMARU, *Proc. Intern. Congress Catalysis Amsterdam* **1964**, 664 (1965).

H. TORKAR and H. T. SPATH, *Monats. Chem.*, **98**, 2020 (1967).

B. M. W. TRAPNELL, *Proc. Roy. Soc. (London) A*, **218**, 563 (1953).

R. van HARDEVELD and A. van MONTFORT, *Surface Science*, **4**, 396 (1966).

E. E. van TAMELEN, R. S. DEWEY, M. F. LEASE and W. H. PIRKLE, *J. Am. Chem. Soc.*, **83**, 4302 (1961).

E. E. van TAMELEN, G. BOCHE, S. W. ELA and R. B. FECHTER, *J. Am. Chem. Soc.*, **89**, 5707 (1967).

E. E. van TAMELEN, G. BOCHE and R. GREELY, *J. Am. Chem. Soc.*, **90**, 1677 (1968).

E. E. van TAMELEN and B. AKERMARK, *J. Am. Chem. Soc.*, **90**, 4492 (1968).

E. E. van TAMELEN and D. A. SEELEY, *J. Am. Chem. Soc.*, **91**, 5194 (1969).

E. E. van TAMELEN, R. B. FECHTER, S. W. SCHNELLER, G. BOCHE, R. H. GREELY and B. AKERMARK, *J. Am. Chem. Soc.*, **91**, 1551 (1969a).

E. E. van TAMELEN, R. B. FECHTER and S. W. SCHNELLER, *J. Am. Chem. Soc.*, **91**, 7196 (1969b).

M. E. VOLPIN, V. B. SHUR and L. P. BICHIN, *Izv. Akad Nauk S.S.S.R., Otd. Khim. Nauk*, 721 (1965a).

M. E. VOLPIN, M. A. ILATOVSKAYA, E. I. LARIKOV, M. L. KHIDEKEL, Yu. A. SHVETSOV and V. B. SHUR, *Dokl. Akad. Nauk S.S.S.R.*, **164**, 331 (1965b).

M. E. VOLPIN and V. B. SHUR, *Nature*, **209**, 1236 (1966a).

M. E. VOLPIN and V. B. SHUR, *Izv. Akad Nauk S.S.S.R., Otd. Khim. Nauk*, 1873 (1966b).

M. E. VOLPIN, V. B. SHUR, V. N. LATYAEVA, L. I. VISHINSKAYA and L. A. SHULGAITSER, *Izv. Akad. Nauk S.S.S.R., Otd. Khim. Nauk*, 385 (1966).

M. E. VOLPIN, M. A. ILATOVSKAYA, L. V. KOSYAKOVA and V. B. SHUR, *Chem. Commun.*, 1074 (1968a).

M. E. VOLPIN, M. A. ILATOVSKAYA, L. V. KOSYAKOVA and V. B. SHUR, *Dokl. Akad. Nauk S.S.S.R.*, **180**, 103 (1968b).

M. E. VOLPIN, V. B. SHUR, R. V. KUDRYAVTSEV and L. A. PRODAYKO, *Chem. Commun.*, 1038 (1968c).

M. E. VOLPIN, A. A. BELII and N. A. KATKOV, *Izv. Akad. Nauk S.S.S.R., Otd. Khim. Nauk*, 2858 (1969).

D. W. WYLIE, A. J. SHUSKUS, G. C. YOUNG, O. R. GILLIAM and P. W. LEVY, *Phys. Rev.*, **125**, 451 (1962).

A. YAMAMOTO, M. OKAWA and S. IKEDA, *Chem. Commun.*, 841 (1969).

K. B. YATSIMIRSKII and V. K. PAVLOVA, *Dokl. Akad. Nauk S.S.S.R.*, **165**, 130 (1966).

CHAPTER 3

Dinitrogen Complexes and Nitrogen Fixation

J. CHATT and R. L. RICHARDS

A.R.C. Unit of Nitrogen Fixation, University of Sussex,
Falmer, Sussex, BN1 9QJ, England

3.1 INTRODUCTION

3.1.1 *History and relevance*

The formation of dinitrogen complexes from nitrogen gas is the only
definite reaction of molecular nitrogen which is known to occur at

ordinary temperature and pressure with reagents sufficiently mild to exist in water and other protic media. This reaction thus provides at present the only reasonable chemical model for the initial uptake of dinitrogen by nitrogenase; hence its relevance, and the relevance of dinitrogen complexes to our subject.

Dinitrogen complexes are substances with dinitrogen directly attached as a ligand to a metal. The first was discovered by Allen and Senoff (December, 1965). It is a ruthenium complex, $[Ru(NH_3)_5(N_2)]^{2+}$, similar to the hexaamine, $[Ru(NH_3)_6]^{2+}$, but with one molecule of ammonia replaced by dinitrogen. Already (December, 1970) mononuclear dinitrogen complexes derived from ten of the fifteen metallic elements of Groups VI, VII and VIII are known, and dinuclear complexes promise to involve acceptor atoms from an even wider area of the Periodic Table. Bortels's (1930) discovery that molybdenum is essential for the metabolism of dinitrogen by bacteria, but not essential for their growth on ammoniacal nitrogen, set chemists thinking in terms of dinitrogen complexes of molybdenum even in the 1930's. Sporadic attempts to obtain them followed, especially by compressing dinitrogen onto molybdenum compounds, and later, because it has a formal resemblance to acetylene, $HC\equiv CH$, onto metal complexes known to form π-complexes with olefins and acetylenes, such as complexes of platinum(II). These attempts, often unpublished, gave no hint of the possible existence of dinitrogen complexes. During the 1940's, and 1950's the development of the theories of bonding of carbon monoxide, (Pauling, 1939) and of unsaturated hydrocarbons (Dewar, 1951; Chatt and Duncanson, 1953) in transition metal complexes, and the measurement of the molecular-orbital energy-levels in the dinitrogen molecule (see Mulliken, 1959), provided good reason to suppose that dinitrogen complexes might not exist as stable entities. This was the position about 1960. It is remarkable therefore that Allen and Senoff's (1965) accidental discovery of $[Ru(NH_3)_5(N_2)]^{2+}$ by the easy reaction of hydrazine with ruthenium trichloride in aqueous solution at 25° should have been followed within little more than a year by a series of similar accidental discoveries of complexes formulated trans-$[IrCl(N_2)(PPh_3)_2]$ (Collman and Kang, 1966), $[Co(N_2)(PPh_3)_3]$ (Yamamoto et al., 1967(a)) and $[CoH(N_2)(PPh_3)_3]$ (Sacco and Rossi, 1967; Misono et al., 1967), providing three new preparative methods for dinitrogen complexes. At about the same time Shilov, Shilova and Borod'ko (1966) showed that the reduction of ruthenium trichloride in tetrahydrofuran under dinitrogen by zinc amalgam yielded a tetrahydrofuran complex containing dinitrogen attached to ruthenium. This was the first evidence that dinitrogen complexes could be obtained directly from dinitrogen, although both of the above cobalt compounds were obtained in good yield from dinitrogen within a year of Shilov's discovery.

The discovery of new types of dinitrogen complexes has been largely a matter of chance and even now a greater number of stable dinitrogen complexes has been obtained by accident than by design. Further, the knowledge that a dinitrogen complex exists does not necessarily lead to the production of analogues and homologues, because a very slight change in the composition or structures of the co-ligands with dinitrogen is often sufficient to render the expected analogue or homologue unstable. The complex $trans$-$[IrCl(N_2)$-$(PPh_3)_2]$ provides a good example. It is readily obtained by reaction (3.1) in reagent grade chloroform at $0°$. The carbonyl complexes of type (I), available with a great variety of tertiary phosphines, show

$$[IrCl(CO)(PR_3)_2] + PhCON_3 \rightarrow [IrCl(N_2)(PR_3)_2] + PhCONCO \quad (3.1)$$
$$\text{(I)} \qquad\qquad\qquad\qquad \text{(II)}$$

no sign of instability depending upon the phosphine. Nevertheless the triphenylphosphine dinitrogen complex (II, R = Ph) is the only dinitrogen complex of this series to have been isolated pure, even tri-p-tolylphosphine (R = $p$tolyl) and methyldiphenylphosphine (R = MePh$_2$) do not give stable analogues of (II) by the above reaction (Chatt, Melville and Richards, 1969). Thus the discovery of a stable dinitrogen complex often leads to only one or two specific examples, rather than provides a series of complexes, and only chance could have thrown up the first examples.

Initially it appeared that some of the dinitrogen in $[Ru(NH_3)_5$-$(N_2)]^{2+}$ was reduced to ammonia by aqueous sodium borohydride (Allen and Senoff, 1965; Allen et $al.$, 1967). Thus an aqueous reduction of dinitrogen to ammonia depended only on the preparation of $[Ru(NH_3)_5(N_2)]^{2+}$ in good yield from dinitrogen in aqueous solution. Harrison and Taube (1967) provided this link, when they showed that dinitrogen displaces water from the aqua-ion, $[Ru(NH_3)_5(H_2O)]^{2+}$ (III, reaction 3.2), which is produced by the

$$2[Ru(NH_3)_5(H_2O)]^{2+} + N_2 \rightarrow$$
$$\text{(III)}$$

$$[(NH_3)_5Ru \cdots N\equiv N \cdots Ru(NH_3)_5]^{4+} + 2H_2O \quad (3.2)$$
$$\text{(IV)}$$

reduction of $[Ru(NH_3)_5Cl]^{2+}$ with amalgamated zinc in 0.1 M aqueous sulphuric acid. The immediate product is a dinuclear complex (IV) (Harrison, Weissberger and Taube, 1968), the first complex with bridging dinitrogen to be characterized, and this is split by ammonia (reaction 3.3).

$$[(NH_3)_5Ru \cdots N{\equiv}N \cdots Ru(NH_3)_5]^{2+} + NH_3 \rightarrow$$
$$\text{(IV)}$$

$$[Ru(NH_3)_6]^{2+} + [Ru(NH_3)_5(N_2)]^{2+} \quad (3.3)$$
$$\text{(V)}$$

Although it was later shown that dinitrogen in neither the compound (IV) nor (V) can be reduced to ammonia (Borod'ko, Shilova and Shilov, 1970; Chatt *et al.*, 1968, 1970) reaction (3.2) is very important because it shows that dinitrogen can displace water from a metal site typified by the sixth coordination position on the moiety $Ru(NH_3)_5{}^{2+}$. Reaction (3.2) is the only known reaction of its type, and demonstrates that in the nitrogenase system there could exist a metal site where dinitrogen could compete successfully with water, contrary to all previous experience in complex chemistry.

Reactions (3.4) and (3.5) are also important in the history of dinitrogen complexes. The first (Sacco and Rossi, 1967) provides a model for the competitive inhibition of nitrogenase activity by dihydrogen, both dinitrogen and dihydrogen competing for the same site. The second (Yamamoto *et al.*, 1967(b)) demonstrates that ammonia does not necessarily have a strong affinity for some dinitrogen-holding sites, an essential condition for the final elimination of ammonia if dinitrogen were reduced on a metal site.

$$[CoH_3(PPh_3)_3] + N_2 \rightleftharpoons [CoH(N_2)(PPh_3)_3] + H_2 \quad (3.4)$$

$$[CoH(NH_3)(PPh_3)_3] + N_2 \rightleftharpoons [CoH(N_2)(PPh_3)_3] + NH_3 \quad (3.5)$$

The type of metal site which binds dinitrogen is one which all coordination chemists will recognize as electronically suited for strong binding to carbon monoxide, acetylene, organic isonitriles and all those 'nitrogen analogues' which inhibit the action of the nitrogenase system or are reduced by it (see Chapter 4). Thus the dinitrogen complexes collectively provide models for most of the difficult chemical steps in the supposed action of nitrogenase, except the highly important one of reduction. So far no dinitrogen in any complex has been reduced to ammonia in water or other protic medium, nor in an aprotic medium, by reducing agents sufficiently mild to exist in aqueous solution.

The mononuclear dinitrogen complexes, with very few exceptions, are diamagnetic 'closed electron shell' complexes, all are characterized by a strong sharp band or bands in the 1900 to 2220 cm^{-1} region of their infrared spectra. This is caused by the N—N stretching mode of vibration $[\nu(N_2)]$ which absorbs at 2331 cm^{-1} in the raman spectrum of the gas, and the difference between 2331 and the value in the complex is a crude measure of the weakening of the N—N bond when dinitrogen enters the complex. The high intensity of infrared absorption, and the fact that the nitrogen atoms in at least one dinitrogen complex are resolved in its X-ray induced emission spectrum indicates that the dinitrogen molecule has become very electrically asymmetrical in the mononuclear complexes (see 3.2.6). Nevertheless the asymmetry has not activated the dinitrogen to reduction, but Chatt, Dilworth, Richards and Sanders (1969) have found that 'closed shell' dinitrogen complexes of very low $\nu(N_2)$, particularly $[ReCl(N_2)(PMe_2Ph)_4]$, react with 'open shell', electron-acceptor, transition-metal complexes, e.g. $[MoCl_4(Et_2O)_2]$, apparently by addition of an acceptor-containing fragment to the free end of the complexed dinitrogen, to cause a great lowering of $\nu(N_2)$, even to 1680 cm^{-1} in one example. Since the N—N double bond in azobenzene absorbs in the raman spectrum at 1510 cm^{-1} it is evident that in these asymmetrical dinuclear complexes the N—N strength is approaching that of a double bond. In nitrogenase, it may be that iron(II) plays the closed shell role of rhenium(I) and molybdenum that of $[MoCl_4(Et_2O)_2]$ in the above adducts (see 3.3.6). The function of the two kinds of metal ions, iron and molybdenum, could then be to weaken the N—N bond as in the above-mentioned adducts, but to such an extent that it reduces easily, or even falls apart to leave nitride ligands which are then reduced to ammonia, on each metal atom. It is too early to assess definitely the relevance, if any, of these latest binuclear dinitrogen complexes to the nitrogenase problem.

There have been many claims over a very long period that dinitrogen is reduced to ammonia at ordinary temperature in an aqueous medium by a variety of reducing agents or by hydrogen in the presence of catalysts, especially molybdenum salts, platinum black, or colloidal noble metals (e.g. Prideaux and Lambourne, 1928; Glemser, 1954). These have never been substantiated but they are difficult to confirm or refute without recourse to ^{15}N labelling which became available only comparatively recently; such claims still continue (e.g. see Haight and Scott, 1964). Two of the most recent do however appear to have substance and may involve reduction of dinitrogen complexes. Schrauzer (1970) claims that some ammonia is produced by the passage of dinitrogen through an aqueous solution containing molybdenum (e.g. Na_2MoO_4) and a thiol (e.g. 1-thioglycerol), in equimolecular proportions and a reducing agent (e.g. $Na_2S_2O_4$); this mixture is based on our scanty knowledge of the chemical nature of the biological system. Such mixtures also reduce acetylene to ethylene and simulate the reduction of alkyl isocyanides and azide ion by the biological system. Denisov, et al. (1970) claim that dinitrogen is reduced to hydrazine by a large excess of alkaline titanous chloride in the presence of a solution obtained from molybdenum(V) trichloride oxide or vanadium(III) chloride oxide and magnesium chloride in aqueous solution of pH just greater than 10.5. Our colleague Dr. R. Hill (personal communication) has recently reproduced these results, obtaining only small amounts of ammonia, enhanced by addition of ferric chloride to Schrauzer's mixture, and of hydrazine as described by Denisov et al., respectively, but no confirmation using $^{15}N_2$ has been reported. Small amounts of reduction to ammonia without such confirmation are always suspect because ammonia itself, and ammonium salts, nitrogen oxides, and nitrogen oxysalts, are common impurities, from which ammonia can be derived. Hydrazine and its possible precursors are not common impurities, and the reported production of hydrazine from dinitrogen is probably correct.

Evidence is just becoming available that dinitrogen may form complexes with titanium(III) chloride in dimethylsulphoxide or in propylene carbonate (Franklin and Byrd, 1970). This d^1 system is completely different from any transition metal system which is known to give stable dinitrogen complexes. The metal has an open electron shell structure, and the bonding of the dinitrogen to the metal must be different from that in the previously known complexes, if only because there are fewer electrons to fill bonding orbitals. Such complexes could be important if the reduction of dinitrogen to ammonia occurs through a nitriding process. Open electron shell complexes would be ideal intermediates provided the metal can easily undergo a change of three in its oxidation number. Molybdenum is one of the few metals likely to do this (see 3.2.9 and 3.3.6).

3.1.2 Purpose

The purpose of this chapter is not to present a detailed review of recent developments in the field of dinitrogen complex chemistry. It is too recent, and developing so rapidly, that any review would be out of date before it could be printed. Rather, the aim is to present coherently that part of the chemistry which seems reasonably well-established, but unusually extensive revision may still be necessary because much is taken from preliminary communications. A detailed list of references cannot be provided as it would be too long, but recent references providing a lead into the literature are given, and imply no priority. For the detailed chemistry and lists of references, the reader is directed to reviews. The first, a lecture review by Chatt (1969), concerns bonding in dinitrogen and its complexes and is directed to biologists. An account of early history is given by Allen and Bottomley (1968) and a very useful comprehensive review which contains 172 references on the activation of the nitrogen molecule from the chemical standpoint is given by Murray and Smith (1968). Most recently we have an excellent short review (122 references) specifically on dinitrogen complexes by Borod'ko and Shilov (1969) and another lecture review by Chatt (1971) given at the XIIth Conference on Coordination Chemistry, Sydney, 1969; also a short review by Fergusson and Love (1970).

3.1.3 Nomenclature

The ligand N_2 has been given various names such as 'nitrogeno', 'nitrogenyl' and 'nitrogen', but according to the International Union of Pure and Applied Chemistry (Nomenclature of Inorganic Chemistry, 1957) the molecule N_2 is named dinitrogen and this name should be used without change when the N_2 molecule occurs in a complex compound as a formally unchanged ligand (rule 7.32). This nomenclature has been adopted.

3.1.4 Bonding possibilities of dinitrogen

Dinitrogen could conceivably combine with metal atoms as shown by formulae (VI) to (XI). These are: end-on, as an analogue of carbon monoxide (VI); side-on, as an analogue of acetylene (VII); bridging through one nitrogen atom (VIII); bridging, one nitrogen atom to each metal (IX) or (X); or bridging through both nitrogen atoms (XI).

The types (VI), (VIII) and (IX) are characteristic of carbon monoxide in its complexes, although type (IX) is rare. The types (VII) and (X) and (XI) are characteristic of acetylenes, both (X) and (XI) being rare. So far only types (VI) and (IX) have been positively identified in dinitrogen complexes. Others such as (VII) (e.g. Johnson and Beveridge, 1967) and (XI) (e.g. Brintzinger, 1966) have

$$M\cdots\cdots N\equiv N$$

(VI)

$$M\cdots\cdots\overset{\displaystyle N}{\underset{\displaystyle N}{\|\|}}$$

(VII)

$$\overset{\displaystyle N^-}{\underset{M\diagup\quad\diagdown M'}{\overset{\|}{N^+}}}$$

(VIII)

$$M\cdots N\equiv N\cdots M'$$

(IX)

$$\overset{\textstyle M}{\underset{\textstyle M'}{\diagdown}}\ \overset{N}{\underset{N}{\|}}\diagup$$

cis

$$\overset{\textstyle M}{\diagdown}\ \overset{N}{\underset{N}{\|}}\ \underset{\textstyle M'}{\diagdown}$$

trans

(X)

$$M\vdots\cdots\overset{\displaystyle\cdot\cdot N}{\diagup}\cdots\vdots M'$$

(XI)

been postulated but never substantiated, and a protonated form of (VIII) has been found in the platinum(II) complex (XII) obtained by the reaction of cis-$[PtCl_2(PPh_3)_2]$ with hydrazine (Dobinson et al., 1967).

It should be noted that N=NH is an analogue of nitrogen oxide, but whereas a great variety of mononuclear nitrosyl complexes is known, (XII) is the only known complex of N=NH. Nevertheless N=NH as a ligand is interesting because it is a probable intermediate in the reduction of dinitrogen complexes to ammonia in protic media, if such reduction ever occurs.

$$\left[\ \begin{matrix} & & \overset{\textstyle H}{\underset{}{N}} & & \\ & & \| & & \\ \text{Ph}_3\text{P} & & N & & \text{PPh}_3 \\ \diagdown & \text{Pt} & & \text{Pt} & \diagup \\ \text{Ph}_3\text{P}\diagup & & N & & \diagdown\text{PPh}_3 \\ & & \| & & \\ & & N & & \\ & & H & & \end{matrix}\ \right]^{2+}$$

(XII)

3.1.5 Classes of dinitrogen complexes

Only two major classes of compound having dinitrogen directly attached to a metal atom or metal atoms have been established.

These correspond to the bonding possibilities, terminal (VI) and bridging (IX). The terminally-bonded dinitrogen class was discovered first, and its chemistry is by far the more developed. The chemistry of the bridging-dinitrogen class is only just beginning but promises to be the more versatile All members of the second class must be di- or poly-nuclear. All known members of the first class are mono-nuclear.

Related to the terminal-dinitrogen complexes are the well-known series of substances, dinitrogen oxide, N_2O, hydrazoic acid, N_2NH, and diazomethane, N_2CH_2, in which the N_2 molecule is attached at one end by a multiple bond to a non-metal, and also the unstable diazonium ions, e.g. $C_6H_5N_2^+$, the inner diazonium salts, e.g. $^-O_3SC_6H_4N_2^+$, and substances $1,10\text{-}B_{10}H_8(N_2)_2$ and related species, which are essentially inner diazonium salts being derived directly by diazotization, e.g. of $1,10\text{-}B_{10}H_8(NH_2)_2^{2+}$ (Knoth, 1966). All these non-metal dinitrogen derivatives show an absorption band or bands in their infrared spectra in the region of the $N{\equiv}N$ stretching vibration in the metal complexes. They are related to the terminal-dinitrogen complexes of the metals but have not yet been prepared from dinitrogen itself under mild reaction conditions, although $CH_2{}^{14}N_2$ photolytically or at its decomposition temperature undergoes exchange with $^{15}N_2$ (Borod'ko, Shilov and Shteinman, 1966). These non-metal diazo-compounds will not be considered further in this chapter, nor will such substances as azobenzene $Ph{-}N = N{-}Ph$, which are only superficially related to bridging-dinitrogen complexes (see 3.3).

3.2   TERMINAL-DINITROGEN COMPLEXES

3.2.1   Occurrence

Terminal-dinitrogen complexes are formed only by transition metals in low oxidation states, often with hydrogen as a co-ligand. This means that they are all reduced complexes. The transition metals are set out in Table 3.1, and those which have yielded reasonably

TABLE 3.1   Transition metals. Mononuclear dinitrogen complexes have been obtained from those in italics.

Sc	Ti	V	Cr	Mn	*Fe*	*Co*	*Ni*	Cu
Y	Zr	Nb	*Mo*	Tc	*Ru*	*Rh*	Pd	Ag
La	Hf	Ta	*W*	*Re*	*Os*	*Ir*	Pt	Au

characterized dinitrogen complexes are shown in italics. They lie about a line joining tungsten to nickel.

3.2.2   Co-ligands with dinitrogen

The stabilities of dinitrogen complexes are very dependent on the co-ligands, of which only four types are commonly found, organic phosphines, hydride, chloride, and ammonia. The most common are tertiary organic phosphines, usually triphenylphosphine or the lower alkaryl-phosphines, often together with hydride ion in the complexes of iron and cobalt, or with one or two chloride ions in the complexes of the metals of the third period. Ruthenium(II) and osmium(II) are still unique in forming very stable, well-characterized pentaammine dinitrogen complexes and a few related substances. All others contain tertiary phosphines and, because nitrogenase is rich in sulphur protein, it is worth noting that no complex having sulphur as coordinating atom together with dinitrogen has yet been isolated.

3.2.3   Types of terminal-dinitrogen complexes

These are presented in Table 3.2, listed in the order of the metal atoms in the Periodic Table. Many types are represented by only the one example listed; homologues and related series are somewhat rare. Even some of the examples listed are not completely characterized, but all are crystalline. Oily products showing 'dinitrogen bands' in their infrared spectra ('ghosts', see p. 79) are ignored. Table 3.2 also provides one recent reference serving as a lead into the literature on each type of complex, concerning particularly the methods of preparation and the reactions, which are discussed in 3.2.4 and 3.2.5.

3.2.4   Preparation of dinitrogen complexes

There are three methods of putting dinitrogen onto a metal, but none is completely general. They are, in order of decreasing generality:—

(a) by direct reaction of dinitrogen with a suitable complex of the metal,

(b) by degradation of a nitrogen chain ligand, e.g. N_2H_4 or N_3^-,

(c) by building up dinitrogen from a mononitrogen ligand such as NH_3 by reaction with a nitrogenous reagent.

(a) *Direct method.* An appropriate complex of the metal is reduced in the presence of an excess of the co-ligand and of dinitrogen as in the following typical examples, listed (i)-(iii) according to the type of reducing agent. Occasionally the reduced complex may be isolated and treated directly with dinitrogen [e.g. reactions under (ii) and reaction 3.18] (for references see Table 3.2).

(i) *Metal, metal amalgam, e.g. Mg, Na/Hg or amalgamated Zn.* (THF = tetrahydrofuran).

$$[WCl_4(PMe_2Ph)_2] + Na/Hg \xrightarrow[THF]{N_2} [W(N_2)_2(PMe_2Ph)_4] \qquad (3.6)$$

$$[OsCl_3(PR_3)_3] + Hg/Zn \xrightarrow[THF]{N_2} [OsCl_2(N_2)(PR_3)_3] \qquad (3.7)$$

$$[Ru(NH_3)_5Cl]^{2+} + Hg/Zn \xrightarrow[0.1 \ M \ H_2SO_4]{N_2, \ 90 \ atm.} [Ru(NH_3)_5(N_2)]^{2+} \qquad (3.8)$$

These reactions may all involve intermediates directly related to the final products. Reaction 3.8 is known to go through $[Ru(NH_3)_5-(H_2O)]^+$, from which the aqua-ligand is readily replaced by dinitrogen. This is the only formation reaction to have received kinetic study (Elson, Itzkovitch and Page, 1970). Reaction 3.7 often shows evidence of a green intermediate, possibly $[OsCl_2(THF)(PR_3)_3]$, but it is not always obvious, as when $PR_3 = PMe_2Ph$.

(ii) *Complex metal hydride.* These reducing agents usually involve unstable hydride intermediates, many of which have been isolated. The dinitrogen then reacts by simple addition or with displacement of dihydrogen, and occasionally by other replacements; as in the following three pairs of equations.

Reactions such as 3.12 involving dihydrogen and dinitrogen are usually reversible.

$$FeCl_2 \cdot 2H_2O + NaBH_4 \xrightarrow[PEtPh_2]{EtOH} [FeH_2(PEtPh_2)_3] \qquad (3.9)$$

$$[FeH_2(PEtPh_2)_3] + N_2 \rightarrow [FeH_2(N_2)(PEtPh_2)_3] \qquad (3.10)$$

$$[CoCl_2(PPh_3)_2] + PPh_3 + NaBH_4 \xrightarrow[EtOH]{N_2} [CoH_3(PPh_3)_3] \qquad (3.11)$$

$$[CoH_3(PPh_3)_3] + N_2 \xrightarrow{EtOH} [CoH(N_2)(PPh_3)_3] + H_2 \qquad (3.12)$$

$$[FeCl_2(depe)_2] + LiAlH_4 \rightarrow [FeHCl(depe)_2] \qquad (3.13)$$

$$[FeHCl(depe)_2] + NaBPh_4 + N_2 \xrightarrow{Me_2CO}$$

$$[FeH(N_2)(depe)_2]BPh_4 + NaCl \qquad (3.14)$$

$$(depe = Et_2PCH_2CH_2PEt_2)$$

TABLE 3.2 Terminal-dinitrogen complexes

	Oxidation number	Colour	Configuration	$\nu(N\equiv N)$	Preparation*	Reference
Group VI (No Cr)						
trans-$[Mo(N_2)_2(P-P)_2]$	0	orange-yellow	d^6 oct.	2220, 1970	a (iii)	1
$[Mo(N_2)(toluene)(PPh_3)_2]$	0	orange	d^6	2005	a (iii)	1
cis-$[W(N_2)_2(PMe_2Ph)_4]$	0	yellow	d^6 oct.	1931^e, 1998^e	a (i)	2
trans-$[W(N_2)_2(P-P)_2]$	0	orange	d^6 oct.	1953^e	a (i)	2
Group VII (No Mn or Tc)						
trans-$[ReCl(N_2)(PMe_2Ph)_4]$	I	pale-yellow	d^6 oct.	1925^d	a (i), b (i)	3
trans-$[ReCl(N_2)(P-P)_2]$**	I	pale-yellow	d^6 oct.	1980^d	b (i)	3
$[ReCl(CO)_2(N_2)(PPh_3)_2]$	I	yellow	d^6 oct.	2060^{e}†	b (i)	4
trans-$[ReCl(N_2)(P-P)_2]^+$	II	purple and green isomers	d^5 oct.	2060^{c}‡		5
Group VIIIa						
$[FeH_2(N_2)(PEtPh_2)_2]$	II		d^6 bipy.	1989^n	a (ii)	6
$[FeH_2(N_2)(PEtPh_2)_3]$	II	yellow	d^6 oct.	ca 2057^n	a (ii)	6
$[FeH(C_6H_4PEtPh)(N_2)(PEtPh_2)_2]$	II		d^6		∅	6
trans-$[FeH(N_2)(depe)_2]BPh_4$	II	orange	d^6 oct.	2090	a (ii)	7
$[RuH_2(N_2)(PPh_3)_3]$	II	white	d^6 oct.	2147^n	a (iii)	8
$[RuCl(N_2)(das)_2]^+$	II	white	d^6 oct.	2130	b (ii)	9
$[Ru(NH_3)_5(N_2)]Br_2$	II	pale-yellow	d^6 oct.	2114^n	b (i), b (ii)	10
cis-$[Ru(en)_2(H_2O)(N_2)][BPh_4]_2$	II	pale-yellow	d^6 oct.	2130^n	b (ii)	11

Complex	Oxidation state	Color	d^n / geometry	$\nu(N{\equiv}N)$	Prep.	Ref.
cis-[Ru(en)$_2$(N$_2$)$_2$]$^{2+}$	II	pale-yellow	d^6 oct.	2220^n, 2190^n	b (ii)	11
[RuCl$_2$(N$_2$)(H$_2$O)$_2$(THF)]**	II	white	d^6 oct.	2153^n	a (i)	12
OsH$_2$(N$_2$)(PEtPh$_2$)$_3$	II	white	d^6 oct.	2085^n	b (ii)	13
mer-[OsCl$_2$(N$_2$)(PMe$_2$Ph)$_3$]**	II	white	d^6 oct.	2082^e	a (i)	13
mer-[OsHCl(N$_2$)(PMe$_2$Ph)$_3$]**	II	white	d^6 oct.	2057^e	a (ii)	13
[Os(NH$_3$)$_5$(N$_2$)]Br$_2$	II	pale-yellow	d^6 oct.	2028^n	b (i)	14
cis-[Os(NH$_3$)$_4$(N$_2$)$_2$]Cl$_2$	II	pale-yellow	d^6 oct.	2120, 2175	c	15
Group VIIIb						
[Co(N$_2$)(PPh$_3$)$_3$]	0	red	d^9	2093^t	a (iii)	16
[CoH(N$_2$)(PPh$_3$)$_3$]**	I	red-orange	d^8 dipy.	2090^e	a (iii)	17
[RhCl(N$_2$)(PPh$_3$)$_2$]	I	light-yellow	d^8 plan.	2152^n	b (ii)	18
[IrCl(N$_2$)(PPh$_3$)$_2$]	I	yellow	d^8 plan.	2105^d	b (ii)	19
[IrCl(N$_2$)(dmm)(PPh$_3$)$_2$]	III?	pale-yellow	d^6?	2190^e	b (ii)	19
Group VIIIc (No Pd or Pt)						
[NiH(N$_2$)(PEt$_3$)$_2$]	I	orange yellow	d^9	2076^e	a (iii)	20

* Preparation, for methods a (iii), etc., see 3.2.4.

** A series of homologues or related complexes of that particular element has been isolated.

† Doubtless coupled with $\nu(C{\equiv}O)$.   ‡ By mild oxidation of *trans*-[ReCl(N$_2$)(P–P)$_2$].

∅ By loss of H$_2$ from [ReH$_2$(N$_2$)(PEtPh$_2$)$_3$] in sunlight.

References to Table

d in CHCl$_3$ *e* in benzene   *n* Nujol mull   *t* in tetrahydrofuran

P–P = Ph$_2$PCH$_2$CH$_2$PPh$_2$ depe = Et$_2$PCH$_2$CH$_2$PEt$_2$ dmm = dimethylmaleate.

das = o-phenylene bis(dimethylarsine)

TABLE 3.2 (*continued*)

1. M. HIDAI, K. TOMINARI, Y. UCHIDA and A. MISONO, *Chem. Comm.*, 1392 (1969).
2. B. BELL, J. CHATT and G. J. LEIGH, *Chem. Comm.*, 842 (1970).
3. J. CHATT, J. R. DILWORTH and G. J. LEIGH, *Chem. Comm.*, 687 (1969).
4. J. CHATT, J. R. DILWORTH and G. J. LEIGH, *J. Organometallic Chem.*, 21, P49 (1970).
5. J. CHATT, J. R. DILWORTH, H. P. GUNZ, G. J. LEIGH and J. R. SANDERS, *Chem. Comm.*, 90 (1970).
6. A. SACCO and M. ARESTA, *Chem. Comm.*, 1223 (1968).
7. G. M. BANCROFT, M. J. MAYS and B. E. PRATER, *Chem. Comm.*, 589 (1969).
8. W. H. KNOTH, *J. Amer. Chem. Soc.*, 90, 7172 (1968).
9. P. G. DOUGLAS, R. D. FELTHAM and H. G. METZGER, *Chem. Comm.*, 889 (1970).
10. A. D. ALLEN, F. BOTTOMLEY, R. O. HARRIS, V. P. REINSALU and G. V. SENOFF, *J. Amer. Chem. Soc.*, 89, 5595 (1967).
11. L. A. P. KANE-MAGUIRE, P. F. SHERIDAN, F. BASOLO and R. G. PEARSON, *J. Amer. Chem. Soc.*, 90, 5295 (1968).
12. Y. G. BOROD'KO, A. K. SHILOVA and A. E. SHILOV, *Doklady Akad. Nauk S.S.S.R.*, 176, 1297 (1967).
13. B. BELL, J. CHATT and G. J. LEIGH, *Chem. Comm.*, 576 (1970); J. CHATT, G. J. LEIGH and R. L. RICHARDS, *J. Chem. Soc.(A)*, 2243 (1970); J. CHATT, D. P. MELVILLE and R. L. RICHARDS, *J. Chem. Soc.(A)*, 895 (1971).
14. A. D. ALLEN and J. R. STEVENS, *Chem. Comm.*, 1147 (1967).
15. H. A. SCHEIDEGGER, J. N. ARMOR and H. TAUBE, *J. Amer. Chem. Soc.*, 90, 3263 (1968).
16. G. SPEIER and L. MARKO, *Inorg. Chim. Acta*, 3, 126 (1969).
17. A. SACCO and M. ROSSI, *Chem. Comm.*, 471 (1969).
18. L. Y. UKHIN, A. Y. SHVETSOV and M. L. KHIDEKEL, *Invest. Akad. Nauk S.S.S.R., Ser. Khim.*, 957 (1967).
19. J. P. COLLMAN, M. KUBOTA, F. D. VASTINE, J. Y. SUN and J. W. KANG, *J. Amer. Chem. Soc.*, 90, 5430 (1968).
20. S. C. SRIVASTAVA and M. BIGORGNE, *J. Organometallic Chem.*, 18, P30 (1969).

(iii) *Metal alkyls.* Those most used to obtain dinitrogen complexes are diethylaluminium ethoxide, triethylaluminium, diisobutyl-aluminium hydride and triisobutylaluminium; trimethylaluminium tends to give methyl complexes. The 2,4-pentanedionate (acetyl-acetonate, acac.) of the metal, or other suitable metal complex, is usually suspended or dissolved in ether or toluene, and the alkyl aluminium added dropwise under dinitrogen with stirring or with a good stream of dinitrogen passing through the mixture.

$$[Co(acac)_2] + 3PPh_3 + Et_2AlOEt \xrightarrow[\text{toluene}]{N_2}$$

$$[Co(N_2)(PPh_3)_3] + [CoH(N_2)(PPh_3)_3] \qquad (3.15)$$

$$[Mo(acac)_3] + 2Ph_2PCH_2CH_2PPh_2 + AlBu^i_3 \xrightarrow[\text{toluene}]{N_2}$$

$$trans\text{-}[Mo(N_2)_2(Ph_2PCH_2CH_2PPh_2)_2] \qquad (3.16)$$

$$[RuHCl(PPh_3)_3] + AlEt_3 \xrightarrow[\text{ether}]{N_2} [RuH_2(N_2)(PPh_3)_3] \qquad (3.17)$$

Reactions (3.15)-(3.17) probably proceed through a hydride intermediate produced from the transition metal (M) alkyl with elimination of olefin (e.g. $M-CH_2-CH_2R \rightarrow M-H + CH_2 = CHR$). Organic groups may also be replaceable by dinitrogen as in the few cases where trimethyl aluminium has been used successfully, and in the series of reactions (overall as 3.18) involving a hydrocarbon complex which was isolated. This reaction is particularly useful for the preparation of alkarylphosphine complexes of the type $[CoH(N_2)(PR_3)_3]$ (PR_3 = alkarylphosphine).

$$[Co(C_8H_{13})(C_8H_{12})] + PR_3 \xrightarrow{N_2}$$

$$[CoH(N_2)(PR_3)_3] + 2C_8H_{12} \qquad (3.18)$$

(C_8H_{13} = cyclo-octenyl, C_8H_{12} = 1,5 cyclo-octadiene)

(b) *From a nitrogen chain.* This method has limited application, but is excellent when it works. The first dinitrogen complex was obtained during the catalytic degradation of hydrazine to dinitrogen, dihydrogen and ammonia by aqueous ruthenium trichloride, and the second by degradation of organic acid azides on an iridium(I) carbonyl complex, i.e. degradation of N_2 and N_3 compounds respectively (reactions 3.19 and 3.22). The degradation may be spontaneous (reactions 3.19 and 3.22), require an external reagent (reactions 3.21 and 3.24) or involve other ligands (reaction 3.20).

(i) N_2 chains.

$$RuCl_3 + N_2H_4 \xrightarrow[H_2O]{25°} [Ru(NH_3)_5(N_2)]^{2+} \qquad (3.19)$$

The osmium analogue is prepared similarly from $[OsCl_6]^{2-}$

$$[(PPh_3)_2Cl_2\overset{\overset{\displaystyle O}{|\quad\quad|}}{Re-N=N-C}-Ph] + 4PR_3 \xrightarrow[reflux]{MeOH}$$

$$trans\text{-}[ReCl(N_2)(PR_3)_4] + HCl + 2PPh_3 + C_6H_5COOMe \qquad (3.20)$$

(ii) N_3 chains.

$$trans\text{-}[IrCl(CO)(PPh_3)_2] + PhCON_3 \xrightarrow[CHCl_3]{0°}$$

$$trans\text{-}[IrCl(N_2)(PPh_3)_2] + PhCONCO \qquad (3.21)$$

$$[Ru(NH_3)_5Cl]^{2+} + N_3^- \rightarrow [Ru(NH_3)_5(N_3)]^{2+} \rightarrow$$

$$[Ru(NH_3)_5(N_2)]^{2+} + \tfrac{1}{2}N_2 \qquad (3.22)$$

$$cis\text{-}[Ru(en)_2(N_2)(N_3)]^+ + HNO_2 \xrightarrow[H_2O]{0°}$$

$$cis\text{-}[Ru(en)_2(N_2)_2]^{2+} + N_2O + OH^- \qquad (3.23)$$

$$[RuCl(N_3)(das)_2]PF_6 + NOPF_6 \rightarrow [RuCl(N_2)(das)_2]PF_6 \qquad (3.24)$$

$$[OsH_4(PEtPh_2)_3] + p\text{-tolSO}_2N_3 \rightarrow$$

$$[OsH_2(N_2)(PEtPh_2)_3] + p\text{-tolSO}_2NH_2 \qquad (3.25)$$

(c) *From a mononitrogen ligand.*   Ammines can react with nitrous acid to form dinitrogen complexes, but this method has very limited applicability (reaction 3.26).

$$[Os(NH_3)_5(N_2)]^{2+} + HNO_2 \rightarrow \textit{cis-}[Os(NH_3)_4(N_2)_2]^{2+} + 2H_2O \quad (3.26)$$

There is also the remarkable reaction whereby $RuCl_3$ is converted to the dinitrogen complex in liquid ammonia by reduction with zinc dust followed by hydrolysis of the product. This causes considerable effervescence and gives a poor yield of $[Ru(NH_3)_5(N_2)]^{2+}$ (Chatt and Fergusson, 1968).

3.2.5   Molecular structure

Theoretical considerations (3.2.6) indicate that dinitrogen should be bound in an essentially linear metal-N–N system ((VI), 3.1.4). The strong absorption in the infrared spectrum, assigned to the N–N stretching mode of vibration, is evidence of this arrangement in all the characterized terminal-dinitrogen complexes; side-on bonding of dinitrogen would give only a very weak absorption. End-on bonding has been confirmed in the X-ray structures of three complexes. In the pentaammine complex $[Ru(NH_3)_5(N_2)]Cl_2$ the ruthenium atom is octahedrally coordinated but, due to random orientation of the cation in the structure, the important bond lengths associated with the dinitrogen ligand cannot be accurately determined. There is similar disorder in octahedral $\textit{trans-}[RuCl(N_2)(PMe_2Ph)_4]$. To relieve steric strain in this latter structure the four phosphorus atoms lie in a puckered ring round the metal. The structure of $[CoH(N_2)(PPh_3)_3]$ is reproduced in Fig. 3.1 and the N–N bond distances are listed in Table 3.3 together with those of related molecules.

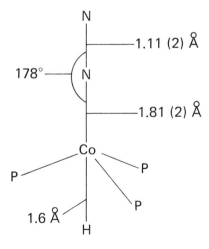

P = PPh$_3$

Fig. 3.1. Structure of $[CoH(N_2)(PPh_3)_3]$.

TABLE 3.3.   Bond distances (Å) in end-bonded dinitrogen compounds.

Substance	M—N*	N—N	Reference
N_2		1.0976 (2)†	‡
$[Ru(NH_3)_5(N_2)]Cl_2$	2.10 (1)	1.12 (8)	1
$[CoH(N_2)(PPh_3)_3]$	1.81 (2)	1.11 (2)	2
$[Ru(N_3)(N_2)(H_2NCH_2CH_2NH_2)_2]PF_6$	1.894 (9)	1.106 (11)	3
$[ReCl(N_2)(PMe_2Ph)_4]$	1.97 (2)	1.06 (3)	4
H_2CN_2	1.34 (5)	1.13 (4)	‡
HNN_2	1.240 (5)	1.134 (5)	‡
ON_2	1.1875 (3)	1.1286 (2)	‡

* M = metal or C, N or O as appropriate.

† Figure in parentheses is estimated error in final figure of distance.

‡ 'Tables of Interatomic Distances etc.', Special Publications Nos. 11 and 18 of the Chemical Society, London (1958), (1965).

1. F. BOTTOMLEY and S. C. NYBURG, *Acta Cryst. B.*, **24**, 1289 (1968).
2. B. R. DAVIS, N. C. PAYNE and J. A. IBERS, *Inorg. Chem.*, **8**, 2719 (1969).
3. B. R. DAVIS and J. A. IBERS, *Inorg. Chem.*, **9**, 2768 (1970).
4. B. R. DAVIS and J. A. IBERS, personal communication.

The N—N distances are generally longer than in free dinitrogen, and in the complexes the margins of error are high. Nevertheless, the distances do not appear to be very different from the N—N distance in dinitrogen, and nowhere near that of the single N—N bond in hydrazine [1.451(5) Å] or of the double N—N bond in azomethane [1.25(2) Å] (Tables of Interatomic Distances etc., 1965).

The structures of the dinitrogen complexes are closely similar to those expected for the corresponding carbonyl complexes, which have the isoelectronic carbon monoxide molecule in place of dinitrogen, so indicating closely similar electronic structures. Since acetylene is reduced to ethylene by the nitrogenase system it is worth noting, too, that the structures of the dinitrogen complexes show no similarity to those of acetylene complexes, where the two carbon atoms are equally bonded to the metal; rather they resemble the acetylide complexes where the metal takes the place of one hydrogen atom in the acetylene.

3.2.6   *Electronic structure*

In basic terms, the bond between dinitrogen and a metal is a donor-acceptor bond, the dinitrogen molecule donating an electron pair into a vacant orbital on the metal. At first sight the nitrogen lone-pair electrons might seem suitable with the bonding scheme shown in (XIII), but the lone-pair electrons are essentially in

$$M \leftarrow: N\equiv N:$$
(XIII)

s-orbitals and of such low energy that donation to form a strong bond is quite impossible. It is necessary to consider the molecular orbital structure of the dinitrogen molecule and how molecular orbitals can interact with the metal orbital system. The correlation diagram (see Chapter 2, Fig. 2.3) shows the derivation of the energy levels in the dinitrogen molecule from the atomic orbitals, and Fig. 3.2 shows the general form of the atomic orbitals. The dinitrogen molecule has no orbitals in the middle range of energy levels, where most reactive molecules have filled or vacant orbitals which allow chemical attack, hence its inert character. Its lowest-energy vacant orbital, the $1\pi_g$, is so high in energy (-7 eV) that only alkali and other very electropositive metals could possibly put electrons into it, and its highest-energy occupied orbital is so low in energy (-15.6 eV) that it is almost as difficult to remove electrons from it as from argon (-15.75 eV). However, the electronic energy levels are closely similar to those in carbon monoxide (see Chapter 2, 2.1) except that the highest-energy occupied orbital in carbon monoxide is essentially non-bonding on carbon, in fact weakly anti-bonding, whereas in dinitrogen the electrons are in a weakly bonding σ-orbital ($3\sigma_g$). However, the bonding of carbon monoxide and of dinitrogen to a metal are essentially the same and very similar bonding schemes can be given for both as shown very approximately in Fig. 3.3. The essential features of the bonding are the donation of electrons mainly from the $3\sigma_g$-orbital of dinitrogen into a metal orbital to form a σ-bond and back donation from the metal non-bonding d-orbitals into the anti-bonding $1\pi_g$-orbital of the dinitrogen molecule. Thus the bond is formed by a concerted electron donor and electron acceptor action of the dinitrogen molecule. The bond between the metal and the dinitrogen molecule is formed at the expense of the bond between the two nitrogen atoms. The metal receives electrons from the bonding $3\sigma_g$-orbital, a process which on first sight should weaken the N—N bond, but owing to decreased electron repulsion in the volume between the nitrogen nucleii is in fact N—N σ-bond strengthening (see Purcell and Drago, 1966; Purcell, 1969; Caulton, De Kock and Fenske, 1970). The metal also donates electrons into the $1\pi_g$-orbitals of dinitrogen, a process which is definitely bond weakening. It so outweighs the σ-bond strengthening

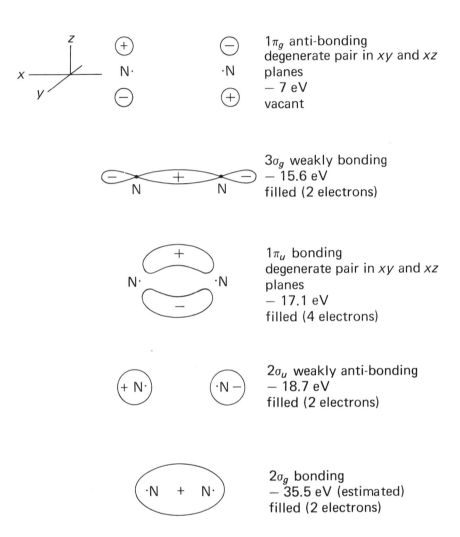

Fig. 3.2. Schematic molecular orbital diagram for N_2.

that the N—N bond is considerably weakened when dinitrogen enters a complex. A similar balance of effects occurs in the binding of carbon monoxide in all of its complexes except H_3BCO, but not generally in the otherwise closely related organonitrile or isonitrile complexes where the strengthening of the C—N σ-bond is usually greater than the weakening in the π-bond. However, when the nitrile ligands take the place of dinitrogen in a complex slight overall C—N bond weakening is usual, thus demonstrating the very strong tendency for metal sites which have the capacity to hold dinitrogen as a ligand to push electrons into the π*-orbitals of the ligands.

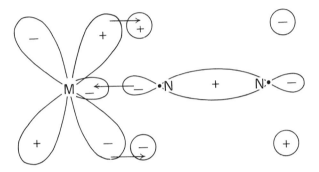

Fig. 3.3. Metal-dinitrogen bonding scheme.

The great reduction in $\nu(N_2)$ from 2331 cm^{-1} in the gas, (raman spectrum) when dinitrogen attaches itself to a metal is sufficient evidence of weakening of the N—N bond. In its complexes the dinitrogen molecule is also rendered very asymmetrical electrically as shown by the strong $\nu(N_2)$ absorption bands in their infrared spectra and resolution of the nitrogen atoms in the X-ray induced electron emission spectrum of trans-$[ReCl(N_2)(Ph_2PCH_2CH_2PPh_2)_2]$ and near resolution in the positive ion of exactly the same formula (Leigh, Murrell, Bremser and Proctor, 1970). In the neutral complex two nitrogen atoms gave emissions from the $1s$ shell corresponding to electron binding energies of 397.9 eV and 399.9 eV relative to the sharp unresolved carbon peak at 284 eV. In the cation the nitrogen atoms do not give a resolved spectrum, but a broad peak at 399.6 eV. These can be compared with 411 eV in the dinitrogen molecule, 398 eV in benzonitrile and amines generally and an estimated 399 eV in the neutral nitrogen atom.

The authors have estimated that in the neutral complex, the charge difference between the nitrogen atoms is about $0.4e$, and in the charged complex only $0.2e$. It is probable from the bonding scheme (Fig. 3.3) that the terminal nitrogen atom in the above complexes is the more negatively charged; also that the lower $\nu(N_2)$ in a complex, i.e. the weaker the dinitrogen molecule, the greater the electrical asymmetry, and the more negative the terminal nitrogen atom.

Evidence concerning the type of metal site which binds dinitrogen has come mainly from comparative studies of corresponding dinitrogen and carbonyl complexes. Their resemblances are much more striking than their differences, but carbon monoxide is the more versatile in the great range of stable complexes which it can form. It is easier to produce a metal-containing entity with orbitals energetically matched to those on carbon monoxide with its rather smaller energy difference between the donor $\sigma_g 2p$ and acceptor $\pi_g 2p$ orbitals (Chapter 2, Fig. 2.1) than to those on dinitrogen (Chatt, 1971). Purcell (1969) has calculated, by comparing the simpler analogous sytems NCO^- and NNN^-, that dinitrogen donates in the σ-bond about a $\frac{1}{4}e$ less charge and receives into its π^* bond about $\frac{1}{4}e$ less charge, thus he has quantified qualitative experimental observations from different data from a number of groups of workers (e.g. Bancroft, Mays and Prater, 1969) that dinitrogen is both the weaker electron donor and weaker electron acceptor in analogous complexes. The influence of carbon monoxide on the metal is also very similar to that of dinitrogen, thus both ligands have the same *trans*-influence on the Ir–X bond in the complex *trans*-$[IrX(Y)(PPh_3)_2]$ (X = Cl or Br, Y = N_2 or CO) (Chatt, Melville and Richards, 1969).

Most molecules when they enter complexes as ligands acquire considerable positive charge, but carbon monoxide remains essentially uncharged, probably acquiring either slight overall positive or negative charge according to the electronic character of the metal site which it occupies. If follows from Purcell's calculations and unpublished comparisons by ourselves and our colleague Mr. Crabtree of dipole moments of analogous carbonyl and dinitrogen complexes that dinitrogen as a ligand must also be essentially uncharged. Nevertheless, it is polarized so that the two dinitrogen atoms develop a difference in charge, and the terminal nitrogen atom in complexes of very low $\nu(N_2)$ may acquire considerable negative charge.

The strength of the bond between the nitrogen atoms in

dinitrogen resides mainly in the $1\pi_u$-orbitals, and its strength in the complexes is largely controlled by the extent of electron drift from the non-bonding d-orbitals of the metal into the $1\pi_g$-orbitals of the dinitrogen ligand, usually known as 'back donation'. The electron drift in the π-system is thus of great importance in determining $\nu(N_2)$ but it does not appear to be decisively important in the stabilizing of the metal to nitrogen bond. By the use of more electropositive co-ligands with dinitrogen it is possible to increase back donation in dinitrogen complexes, as when dimethylphenylphosphine is used in place of triphenylphosphine in $[IrCl(N_2)(PPh_3)_2]$; $\nu(N_2)$ is lowered, but so is the stability of the complex (Chatt, Melville and Richards, 1969), presumably because dimethylphenylphosphine, in raising the energy of the non-bonding d-orbitals in the metal, has also raised its σ-orbital energies so that mixing with the $3\sigma_g$-orbital of dinitrogen becomes too poor (see also section 3.2.8).

3.2.7 Chemical properties of terminal-dinitrogen complexes

In general, dinitrogen complexes have low thermal stability and, as might be expected, nearly all are of the normally very stable, octahedral d^6-electronic configuration; the remaining few, with minor exception, are planar d^8 or trigonal bipyrimidal d^8 (Table 3.2). All except the two rhenium d^5 and cobalt d^9-complexes are diamagnetic.

The range of stability over the whole field of dinitrogen complexes is wide. Those formed from the metals towards the tungsten end of Table 3.1 are the most stable. The heavy elements in any one Group also tend to form the more stable complexes, except when they are of the hydridic type, then the light elements form the more stable complexes. Nevertheless, the heavy elements platinum and palladium have not yielded stable dinitrogen complexes. Complexes containing two terminal-dinitrogen ligands are less stable than those containing one, and all the *bis*-dinitrogen complexes are formed by elements around the tungsten end of the line of Table 3.1 (W, Mo, Os and Ru).

Many dinitrogen complexes are too unstable to be isolated, but they can be recognized by the characteristic N—N stretching band in the infrared spectra of crude preparations. These have been dubbed 'ghosts' (Chatt, 1969), because they disappear on attempted isolation.

In general dinitrogen complexes, being thermally unstable, are chemically very reactive, and the reactivity normally parallels their thermal stability. They usually react to lose dinitrogen and leave a chemically sensitive metal-containing entity, often a highly reactive olefin polymerization catalyst. In this sense the reactions resemble those of the aliphatic diazo-compounds which lose dinitrogen and leave the highly reative carbene-based fragment.

Dinitrogen complexes are particularly sensitive to oxidation with loss of dinitrogen, so strong oxidation has often been used to estimate dinitrogen quantitatively (e.g. Harrison and Taube, 1967; Chatt, Leigh and Richards, 1970).

$$[OsCl_2(N_2)(PEt_2Ph)_3] + \tfrac{1}{2}Cl_2 \rightarrow [OsCl_3(PEt_2Ph)_3] + N_2 \qquad (3.27)$$

Some very stable complexes can be oxidized by at least one unit (e.g. reaction 3.28) without loss of dinitrogen. The oxidized species is much less stable and the metal-dinitrogen bond evidently much weakened. The N–N stretching frequency is raised by about 80-100 cm^{-1} as expected, owing to weaker 'back donation' of electrons from the more positive metal of the oxidized species.

$$[ReCl(N_2)(Ph_2PCH_2CH_2PPh_2)_2] + \tfrac{1}{2}Cl_2 \rightarrow$$

$$[ReCl(N_2)(Ph_2PCH_2CH_2PPh_2)]Cl \qquad (3.28)$$

Electrochemical oxidation of $[Os(NH_3)_5(N_2)]^{2+}$ produces $[Os(NH_3)_5(N_2)]^{3+}$ as a transient species with a decomposition rate constant of 0.02 sec^{-1} (Elson et al., 1970). Hydrogen chloride oxidizes yellow, trans-$[IrCl(N_2)(PPh_3)_2]$ via a transient, colourless intermediate which is probably an iridium(III) dinitrogen complex produced by oxidative addition, which very quickly loses dinitrogen (reaction (3.29)) (Chatt, Richards, Sanders and Fergusson, 1969).

$$trans\text{-}[IrCl(N_2)(PPh_3)_2] + HCl \rightarrow \left[\substack{\text{colourless} \\ \text{intermediate}}\right] \rightarrow$$

$$[IrHCl_2(PPh_3)_2] + N_2 \qquad (3.29)$$

Attempts to reduce coordinated dinitrogen to ammonia usually cause liberation of dinitrogen from the complex and there has been little investigation of the metal-containing products of such reactions. However, Speier and Markó (1970) have reduced the complexes $[Co(N_2)(PR_3)_3]$ (R = Ph or Bu^n) by sodium amalgam or by n-butyl lithium. From the triphenylphosphine complex they obtain products which they formulated $[Co(PPh_2)(N_2)(PPh_3)_2]$ and $[Co(PPh_2)_2(N_2)(PPh_3)]^{2-}$. These have N–N stretching frequencies up to 200 cm^{-1} lower than those of the starting material, the lowering being caused by the accumulation of negative charge on the complex, which increases the 'back-donation' of electrons from the metal into the anti-bonding $1\pi_g$-orbitals of the dinitrogen. These are the only anionic dinitrogen complexes reported.

In the above reductions some of the triphenylphosphine ligand was cleaved by the reducing agent to form PPh_2^- which stayed on the metal and also, presumably, Ph^- which was lost. In another example (reaction (3.30)) chloride on the metal is substituted by hydride (Chatt, Melville and Richards, 1971).

$$mer\text{-}[OsCl_2(N_2)(PMe_2Ph)_3] + NaBH_4 \rightarrow$$

$$mer\text{-}[OsHCl(N_2)(PMe_2Ph)_3] \qquad (3.30)$$

Some other ligand exchange and displacement reactions without loss of dinitrogen are known. We have already mentioned that coordinated ammonia is converted to dinitrogen coordinated to the metal by a diazotization type of reaction (reaction (3.26)), and all the dimethylphenylphosphine ligands of $[ReCl(N_2)(PMe_2Ph)_4]$ may be displaced by $Ph_2PCH_2CH_2PPh_2$ in boiling toluene (reaction (3.31)). Only the more stable dinitrogen complexes behave in this way.

$$trans\text{-}[ReCl(N_2)(PMe_2Ph)_4] + 2Ph_2PCH_2CH_2PPh_2 \rightarrow$$

$$trans\text{-}[ReCl(N_2)(Ph_2PCH_2CH_2PPh_2)_2] + 4PMe_2Ph \qquad (3.31)$$

In general the less stable dinitrogen complexes react with other two-electron donor molecules with displacement of dinitrogen, and the ease of displacement usually follows the order of thermal stabilities of the dinitrogen complexes. Thus dinitrogen is displaced rapidly and quantitatively from $[CoH(N_2)(PPh_3)_3]$ by reaction with $Ph_2PCH_2CH_2PPh_2$ at $20°$ (Sacco and Rossi, 1967), but the dinitrogen is displaced completely from $[OsCl_2(N_2)(PEt_2Ph)_3]$ only after 72 hrs. in boiling toluene with a large excess of $Ph_2PCH_2CH_2PPh_2$, which displaces the monophosphines as well (Chatt, Leigh and Richards, 1970). Although displacement of dinitrogen from the complex $[CoH(N_2)(PPh_3)_3]$ is easy, some donors such as ammonia and hydrogen set up an equilibrium as shown in reactions (3.4) and (3.5). The equilibrium with dihydrogen is complex because dideuterium, in addition to displacing the dinitrogen, exchanges with all eighteen ortho-hydrogen atoms in the phenyl groups of the triphenylphosphine ligands. The mechanism proposed for this exchange, shown in the series of reactions (3.32), involves the formation of an intramolecular metal-carbon bond (Parshall, 1968), a common type of reaction where aromatic groups, and on rare occasions aliphatic groups, are in close contact with reduced transition metal complexes (Chatt and Davidson, 1965).

(3.32)

The related complex $[RuH_2(N_2)(PPh_3)_3]$ undergoes similar reversible displacement of dinitrogen and exchange of hydrogen (Knoth, 1968). An exception to the general pattern of displacement of dinitrogen by donor ligands is provided by the simple addition of diethyl maleate to the coordinatively unsaturated complex *trans*-$[IrCl(N_2)(PPh_3)_2]$, increasing the coordination number from four to five, assuming that the double bond occupies the fifth coordination position (Collman and Kang, 1966).

$$[IrCl(N_2)(PPh_3)_2] + EtO_2C(H){=}(H)CO_2Et \rightarrow$$

$$[IrCl(N_2)(EtO_2C(H){=}(H)CO_2Et)(PPh_3)_2] \qquad (3.33)$$

Coordination of maleate raises in N–N stretching frequency by 95 cm^{-1} since dinitrogen and the double bond are now competing for the non-bonding d-electrons on the metal. Acetylenes irreversibly displace dinitrogen from the iridium complex (Collman and Kang, 1967) (reaction (3.34)).

$$[IrCl(N_2)(PPh_3)_2] + RCCR' \rightarrow [IrCl(RCCR')(PPh_3)_2] + N_2 \qquad (3.34)$$

(R or R' = MeO_2C, PhCO, Ph, EtCO, or $p\text{-}O_2NC_6H_4$).

Although the above reactions potentially can provide important new synthetic routes to dinitrogen complexes, and to new catalysts for olefin polymerizations, the properties and reactions of coordinated dinitrogen are much more relevant to nitrogen fixation and they are discussed below.

The dinitrogen in $[Ru(NH_3)_5(N_2)]^{2+}$ undergoes end-over-end rotation, as shown by the isomerization of $[(NH_3)_5Ru{-}^{15}N{\equiv}^{14}N]^{2+}$ to $[(NH_3)_5Ru{-}^{14}N{\equiv}^{15}N]^{2+}$, which occurs with a half life of about 2 h at 25° in aqueous solution, and some forty-five times faster than reaction (3.35). Thus the isomerization occurs without dissociation

$$[Ru(NH_3)_5(N_2)]^{2+} + H_2O \rightleftharpoons [Ru(NH_3)_5(H_2O)]^{2+} + N_2 \qquad (3.35)$$

of dinitrogen from the metal and so passes through the side-on or *dihapto* structure (3.1.4, (VII)). The energetics of the isomerization and dissociation indicate that the activation energy of dissociation from the side-on position in aqueous solution is only 7 kcal/mole as compared with 29 kcal/mole from the end-on position (3.1.4, (VI)), i.e. side-on bonding is very weak (Armor and Taube, 1970).

The terminal nitrogen atom of the coordinated dinitrogen is also an electron donor in the ion $[Ru(NH_3)_5(N_2)]^{2+}$ which reacts, in aqueous solution at 25°, with the ion $[Ru(NH_3)_5(H_2O)]^{2+}$ to displace the water (reaction (3.36)) (Harrison, Weissberger and Taube, 1968).

$$[Ru(NH_3)_5(N_2)]^{2+} + [Ru(NH_3)_5(H_2O)]^{2+} \rightleftharpoons$$

$$[(NH_3)_5RuN_2Ru(NH_3)_5]^{4+} + H_2O \qquad (3.36)$$

The symmetrical binuclear ion thus formed is discussed in section 3.3. Although the complex $[Ru(NH_3)(N_2)]^{2+}$ is not known to react with any other acceptor molecule, the complex $[ReCl(N_2)\text{-}(PMe_2Ph)_4]$ which has a particularly low N—N stretching frequency of 1925 cm^{-1}, reacts with a remarkably wide variety of acceptor molecules, apparently by addition of an electron-acceptor entity to the terminal nitrogen atom. All types of acceptor molecules react, Main Group metal and non-metal, and open shell, half-filled and closed-shell transition metal, but not the proton (Table 3.4). The

TABLE 3.4   Colours and $\nu(N_2)(cm^{-1})$ of products from $[ReCl(N_2)(PMe_2Ph)_4]$ with a selection of electron acceptor molecules.

Acceptor class (1)	Colour	$\nu(N_2)$	Ratio†
TiCl$_3$(THF)$_3$ ⎫	blue	1805	1:1
TiCl$_3$ ⎬	red	n.a.	1:2
TaCl$_5$	brown	1695	1:1
[MoCl$_4$(Et$_2$O)$_2$]	emerald green	1795	1:1
[MoCl$_4$(Et$_2$O)$_2$] ‡	blue	1680	1:2
[MoCl$_4$(PMePh$_2$)$_2$] ‡	purple	n.a.	1:2
PF$_5$	pale-pink	1640	1:1
Class (II)			
[CrCl$_3$(THF)$_3$]	purple	1875	1:1
CoCl$_2$(THF)$_{1.5}$	blue	1855	§
[Pt$_2$Cl$_4$(PEt$_3$)$_2$]	yellow	1890	§
AlEt$_3$	pale-yellow	1890	§

† Ratio of Re to acceptor atom in the product.
‡ An excess of acceptor.
§ Ratio not determined.
n.a. not assignable.

proton appears instead to attack the nucleophilic rhenium centre so causing evolution of dinitrogen. The types of acceptors which react with the rhenium complex are shown in Table 3.4 and since all the products are binuclear complexes they are discussed in section 3.3.

3.2.8 Complexes of dinitrogen analogues, the type of site which binds dinitrogen

Some substances having triply-bonded ligand atoms, such as carbon monoxide, organic cyanides, organic isocyanides, dinitrogen oxide, and azide ion, are known as dinitrogen analogues. They are isoelectronic with, or otherwise closely similar to dinitrogen, and bind in its place to form complexes exactly analogous to dinitrogen complexes. However, because they are more reactive than dinitrogen they form many more complexes, and by extensive studies, usually spectroscopic, of the behaviour of these ligands in a great variety of complexes, it is possible to deduce something of the character of sites which bind dinitrogen.

The complex $[Ru(NH_3)_5(N_2)]^{2+}$ is the most thoroughly investigated system. It takes, in place of dinitrogen, organic cyanides (Clarke and Ford, 1970) organic isocyanides (Chatt, Leigh and Thankarajan, 1970), dinitrogen oxide (Diamantis and Sparrow, 1969; Armor and Taube, 1969) and carbon monoxide (Allen *et al.*, 1969). Substituted acetylenes, $RC{\equiv}CR'$, do not appear to bind in place of dinitrogen in the ruthenium complex, and acetylene and terminal acetylenes, $RC{\equiv}CH$, give intractable products. Dinitrogen oxide complexes are so easily reduced to dinitrogen complexes, that little else is known about them (Armor and Taube, 1969). The infrared stretching frequencies of the triple bonds in all the dinitrogen analogues bound to the ruthenium pentaammine unit in the solid salts are dependent on the anion (X) as is $\nu(N_2)$ in the salts of $[Ru(NH_3)_5(N_2)]X_2$.

In the salts of $[Ru(NH_3)_5(CO)]^{2+}$, $\nu(CO)$ occurs amongst the lowest frequencies shown by terminal-carbonyl complexes. Similarly the stretching frequencies, $\nu(CN)$, of the salts of $[Ru(NH_3)_5(NCR)]^{2+}$ are generally lower than those in the free nitrile, although in the commonly occurring organic cyanide complexes they are higher than in the free nitrile. These spectral features are evidence of a strong release of electrons from the filled non-bonding d-orbitals of the ruthenium atom into the anti-bonding π^*-orbitals on the ligand (3.2.6). It is probably this quality of strong π-type electron release which gives the site its poor ability to bind ligands which have no π-electron acceptor capacities, such as H_2O and NH_3. Indeed the d-electron density is likely to repel those ligands, and the fact that they bind at all is evidence that the dinitrogen binding site has a strong σ-electron acceptor capacity.

The related series of compounds $[OsX_2(Y{\equiv}Z)(PR_3)_3]$ (X = Cl or Br; $Y{\equiv}Z$ = CO, N_2, RNC or RCN; PR_3 = tertiary phosphine) show the same tendency to low $\nu(Y{\equiv}Z)$, with the exception that CH_3CN shows an increased $C{\equiv}N$ stretching frequency in the complex, nevertheless PhCN has a lowered frequency. Evidently in these organic cyanide complexes the strengthening in the σ-(C—N) bond caused by coordination of the nitrogen is counter-balanced almost exactly by the weakening of the π-bond caused by back donation from the metal. Two isomeric series of the osmium complexes have been isolated (XIV) and (XV), but only the (XV) isomer exists when L = N_2 (Chatt, Melville and Richards, 1971).

(XIV)   (XV)

(Q = tertiary phosphine or arsine).

In the series (XIV) the dinitrogen analogue occupies a position *trans* to a tertiary phosphine which has a high *trans*-influence, i.e. in this case, tends to repel the σ-bonding electrons of the ligand *trans* to it onto the ligand, so that the *trans*-ligand-metal bond is weakened (Basolo *et al.*, 1961; Pidcock, Richards and Venanzi, 1966). Thus the $OsCl_2(PR_3)_3$ site of the (XIV) configuration is the poorer σ-electron acceptor, presumably too poor to bind dinitrogen.

We have already seen that transition metals which bind dinitrogen are in low oxidation states, and that oxidation merely weakens the metal-dinitrogen bond, also the site which binds dinitrogen in the manner of Fig. 3.3 must be a good π-type donor of electrons. But equally important the metal site must be a good σ-electron acceptor. These necessary properties of the metal site were thought only a few years ago to be mutually exclusive, but they follow from the weakness of dinitrogen both as a σ-donor and as a π-acceptor discussed earlier. It seems likely that the σ-electron acceptor function is the more difficult to obtain on a metal site, although both σ- and π-effects are important. The prime importance of the acceptor function would explain why the first stable dinitrogen complexes to be discovered were neutral or cationic, whereas in the early development of carbonyl chemistry the first complexes were neutral or anionic. The necessity for good π-donation is shown by the instability of oxidized dinitrogen complexes, and of the hypothetical $[Co(NH_3)_5(N_2)]^{3+}$ which is isoelectronic with $[Ru(NH_3)_5(N_2)]^{2+}$ (Jordan, Sargeson and Taube, 1966).

The electronic conditions for the attachment of dinitrogen to a metal site are difficult to obtain and depend crucially on the position of the metal in the Periodic Table, the co-ligands with dinitrogen and their relative orientations with respect to each other and to dinitrogen. Minor changes in any of these may render the site incapable of holding dinitrogen, even so minor as the replacement of triphenyl- by tri-p-tolyl-phosphine as a co-ligand (Chatt, Melville and Richards, 1969).

Chatt (1971) discussed possible reasons for the sensitivity of dinitrogen complexes to the above factors. Essentially he argued that to achieve good bonding to dinitrogen the energy level of the e_g-orbitals must be low, and the energy separation between the e_g- and t_{2g}-orbitals must be large in octahedral complexes, and similarly the corresponding orbitals in other complexes. This is necessary to attain good mixing with the widely separated $3\sigma_g$- and $1\pi_g$-orbials of dinitrogen (Fig. 3.2) respectively. The energy gap between these orbitals is so wide that only very small upward or downward adjustment in the metal orbitals can be tolerated before either the electron donor ($3\sigma_g$) or the electron acceptor ($1\pi_g$) requirement respectively of the dinitrogen ceases to be met.

3.2.9 Terminal-dinitrogen complexes as possible intermediates in the reduction of dinitrogen to ammonia

This section is concerned with the possible relevance of terminal dinitrogen complexes to the fixation of molecular nitrogen by biological systems. Nitrogenase contains molybdenum and iron as essential elements. Both these elements are capable of forming dinitrogen complexes, the molybdenum in oxidation state zero, and iron in oxidation state two (Table 3.2). The formation of dinitrogen complexes is the only reaction of dinitrogen which is known to occur easily in protic media; the substances known as dinitrogen analogues, which interfere with the natural process, also form complexes where dinitrogen forms complexes. It therefore seems reasonable to suppose that either molybdenum or iron form the active site of the enzyme and that the first stage in the reduction process is the formation of the dinitrogen complex with one of them.

In the resting state the active site may be occupied by such molecules as water which could be replaced by dinitrogen as from the ruthenium complex $[Ru(NH_3)_5(H_2O)]^{2+}$ (reaction (3.2)). Since molybdenum(0) is an uncommon oxidation state it seems likely that if terminal dinitrogen complexes are involved, dinitrogen would attach itself to iron(II), which according to our present knowledge of dinitrogen complexes should be in the low spin d^6-state. It would be attached in an end-on manner (Fe—N≡N) because side-on binding is very weak and the molecule in that state would be very similar to molecular dinitrogen, i.e. more symmetrical and thus more inert chemically. There is plenty of evidence in the complex chemistry of dinitrogen for the above suppositions, but no evidence yet of the reduction of terminally bonded dinitrogen. We have seen that as $\nu(N_2)$ in the complex is lowered the terminal atom acquires greater negative charge until it becomes an electron donor to all sorts of electron acceptors except the proton, presumably because the proton attacks the strongly reduced metal instead, so being reduced to coordinated hydride ion. However, it could be that in the natural system a complex with so negative terminal nitrogen atom is produced, that the proton attack occurs there rather than at the metal. Thus a dimido-complex would result, with the atomic grouping M—N=NH (M = Mo or Fe), isoelectronic with the well-established nitrosyl complexes containing the atomic grouping M—N=O, and equally capable of reduction, but entirely to ammonia. The reduction of the dinitrogen complex could then proceed by feed of electrons through the metal and protons from the aqueous medium according to some sequence such as outlined in the series of reactions (3.37) or (3.38). Other slight modifications of these are possible but the two schemes are essentially different. In these schemes the ligands surrounding the metal are represented by L_5 —five ligands to form an octahedral complex—and the metal is given oxidation state zero in the initial complex (i). This coordination number and oxidation state were chosen arbitrarily and have no other significance than to help in counting electrons and to show immediately changes in coordination number or oxidation state necessitated by the sequence of reactions. Where dotted bonds occur these imply partial π-bond character by overlap of filled non-bonding d-orbitals of the metal with anti-bonding orbitals of the multiply-bonded nitrogen system (as Fig. 3.3), or by overlap of filled non-bonding p-orbitals of the nitrogen atom with a vacant d-orbital on the metal. To what extent such π-bonding occurs will depend upon the exact electronic state of the metal.

The reactions of the sequence (3.37) involve strong multiple bonding between the nitrogen atom and the metal in all stages from the initial dinitrogen complex to the final hydrolysis or reduction of the nitrido-complex (v). A change of oxidation state of the metal of three units is also necessary. Molybdenum could do this easily, $Mo^0 \rightarrow Mo^{III}$, $Mo^{III} \rightarrow Mo^{VI}$, or any corresponding intermediate pair of oxidation states, without any change in the coordination number.

$$[L_5M^\circ \overset{\leftarrow}{\cdot} N\equiv N]^{n+} \xrightarrow{\ H^+\ } [M^1 \overset{..}{-} N = N - H]^{(n+1)+}$$
$$\text{(i)} \qquad\qquad\qquad\qquad \text{(ii)}$$

$$\xrightarrow{\ e\ } \left[L_5M^I \overset{..}{=} \overset{..}{N} \diagdown \underset{\underset{H}{|}}{\overset{..}{N}} \right]^{n+} \quad \text{(iii)}$$

$$\xrightarrow[e]{H^+} \left[L_5M^{II} \equiv \overset{..}{N} \diagdown \underset{\underset{H}{\diagup}}{\overset{\diagup H}{N}} \; \overset{..}{} \right]^{n+} \quad \text{(iv)}$$

$$\xrightarrow[e]{H^+} [L_5M^{III} \equiv N \colon]^{n+} + \colon NH_3$$
$$\text{(v)}$$

$$[L_5M^{III}(O)OH]^{n+} + \colon NH_3 \qquad\qquad \left[L_5M^{II} \overset{..}{=} \overset{..}{N} \diagdown H \right]^{n+} \quad \text{(vi)}$$
$$\text{(ix)}$$

$$\Bigg\downarrow \; \begin{array}{l} 3H^+ \\ 3e \\ N_2 \end{array} \qquad\qquad\qquad H^+ \Bigg| e$$

$$\left[L_5M^I \overset{..}{-} \overset{..}{N} - H \diagdown H \right]^{n+} \quad \text{(vii)}$$

$$\qquad\qquad\qquad\qquad\qquad H^+ \Bigg| N_2$$

$$[L_5M^\circ \overset{\leftarrow}{\cdot} N\equiv N]^{n+} \text{ (i)} \qquad\qquad [L_5M^\circ \overset{\leftarrow}{\cdot} N\equiv N]^{n+} \text{ (i)}$$
$$+ \qquad\qquad\qquad\qquad\qquad +$$
$$2H_2O \qquad\qquad\qquad\qquad\quad \colon NH_3$$

Mode of reduction *via* a nitride complex

$$(3.37)$$

$$[L_5M^0 \rightleftharpoons N\equiv N]^{n+} \xrightarrow{\;H^+\;} [L_5M^I = N=N-H]^{(n+1)^+}$$

$$\text{(i)} \hspace{6cm} \text{(ii)}$$

$$\xrightarrow{\;e\;} \left[L_5M^I = \overset{..}{N} \diagdown \overset{\underset{\displaystyle H}{|}}{\underset{..}{N}} \right]^{n+} \quad \text{(iii)}$$

$$\xrightarrow{\;H^+\;} \left[L_5M^I = \overset{\oplus}{N} \diagup \overset{H}{} \diagdown \overset{\underset{\displaystyle H}{|}}{\underset{..}{N}} \right]^{(n+1)^+} \quad \text{(iv)}$$

$$\xrightarrow{\;e\;} \left[L_5M^0 \rightleftharpoons N \diagup \overset{H}{} \diagdown \overset{\underset{\displaystyle H}{|}}{\underset{..}{N}} \right]^{n+} \quad \text{(v)}$$

$$\xrightarrow{\;H^+\;} \left[L_5M^I = N \diagdown \overset{\displaystyle H}{\underset{\underset{\displaystyle H}{|}}{\overset{\oplus}{N}}} \diagup H \right]^{(n+1)^+} \quad \text{(vi)}$$

$$\xrightarrow{\;e\;} \left[L_5M^I = \underset{..}{N} \diagdown \overset{\displaystyle H}{\underset{\underset{\displaystyle H}{|}}{\underset{..}{N}}} \diagup H \right]^{n+} \quad \text{(vii)}$$

$$\xrightarrow[e]{\;H^+\;} \left[L_5M^0 \leftarrow N \overset{\displaystyle H}{\underset{\displaystyle H}{\diagup}} \quad \underset{H}{\overset{H}{\underset{..}{N}}} \diagdown H \right]^{n+} \quad \text{(viii)}$$

$$\xrightarrow{\;H^+\;} \left[L_5M^0 \rightleftharpoons N \overset{\displaystyle H}{\underset{\displaystyle H}{\diagup}} \quad \underset{H\;\;H}{\overset{\oplus}{N}}-H \right]^{(n+1)^+} \quad \text{(ix)}$$

$$\xrightarrow{\;e\;} \left[L_5M^I - \underset{..}{N} \overset{\diagup H}{\diagdown H} \right]^{n+} + NH_3$$

$$\text{(x)}$$

$$\xrightarrow[e]{\;H^+\;} [L_5M^0 \leftarrow NH_3]^{n+} \xrightarrow{\;N_2\;} [L_5M^0 \rightleftharpoons N\equiv N]^{n+} + NH_3$$

$$\text{(xi)} \hspace{4cm} \text{(i)} \hspace{3cm} (3.38)$$

Mode of reduction *via* diimine and hydrazine

Iron is less likely to do this but could over the range $Fe^0 \rightarrow Fe^{III}$, with change of coordination number from 5 to 6 for example.

The reaction sequence (3.38) involves the rapid degradation of the imido-ligand *via* a diimine complex (v) and hydrazido-complex (vii) to give a hydrazine complex (viii) with complete loss of multiple bond character in the nitrogen-to-metal bond before the N—N is split. Since this sequence of reactions also involves the displacement of ammonia by dinitrogen from (xi) the displacement of hydrazine from (viii) should be equally possible; also if hydrazine was added to the system it should displace dinitrogen from (i) to form (viii) directly, which could then be reduced to ammonia. The fact that hydrazine is neither produced by the natural system, nor reduced by it, strongly suggests that if a terminal dinitrogen complex is produced and reduced without intermediacy of a second metal atom in the natural system, it would be by some scheme such as (3.37) when only ammonia could come from the reduction. Furthermore, the reaction sequence (3.38) provides no advantage over the reduction of gaseous dinitrogen except to assist the first protonation. Reaction sequence (3.37), however, provides for the formation of a strong multiple bond between the metal and its adjacent nitrogen atom as the bond between the nitrogen atoms is being destroyed and the terminal nitrogen atom is being reduced to produce ammonia. If at the end of the reduction dinitrogen is not available to displace the ammonia it could be displaced by two hydride ions produced by protonation of the metal and reduction of the protonated species. Dinitrogen could then displace two hydride ions as dihydrogen, as mentioned above.

These hypothetical schemes assume that dinitrogen is reduced as a terminal dinitrogen complex. So far no terminal dinitrogen complex, not even trans-$[ReCl(N_2)(PMe_2Ph)_4]$ has been reduced by any reagent sufficiently mild to exist in aqueous solution, and it may be that dinitrogen must be attacked by two metal toms as an essential prelude to reduction. This hypothesis is considered in the following section.

3.3   BRIDGING-DINITROGEN COMPLEXES

3.3.1   *Structural possibilities*

These are complexes in which the dinitrogen molecule bonds between two metal atoms. They must therefore be at least dinuclear, and some are polynuclear. There are four possible structures for the dinitrogen bridge (VIII), (IX), (X) and (XI) on p. 64. The few known dinuclear complexes appear to have structures (IX), some might have structure (X), but so far no compounds of structure (VIII) or (XI) have been discovered.

3.3.2 Types of bridging-dinitrogen complexes

These contain the unit MN_2M' and may be classified into two types:

(a) *symmetrical,* where the two units M and M' consisting of a metal and other ligands are identical

(b) *unsymmetrical,* where the metals, the ligands or both are not the same. The members of the latter group, which are by far the most numerous, are mainly derived from one rhenium-containing complex. The symmetrical bridging-dinitrogen complexes are represented by only two members, $[(NH_3)_5Ru-N\equiv N-Ru(NH_3)_5]$ and $[\{(cyclohexyl)_3P\}_2Ni-N\equiv N-Ni\{P(cyclohexyl)_3\}_2]$. The former is well established, the latter no more than characterized (Jolly and Jonas, 1968).

The unsymmetrical bridging-dinitrogen complexes are mainly of the general formula $[(PMe_2Ph)_4ClReN_2M']$ where M' represents a wide variety of acceptor molecules of which Table 3.4 gives a selection (Chatt, Dilworth, Leigh and Richards, 1970), but there is one other. It is $[(NH_3)_5Os-N\equiv N-Ru(NH_3)_5]^{4+}$ closely related to the symmetrical ruthenium complex (Allen, 1970).

3.3.3 Preparation of bridging-dinitrogen complexes

These are prepared either by the reduction of a suitable transition metal complex in the presence of dinitrogen or by the displacement of a labile ligand from a suitable metal complex by certain terminal-dinitrogen complexes. The two methods are probably related. Thus the complex $[(NH_3)_5RuN_2Ru(NH_3)_5]^{4+}$ may be prepared either directly from dinitrogen (reaction (3.39)) (Harrison, Weissberger and Taube, 1968)

$$2[Ru(NH_3)_5Cl]Cl_2 + Zn/Hg \xrightarrow{N_2}$$

$$[(NH_3)_5RuN_2Ru(NH_3)_5]^{4+} \tag{3.39}$$

or by the reaction of $[Ru(NH_3)_5(N_2)]^{2+}$ and $[Ru(NH_3)_5(H_2O)]^{2+}$ with elimination of water. The reaction occurs in aqueous acid, and (3.39) is thus the overall representation of a four-step reaction: reduction, aquation, displacement of water from some of the aqua complex, and condensation of the terminal-dinitrogen complex with the remainder of the aqua complex.

The nickel complex is formed by a complicated reaction of nickel(II) pentane-2,4-dionate with tricyclohexylphosphine and trimethylaluminium in the presence of dinitrogen. The bridging-dinitrogen nickel complex is thought to dissociate in benzene solution (Jolly and Jonas, 1968) (reaction (3.40)).

$$[\{Ni(PCy_3)_2\}_2N_2] \rightleftharpoons [Ni(N_2)(PCy_3)_2] + Ni(PCy_3)_2 \qquad (3.40)$$

$$(Cy = cyclohexyl)$$

The unsymmetrical dinuclear complexes are prepared by the reaction of $[ReCl(N_2)(PMe_2Ph)_4]$ with the acceptor entity. Reaction may occur with ligand displacement as in (3.41) (Chatt, Fay and Richards, 1971) or by direct attachment of a Lewis acid as in reaction (3.42)

$$[ReCl(N_2)(PMe_2Ph)_4] + [CrCl_3(THF)_3] \rightarrow$$
$$[ReCl(PMe_2Ph)_4N_2CrCl_3(THF)_2] + THF \qquad (3.41)$$

$$[ReCl(N_2)(PMe_2Ph)_4] + PF_5 \rightarrow$$
$$[ReCl(PMe_2Ph)_4N_2PF_5] \qquad (3.42)$$

3.3.4   Structures of bridging-dinitrogen complexes

Treital *et al.* (1969) have determined the structure of the ruthenium complex (Fig. 3.4). It is of the linear symmetrical type (IX) as predicted on the basis of infrared and raman spectral studies (Weissberger, Harrison and Taube, 1968; Chatt, Nikolsky *et al.*, 1969). There is no N—N stretching absorption band in its infrared spectrum, and in the raman spectrum of the solid fluoroborate it occurs at 2100 cm^{-1}, close to that of $[Ru(NH_3)_5(N_2)][BF_4]_2$ (2130 cm^{-1}). Similarly, the N—N bond length (Fig. 3.4) of 1.124 Å is not significantly different from that found in the mononuclear complex $[Ru(NH_3)_5(N_2)]^{2+}$ (1.12 ± 0.08 Å) and $[CoH(N_2)(PPh_3)_3]$ (1.11 ± 0.02 Å). Thus the dinitrogen in the terminal complex is altered very little by the attachment of the second metal atom.

The symmetrical nickel complex also appears to be of this linear type, where the N—N bond order is almost three.

The unsymmetrical bridging-dinitrogen complexes have, as expected, a strong N—N stretching absorption band in their infrared spectra. In some complexes its frequency is close to that of the mononuclear rhenium complex from which they were derived, and in others very greatly lowered, depending on the electronic state of the second metal atom which is usually contained in a strong electron acceptor molecule. Some acceptor molecules, together with the N—N stretching frequencies of the derived polynuclear complexes, are given in Table 3.4, where the acceptor molecules are classified

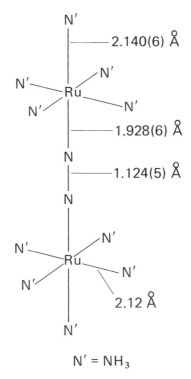

$$N' = NH_3$$

Fig. 3.4. Structure of $[(NH_3)_5 RuN_2 Ru(NH_3)_5]^{4+}$.

according to the frequencies. There are those (Class (i)) which cause a great lowering of the N—N stretching frequency, and others (Class (ii)) which cause little lowering. The acceptor atoms in the first class have vacant d-orbitals on an energy sufficiently low to accept bonding π-electrons from the dinitrogen molecule, and the N—N bond is greatly weakened. The acceptor atoms in the second class are those which have non-bonding d-orbitals which are filled or half-filled with electrons (e.g. platinum(II) or chromium(III) respectively), or are vacant but energetically inaccessible (e.g. AlEt₃). In these, as in the symmetrical type of bridging-dinitrogen complex, the N—N bond strength is scarcely affected by the addition of the second acceptor atom.

The reasons for the difference in bond strength according to the class of acceptor, M, receives ready explanation in terms of π-bonding within a linear $Re^{(I)}$–N–N–M system.

In simplest terms, the d^6-Re(I) system pushes electrons from its t_{2g}-orbitals into the anti-bonding π-orbitals ($1\pi_g$) of the dinitrogen as in Fig. 3.3, and the atom M of Class (i) withdraws electrons into its t_{2g}-orbitals from the bonding π-orbitals ($1\pi_u$) of the dinitrogen, both actions being π-bond weakening on the N–N bond. On the other hand the Class (ii) acceptor cannot receive electron density into its t_{2g}-orbitals either because they contain electrons (platinum(II)) or because they are inaccessibly high in energy so that the N–N bond is little affected by the second metal atom (aluminium(III)) (Chatt, Dilworth et al., 1969).

More exactly, π-bonding in linear M–N–N–M' systems, which is most important in the determination of the bond strength, must be considered in terms of linear combinations of certain d- and p-orbitals on the metal and nitrogen atoms respectively. Treital et al. (1969) have discussed the bonding in symmetrical (D_{4h}) systems, e.g. the $[(NH_3)_5Ru–N–N–Ru(NH_3)_5]^{4+}$ ion, and Chatt, Fay and Richards (1971) the (C_{4v}) system $[(PMe_2Ph)_4ClRe–N–N–CrCl_3-(THF)_2]$.

The π-molecular orbital system produced by linear combination is shown schematically in Fig. 3.5. Four-centre molecular orbitals are constructed from the Md_{xz}-, Np- and $M'd_{xz}$-orbitals. The σ-bond defines the z direction, and there is an equivalent set of π-orbitals in a plane at 90° to the orbitals shown in Fig. 3.5 derived from the metal d_{yz}- and Np_y-orbitals. The $1b_2$- and $2b_2$-molecular orbitals are derived from the metal d_{xy}-orbitals and contribute insignificantly to the bonding. The energy of the orbitals labelled $1e$, $2e$, $3e$ and $4e$ in idealized C_{4v} symmetry increases with the number of nodes. The $1e$- and $4e$-orbitals are predominantly bonding (π) and anti-bonding (π^*) respectively on nitrogen, with a small amount of metal orbitals mixed in, according to the electron affinity of the metal orbitals relative to those on nitrogen. The $2e$- and $3e$-molecular orbitals are mainly metal d_{xz} and d_{yz} in character, and the $2b_2$- and $1b_2$-orbitals are associated mainly with the metal d_{xy}-orbitals. If the energies of the metal orbitals are different, then their contribution to the $2e$, $3e$, $1b_2$ and $2b_2$ levels will differ. Thus in the rhenium-chromium complex, where the low-spin d^6-rhenium(I) atom has the lowest energy d-orbitals, the $2e$- and $1b_2$-orbitals will be mainly rhenium in character and the $3e$- and $2b_2$-orbitals will be mainly chromium in character.

The strength of the N—N bond depends mainly on the number of electrons in the *e*-molecular orbitals. Four electrons are supplied by

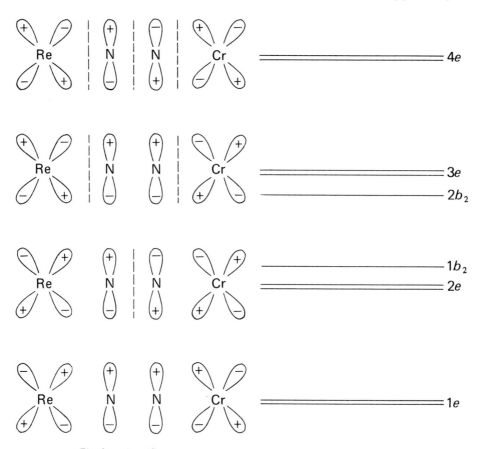

Fig. 3.5. Simplified π-molecular orbital scheme for the Re—N—N—Cr chain in $[(PMe_2Ph)_4ClReN_2CrCl_3(THF)_2]$.

the nitrogen atoms, which fill the 1*e* level. This level is bonding between the nitrogen atoms, but because of its partial metal character, gives a weaker N—N bond than in free dinitrogen. In all the known bridging-dinitrogen complexes at least one metal has the d^6-configuration, i.e. it can contribute six electrons to fill the 2*e* and 1b_2 levels. Since the 2*e* level is anti-bonding between the nitrogen atoms, the N—N bond will be weakened greatly and if no further electrons are available from the other metal, e.g. as in the PF_5 and $TaCl_5$ complexes of Table 3.4, a very low $\nu(N_2)$ is observed.

If the second metal has sufficient electrons to occupy the $2b_2$ and $3e$ levels (e.g. the three electrons from Cr(III) in the rhenium-chromium complex) the occupation of the $3e$ level, which is bonding between the nitrogen atoms, will increase the N–N bond strength, and thus $\nu(N_2)$. The di-ruthenium complex $[(NH_3)_5\text{-}Ru\text{–}N\text{–}N\text{–}Ru(NH_3)_5]^{4+}$ which is symmetrical, contains sufficient electrons to occupy completely all except the $4e$-orbitals, and the N–N bond strength appears to be very little altered from that of the corresponding terminal-dinitrogen complex $[Ru(NH_3)_5(N_2)]^{2+}$, as noted above. The $4e$-orbitals are so high in energy that few reducing agents are likely to be sufficiently strong to put electrons into them. They are anti-bonding on all atoms; if electrons were put into them, all the bonds in the metal-dinitrogen-metal chain would be weakened.

Thus we see that the N–N bond is weakest in those bridging dinitrogen complexes which contain insufficient electrons to fill orbitals on any higher energy than the $2b_2$, and that attempted reduction by the addition of electrons to this system would strengthen the N–N bond, except possibly under circumstances to be discussed in section 3.3.6.

Most of the bridging-dinitrogen complexes with fewer than 16 electrons in the e- and b_2-orbitals are strongly coloured owing to a charge transfer band assignable to the allowed transitions from the $2e$- or $1b_2$-orbitals into the $3e$- or $2b_2$-orbitals. In the complex $[ReCl(PMe_2Ph)_4N_2CrCl_3(THF)]$ a charge transfer band occurs at 18,600 cm^{-1} corresponding to an energy difference of about 2.3 eV between the above orbitals. Charge transfer also occurs in all of the unsymmetrical bridging-dinitrogen complexes from the $3e$ or $2b_2$ into the $4e$-orbitals, and this involves a much higher energy of about 38,000 cm^{-1} (4.7 eV) in the above chromium complex. In the dinuclear ruthenium complex $[(NH_3)_5RuN_2Ru(NH_3)_5]^{4+}$ the corresponding transition occurs at about the same energy, and in the mononuclear complex, $[Ru(NH_3)_5(N_2)]$, it is some 7200 cm^{-1} higher in energy (see Treital *et al.* (1969) for a detailed discussion of these energy levels).

3.3.5   *Reactions of bridging-dinitrogen complexes*

The chemistry of the bridging-dinitrogen complexes is still in its early stages, and the bridging-dinitrogen has not yet been reduced to ammonia by reagents which are capable of existence in water (Chatt, Richards *et al.*, 1969). It can, however, be reduced by such strong reducing agents as lithium naphthalene although not so easily as dinitrogen itself (Chatt and Dilworth, 1971). The most-studied reaction involves the cleavage of one metal-to-nitrogen bond to give the stable mononuclear dinitrogen complex, for example, reaction (3.43) (Harrison and Taube, 1967).

$$[(NH_3)_5 RuN_2 Ru(NH_3)_5]^{4+} + NH_3 \rightarrow$$

$$[Ru(NH_3)_5(N_2)]^{2+} + [Ru(NH_3)_6]^{2+} \qquad (3.43)$$

Many of the acceptor groups which bind to $[ReCl(N_2)(PMe_2Ph)_4]$ in the rhenium series of bridging-dinitrogen complexes are very sensitive to water, and others are destroyed by hydrolysis or aquation, so regenerating the very stable mononuclear-rhenium complex as in reaction (3.44)

$$[(PMe_2Ph)_4ClReN_2CrCl_3(THF)_2] \xrightarrow{H_2O}$$

$$[ReCl(N_2)(PMe_2Ph)_4] + \text{chromium products} \qquad (3.44)$$

Many of the dinuclear bridging-dinitrogen complexes derived from the rhenium(I) complex and an acceptor reagent, such as $TiCl_3 \cdot 3THF$ containing an early transition metal, will react with an excess of the reagent to form a tri- or polynuclear bridging-dinitrogen complex of even lower $\nu(N_2)$ than the corresponding dinuclear complex (Table 3.4).

Strong oxidation causes evolution of dinitrogen from bridging-dinitrogen complexes, but by electrochemically controlled oxidation of the symmetrical ruthenium complex, Elson et $al.$ (1970) observed the transient formation of an oxidized species $[(NH_3)_5 RuN_2 Ru-(NH_3)_5]^{5+}$, which as a decomposition rate constant of 0.10 sec^{-1} in aqueous sulphuric acid. The complex $[(PMe_2Ph)_4ClReN_2CrCl_3-(THF)_2]$ is very sensitive to oxidation by dry air, although neither of its precursors $trans$-$[ReCl(N_2)(PMe_2Ph)_4]$ nor $[CrCl_3(THF)_3]$ is very sensitive. The oxidation leads to a positive mononuclear-dinitrogen complex (reaction (3.45)) (Chatt, Fay and Richards, 1971).

$$[(PMe_2)_4ClReN_2CrCl_3(THF)_2] \xrightarrow{O_2}$$

$$[ReCl(N_2)(PMe_2Ph)]^+ + \text{chromium products} \qquad (3.45)$$

Presumably the oxygen molecule takes an electron from the chromium atom which then withdraws an electron through the Re–N–N–Cr system from the rhenium. Such electron transfer should occur easily in view of the low energy of the rhenium-to-chromium charge-transfer band noted in 3.3.4.

3.3.6 The possible role of bridging-dinitrogen complexes in nitrogen fixation

Long before active extracts of nitrogenase had been obtained from the natural system, or dinitrogen complexes were known, chemists had speculated that the crucial step in nitrogen fixation might involve dinitrogen bridging between metal atoms, particularly between molybdenum atoms, but these speculations were rarely recorded (e.g. Winfield, 1955). Only recently has symmetrical bridging as in $[(NH_3)_5RuN_2Ru(NH_3)_5]^{4+}$ become known and the known compounds provide no evidence that dinitrogen, when bridging between saturated d^6 and similar electronic systems, is in any way activated towards reduction.

The discovery that the nitrogenase proteins contained molybdenum and iron as essential elements led to further speculation along similar lines, but with dinitrogen bridging between molybdenum and iron. The nearest actual analogue to such a system is found in the unsymmetrical bridging-dinitrogen complexes, especially with a bridge of the type $Re^I-N{\equiv}N-Mo^V$. The very low N–N stretching frequencies found in such complexes indicate that the N–N bond has been weakened by some 100 kcal and that perhaps some $Fe^{II}-N{\equiv}N-Mo^V$ system, which would be isoelectronic in the valency shell with the rhenium(I) system, could hold the clue to nitrogen fixation, the N–N bond being so weakened that reduction became possible (Chatt, Dilworth et al., 1969). However, the $Fe^{II}-N{\equiv}N-Mo^V$ system would retain its cylindrical symmetry and the π-molecular orbital system outlined in Fig. 3.5. The $1e$-, $2e$-, and $1b_2$-orbitals would be filled and the $2b_2$ singly occupied, assuming both metals had essentially octahedral coordination. Any attempt at reduction would put electrons into the $3e$-orbital which is bonding between the nitrogen atoms, i.e. reduction would first strengthen the N–N bond. Reduction with weakening of the N–N bond could not occur until electrons entered the $4e$-orbitals, and these are likely to be on an even higher energy level than the $1\pi_g$-orbitals of dinitrogen itself. Thus such a complex would be more difficult to reduce than molecular dinitrogen.

To render a bridging-dinitrogen complex reducible the C_4 symmetry of the molecule should be lowered to C_2, so rendering the e-orbitals degenerate, and forcing the dinitrogen and attached groups towards a non-linear state as in azobenzene, Ph–N=N–Ph, which is easily reducible. This would need great asymmetry in the π-bonding capacity of the ligand or ligands (four coordinating atoms in all) which occupy the plane perpendicular to the metal-nitrogen bond.

Molybdenum might well satisfy these conditions if it had terminal oxygen atoms bonded as Mo=O perpendicular to the Mo—N$_2$ bond. In iron complexes such asymmetry would most probably demand great π-asymmetry in a macrocyclic tetradentate system. Alternatively, a non-linear state might be imposed by the stereochemistry of the metal sites regardless of the metals. There is no firm evidence yet as to the chemical environment of the metal atom in the enzyme on which to support or reject such hypotheses.

Another possibility exists involving bridging-dinitrogen complexes. If they contain metal atoms which are similar electronically and contain sufficient electrons to fill the 1e-, 2e- and b$_2$-orbitals, but not more, then the bridging-dinitrogen molecule could be rendered very weak. If the metal atoms are octahedrally coordinated, this condition requires that at least two, and not more than four, electrons are contained in the non-bonding d-orbitals, i.e. t$_{2g}$-orbitals.

If the metal atoms had high electron affinity, as they would if their oxidation state was high, e.g. molybdenum(III) with three electrons in t$_{2g}$-orbitals, then bonding between the metal and nitrogen would strengthen at the expense of bonding between the nitrogen atoms, and indeed if the M≡N bond strength were greater than about 112 kcal, which is not unreasonable, the molecule would fall apart at the N—N bond. Assuming octahedral coordination and oxidation state (III) on molybdenum, neither of these being a necessary condition but used only by way of illustration to monitor changes of coordination number and oxidation state, the process of N—N bond cleavage would be as illustrated in the scheme (3.46) below. In this process the dinitrogen is reduced by the attached metal. The metal increases its formal oxidation state by three units and the nitrogen molecule becomes formally two coordinated N^{3-} ions. Such nitride complexes are known for elements around molybdenum in the Periodic Table and are capable of reduction or hydrolysis to ammonia as in the final states of reaction scheme (3.37).

To form a nitride complex the metal must provide three electrons from non-bonding d-orbitals or they must be brought into the system from outside as they are used. If the metal and system can provide only two electrons a hydrazide(4−) complex would result, e.g. from molybdenum(IV) in reaction (3.47), and hydrazine could result from hydrolysis or hydrogenation. This could be the origin of hydrazine in the system reported by Denisov et al. (see 3.1.1).

On the basis of chemical evidence and theory at present the most likely mechanism for the reduction of dinitrogen is one as outlined

$$2L_5 Mo^V H_2 + N_2 \rightarrow L_5 Mo^{III} - N \equiv N - Mo^{III} L_5 + 2H_2$$
$$\downarrow$$
$$2L_5 Mo^{VI} \equiv N \qquad\qquad (3.46)$$

$$L_5 Mo^{IV} H_2 + N_2 \rightarrow L_5 Mo^{IV} - N \equiv N\ Mo^{IV} L_5 + 2H_2$$
$$2L_5 Mo^{VI} = N - N = Mo^{VI} L_5 \qquad (3.47)$$

(L_5 = coordinating atoms, 2 neutral and 3 anionic to fit the oxidation states in the above examples).

above, where molybdenum(II) or molybdenum(III) becomes attached to each end of the dinitrogen molecule, and by autoxidation of the metal and reduction of the dinitrogen the latter becomes two coordinated nitride ions, which are then hydrolyzed or hydrogenated to form ammonia. The iron may have the function of storing the electrons which are then transferred to the molybdenum site during reduction of the dinitrogen, or it may even first take up the dinitrogen molecule as a terminal-dinitrogen complex, and pass it on to the two molybdenum atoms. It seems unlikely that iron could function directly to form a reducible bridging-dinitrogen complex because iron(II) already contains six electrons in non-bonding d-orbitals and, as explained earlier, reduction in the presence of so many electrons can only strengthen the nitrogen-nitrogen bond.

REFERENCES

A. D. ALLEN, Section Lecture, *XIIIth Intern. Conf. Coord. Chem.*, *Zakopane* (1970).

A. D. ALLEN and F. BOTTOMLEY, *Accounts of Chem. Res.*, **1**, 360 (1968).

A. D. ALLEN, F. BOTTOMLEY, R. O. HARRIS, V. P. REINSALU and C. V. SENOFF, *J. Amer. Chem. Soc.*, **88**, 5595 (1967).

A. D. ALLEN, T. ELIADES, R. O. HARRIS and V. P. REINSALU, *Cand. J. Chem.*, **47**, 1605 (1969).

A. D. ALLEN and C. V. SENOFF, *Chem. Commun.*, 621 (1965).

J. N. ARMOR and H. TAUBE, *J. Amer. Chem. Soc.*, **91**, 6874 (1969).

J. N. ARMOR and H. TAUBE, *J. Amer. Chem. Soc.*, **92**, 2560 (1970).

G. M. BANCROFT, M. J. MAYS and B. E. PRATER, *Chem. Commum.*, 589 (1969).

F. BASOLO, J. CHATT, H. B. GRAY, R. G. PEARSON and B. L. SHAW, *J. Chem. Soc.*, 2207 (1961).

Y. G. BOROD'KO and A. E. SHILOV, *Russ. Chem. Rev.*, **38**, 355 (1969).

Y. G. BOROD'KO, A. K. SHILOVA and A. E. SHILOV, *Zhur. fiz. Khim.*, **44**, 627 (1970).

Y. G. BOROD'KO, A. E. SHILOV and A. A. SHTEINMAN, *Doklady Akad. Nauk. S.S.S.R.*, **168**, 581 (1966).

H. BORTELS, *Arch. Mikrobiol.*, **1**, 333 (1930).

H. BRINTZINGER, *Biochemistry*, **5**, 3947 (1966).

K. G. CAULTON, R. L. DECOCK and R. F. FENSKE, *J. Amer. Chem. Soc.*, **92**, 515 (1970).

J. CHATT, *Proc. Roy. Soc. B.*, **172**, 327 (1969).

J. CHATT., *Proc. XIIth Internat. Conf. Coord. Chem.*, Sydney, 1969. Butterworths, London (1971).

J. CHATT and J. M. DAVIDSON, *J. Chem. Soc.*, 843 (1965).

J. CHATT, J. R. DILWORTH, G. J. LEIGH and R. L. RICHARDS, *Chem. Commun.*, 955 (1970).

J. CHATT, J. R. DILWORTH, R. L. RICHARDS and J. R. SANDERS, *Nature*, **224**, 5225 (1969).

J. CHATT and L. A. DUNCANSON, *J. Chem. Soc.*, 2939 (1953).

J. CHATT, R. C. FAY and R. L. RICHARDS, *J. Chem. Soc. (A)*, 702 (1971).

J. CHATT and J. E. FERGUSSON, *Chem. Commun.*, 126 (1968).

J. CHATT, J. E. FERGUSSON, J. L. LOVE, A. B. NIKOLSKY, R. L. RICHARDS and J. R. SANDERS, *J. Chem. Soc. (A)*, 1479 (1970).

J. CHATT, G. J. LEIGH and R. L. RICHARDS, *J. Chem. Soc.(A)*, 2243 (1970).

J. CHATT, G. J. LEIGH and N. THANKARAJAN, *J. Organometallic Chem.*, **25**, C77 (1970).

J. CHATT, D. P. MELVILLE and R. L. RICHARDS, *J. Chem. Soc.(A)*, 2841 (1969).

J. CHATT, D. P. MELVILLE and R. L. RICHARDS, *J. Chem. Soc.(A)*, 1169 (1971).

J. CHATT, A. B. NIKOLSKY, R. L. RICHARDS and J. R. SANDERS, *Chem. Commun.*, 154 (1969).

J. CHATT, R. L. RICHARDS, J. E. FERGUSSON and J. L. LOVE, *Chem. Commun.*, 1522 (1968).

J. CHATT, R. L. RICHARDS, J. R. SANDERS and J. E. FERGUSSON, *Nature*, **221**, 551 (1969).

R. E. CLARKE and P. C. FORD, *Inorg. Chem.*, **9**, 495 (1970).

J. P. COLLMAN and J. W. KANG, *J. Amer. Chem. Soc.*, **88**, 3459 (1966).

J. P. COLLMAN and J. W. KANG, *J. Amer. Chem. Soc.*, **89**, 844 (1967).

N. T. DENISOV, Y. F. SHUVALOV, N. I. SHUVALOVA, A. K. SHILOV and A. E. SHILOV, *Kinetika i Kataliz.*, **11**, 819 (1970).

M. J. S. DEWAR, *Bull. Soc. Chim.*, **18**, C79 (1951).

A. A. DIAMANTIS and G. J. SPARROW, *Chem. Commun.*, 469 (1969).

G. C. DOBINSON, R. MASON, G. B. ROBERTSON, R. UGO, F. CONTI, D. MONELLI, S. CENENI and F. BONATI, *Chem. Commun.*, 739 (1967).

C. M. ELSON, J. GULENS, I. J. ITZKOVITCH and J. A. PAGE, *Chem. Commun.*, 895 (1970).

C. M. ELSON, I. J. ITZKOVITCH and J. A. PAGE, *Can. J. Chem.*, **48**, 1639 (1970).

J. E. FERGUSSON and J. L. LOVE, *Rev. Pure and Appl. Chem.*, **20**, 33 (1970).

T. C. FRANKLIN and R. C. BYRD, *Inorg. Chem.*, **9**, 986 (1970).

O. GLEMSER, Ger. Pat. Appl. F15431 IVa/12K (1954).

G. P. HAIGHT and R. SCOTT, *J. Amer. Chem. Soc.*, **86**, 743 (1964).

D. E. HARRISON and H. TAUBE, *J. Amer. Chem. Soc.*, **89**, 5706 (1967).

D. F. HARRISON, E. WEISSBERGER and H. TAUBE, *Science*, **159**, 320 (1968).

G. L. JOHNSON and W. D. BEVERIDGE, *Inorg. and Nucl. Chem. Lett.*, **3**, 323 (1967).

P. W. JOLLY and K. JONAS, *Angew. Chem.*, **80**, 205 (1968).

R. B. JORDAN, A. M. SARGESON and H. TAUBE, *Inorg. Chem.*, **5**, 1091 (1966).

W. H. KNOTH, *J. Amer. Chem. Soc.*, **88**, 935 (1966).

W. H. KNOTH, *J. Amer. Chem. Soc.*, **90**, 7172 (1968).

G. J. LEIGH, J. N. MURREL, W. BREMSER and W. G. PROCTOR, *Chem. Commun.*, 1661 (1970).

A. MISONO, Y. UCHIDA and T. SAITO, *Bull. Chem. Soc. Japan*, **40**, 700 (1967).

R. S. MULLIKEN, *Can. J. Chem.*, **36**, 10 (1959).

R. MURRAY and D. C. SMITH, *Coordination Chem. Rev.*, **3**, 429 (1968).

Nomenclature of Inorganic Chemistry, International Union of Pure and Applied Chemistry, Butterworths Scientific Publications, London (1957).

G. W. PARSHALL, *J. Amer. Chem. Soc.*, **90**, 1669 (1968).

L. PAULING, The Nature of the Chemical Bond, p. 252. Cornell University Press (1939).

A. PIDCOCK, R. E. RICHARDS and L. M. VENANZI, *J. Chem. Soc.(A)*, 1707 (1966).

A. B. R. PRIDEAUX and H. LAMBOURNE, A Textbook of Inorganic Chemistry. (J. Newton Friend ed.) VI. Charles Griffin and Co., London (1928).

K. F. PURCELL, *Inorg. Chim. Acta*, **3**, 540 (1969).

K. F. PURCELL and R. S. DRAGO, *J. Amer. Chem. Soc.*, **88**, 919 (1966).

A. SACCO and M. ROSSI, *Chem. Commun.*, 316 (1967).

A. E. SHILOV, A. K. SHILOVA and Y. G. BOROD'KO, *Kinetika i Kataliz*, **7**, 768 (1966).

G. N. SHRAUZER and G. SCHLESINGER, *J. Amer. Chem. Soc.*, **92**, 1808 (1970).

G. SPEIER and L. MARKO, *J. Organometallic Chem.*, **21**, p. 46 (1970).

Tables of Interatomic Distances and Configurations in Molecules and Ions, Supplement 1956-1959, Special Publication No. 18, Chemical Society, London (1965).

J. M. TREITAL, M. T. FLOOD, R. E. MARSH and H. B. GRAY, *J. Amer. Chem. Soc.*, **91**, 6512 (1969).

M. E. WINFIELD, *Rev. Pure and Appl. Chem.*, **5**, 217 (1955).

A. YAMAMOTO, S. KITAZUME, L. S. PU and S. IKADA, *Chem. Commun.*, 79 (1967a).

A. YAMAMOTO, S. KITAZUME, L. S. PU and S. IKEDA, *J. Amer. Chem. Soc.*, **89**, 3071 (1967b).

CHAPTER 4

Fixation by Free-Living Micro-Organisms: Enzymology

R. H. BURRIS

Department of Biochemistry,
University of Wisconsin, Madison,
Wisconsin 53706, U.S.A.

4.1   INTRODUCTION

4.1.1   *Nomenclature*

N_2 fixation is accomplished by the combined action of two proteins, and this complicates the problem of nomenclature. The problem is particularly acute because it is not clear whether the proteins act independently or as a complex, nor is it clear what specific action each protein effects. Nitrogenase or the N_2-fixing complex refers to the two proteins acting together. The individual proteins of nitrogenase will be termed the Mo-Fe protein and the Fe protein; some workers refer to these as 'protein 1' or 'fraction 1' and 'protein 2' or 'fraction 2' respectively. Although these designations are not completely definitive, because other Fe proteins do exist in the N_2-fixing cells, they seem preferable currently to azoferredoxin and molybdoferredoxin. Mortenson, Morris and Jeng (1967) introduced the term azoferredoxin to designate the N_2-binding protein of the nitrogenase complex, but later work (Bui and Mortenson, 1968) suggested that it is the Mo-Fe protein which binds N_2. When the chemical nature of the Mo-Fe and Fe proteins and their functions are better established a more rational nomenclature can be assigned.

The International Commission on Biochemical Nomenclature has given no name to nitrogenase. Current information could support a designation of the complex such as 1.8.6.2 Ferredoxin: N_2 oxidoreductase (ATP hydrolyzing), but assignment of a name must be left to the commission, and at the moment the commission is not inclined to name complexes.

Hydrogen metabolism in many N_2-fixing agents is confusing, because H_2 has multiple roles. It is a product, an inhibitor and a potential electron donor. Furthermore, it can be liberated both by 'classical hydrogenase' (ATP independendent, CO sensitive) or by nitrogenase *via* the 'H_2-evolving system' (ATP dependent, CO insensitive). To distinguish these entities we will refer to the former merely as hydrogenase and to the latter as the H_2-evolving system.

4.1.2   *Sources of cell-free extracts*

Although numerous cell-free preparations capable of fixing N_2 had been prepared before 1960 (Magee and Burris, 1956; Hoch and

Westlake, 1958; Burris, 1966), fixation by these preparations was weak and was not reproducible. Because of these limitations, the preparations had not been useful in determing the mechanism of N_2 fixation. The discovery by Carnahan, Mortenson, Mower and Castle (1960a, b) of a method to produce cell-free preparations which consistently supported vigorous N_2 fixation was a turning point in studies of the enzymology of N_2 fixation, and this discovery markedly accelerated progress in the field.

Extracts from *Clostridium pasteurianum* have been used most extensively for studies on the mechanism of N_2 fixation, and extracts from *Azotobacter vinelandii* and *Azotobacter chroococcum* have been second in popularity. Active cell-free preparations have been made, but have been used much less extensively, from blue-green algae (Schneider *et al.*, 1960; Smith and Evans, 1970; Cox and Fay, 1967), *Rhodospirillum rubrum* (Schneider *et al.*, 1960; Burns and Bulen, 1966; Munson and Burris, 1969), *Chromatium* sp. (Winter and Arnon, 1970), *Bacillus polymyxa* (Grau and Wilson, 1963; Fisher and Wilson, 1970), *Klebsiella pneumoniae* (Mahl and Wilson, 1968), and *Mycobacterium flavum* (Biggins and Postgate, 1969).

4.1.3 General requirements for reactivity

The early attempts to prepare cell-free extracts from N_2-fixing organisms were concentrated on the azotobacter, because they grow rapidly and fix N_2 vigorously. Bach *et al.* (1934) reported cell-free fixation by extracts from *A. chroococcum,* but Burk was unable to repeat the work in Bach's laboratory, and Roberg (1936) could not reproduce the results. The fact that Bach *et al.* published only a preliminary note on their method, suggests that they were unsuccessful in repeating the work and utilizing cell-free extracts for studying N_2 fixation.

The availability of ^{15}N as a tracer furnished a far more sensitive tool for detecting N_2 fixation than had the determination to total N as employed by Bach *et al.* (1934). Many experiments were performed by exposing bacterial extracts to $^{15}N_2$, but the results summarized in 1943 (Burris, Eppling, Wahlin and Wilson) were predominantly negative. Magee and Burris (1956) summarized more extensive trials, which supplied a variety of substrates primarily to extracts from *A. vinelandii* disrupted by a number of methods. About 27% of the experiments gave positive results, in the sense that the preparations accumulated greater than 0.015 atom-% ^{15}N excess, about 5 times the level detectable with the mass spectrometers employed.

In some of these early experiments it is possible that intact cells remaining in the preparations were responsible for the N_2 fixation observed. However, Hoch and Westlake (1958) exercised extreme care to avoid inclusion of intact cells and obtained readily measure-able fixation of $^{15}N_2$ by extracts from *C. pasteurianum*. In the fall of 1959 Schneider *et al.* (1960) obtained N_2 fixation with cell-free extracts from blue-green algae. Despite numerous demonstrations of N_2 fixation, these early studies had no appreciable impact on investigation of the mechanism of N_2 fixation. Extracts of *C. pasteurianum* employed by Carnahan *et al.* (1960a, b) were capable of metabolizing pyruvate anaerobically to generate reducing power and the energy source needed to support N_2 fixation.

Early cell-free preparations for studying N_2 fixation commonly were kept cold during all steps in processing in an attempt to prevent denaturation of the enzymes; lyophilized preparations also were tested. Carnahan *et al.* (1960a, b) dried *C. pasteurianum* in a rotary vacuum drier; the water bath was kept near 40°, and at no time were the cells allowed to freeze. The N_2-fixing system may be inactivated by cold but is not inactivated in a rotary vacuum drier; however, the permeability of the dried cells is altered so that extraction with buffer will release the soluble enzyme system.

In early studies, a variety of substrates, including sugars and organic acids, were furnished to the cell-free extracts to supply the energy required for N_2 fixation. Although pyruvate was among the substrates tested, it normally was furnished at levels to approximate saturation of the phosphoroclastic enzyme system which produces H_2, CO_2 and acetyl phosphate from pyruvate. Extracts from *C. pasteurianum* degrade pyruvate so rapidly that the formation of gas bubbles is visible. Carnahan and coworkers systematically investi-gated the influence of pyruvate concentration on N_2 fixation, and this proved highly important in achieving consistent cell-free fixation. A clearly defined optimal concentration was observed, and higher and lower levels of pyruvate were much less effective in supporting fixation. Although the optimal level varies among preparations, in general about 0.15 M pyruvate gives the best N_2 fixation. A continuing pyruvate metabolism was crucial in supporting N_2 fixation in the system of Carnahan *et al.* (1960b), because it served both as the source of electrons to reduce N_2 and as an energy source.

4.2 EXTRACTION, PURIFICATION AND PROPERTIES OF THE ENZYMES

4.2.1 Methods of extraction

A relatively simple preparation of an active nitrogenase can be obtained from dried cells of *C. pasteurianum* as described by Carnahan *et al.* (1960a, b). The cells which have been dried in a rotary vacuum drier will retain activity for months when stored in a vacuum or under an inert atmosphere in the refrigerator. When these cells are shaken anaerobically with buffer solution, the nitrogenase is extracted. Centrifugation sediments the cell debris and leaves a preparation whose nitrogenase is soluble in the sense that it is not sedimented by prolonged centrifugation at 150 000 x g. The stability of the stored, dried cells and the ease of preparation of the soluble enzyme by direct extraction recommend this method. In studies of N$_2$ fixation by *C. pasteurianum* extracts subsequent to 1960, this type of preparation has been used more extensively than any other.

In their 1960 investigations Carnahan *et al.* also indicated that they could make active preparations with a Hughes press. The Hughes press is a device which extrudes a frozen paste of bacterial cells through a small orifice by successive blows on a piston with a fly press. As the block of the press is pre-cooled, material extruded through the orifice is immediately re-frozen, and the treatment is relatively mild. The fact that the press has low capacity has limited its use.

Preparations from the azotobacter have generally been made by extrusion of a heavy slurry of cells through a French press (Bulen *et al.*, 1965). The French pressure cell is a heavy cylinder which can be subjected to 10 000-20 000 pounds pressure per sq. inch; the slurry is extruded past a narrow orifice adjusted by a needle valve, and in passing through the orifice the cells are ruptured. The method will handle a reasonable quantity of cells, and at this stage of the preparation it is not necessary to keep the azotobacter anaerobic to preserve activity.

It also is possible to obtain excellent preparations from the azotobacter by osmotic shock. Robrish and Marr (1962) disrupted the azotobacter by osmotic shock during their studies on the structure of the organism. The method was not utilized for preparing nitrogenase until rather recently when Oppenheim *et al.* (1970) described its application. The method is simple and yields a preparation with cell fragments somewhat different from those obtained with the French press.

Grau and Wilson (1963) prepared extracts by lysing *B. polymyxa* with the aid of lysozyme. The extract was active in N_2 fixation when supported by pyruvate metabolism.

Sonic oscillation also has been employed to disrupt cells to extract nitrogenase. This has been the method of choice for preparations from *R. rubrum* (Burns and Bulen, 1966; Munson and Burris, 1969). The disruption is performed in a suspension of the organisms through which inert gas is passed to maintain anaerobic conditions. The cells are disrupted in a vessel cooled with flowing water during the sonication.

4.2.2   Soluble and particulate preparations

The extracts from the dry cells of *C. pasteurianum* are notable because the nitrogenase is soluble. Solubility is a relative term, but in an operational sense vigorous centrifugation does not sediment a particle containing nitrogenase from these extracts. In contrast, when the azotobacter cells are disrupted with the French press, the bulk of the cell debris can be sedimented in a field of about 30 000 x g, and the resulting supernatant will yield the bulk of its nitrogenase activity in a pellet sedimented in 4-6 h at about 150 000 x g. This sedimentation can constitute a mechanical purification as an early step in the purification of nitrogenase. However, it necessitates that steps be taken subsequently to solubilize the protein constituents to effect their further purification. The particles sedimented from the azotobacter extracts are not uniform in structure but appear in the electron microscope as disrupted pieces of the cell membrane. There has been little study of the particulate or soluble nature of preparations from N_2-fixing agents other than the clostridia and the azotobacter.

4.2.3   Purification of the Mo-Fe protein and the Fe protein

Clostridium pasteurianum.   Methods for purification vary among laboratories but all follow many common steps. We will outline a few preparations, but the original papers must be consulted for details.

The preparations all start with the anaerobic extraction of dried *C. pasteurianum* by shaking with buffer for 30-60 min. The cell debris is sedimented leaving a supernatant with 25-40 mg protein/ml. It is advantageous to work with concentrated extracts to avoid unnecessary concentration procedures later. For some preparations a treatment with nuclease to reduce the viscosity by hydrolyzing the nucleic acids is useful.

The initial extracts should have a specific activity of 5 to 25. Specific activity is defined here in its most widely accepted form as nanomoles N_2 reduced per min per mg protein.

For many experiments a simple purification may be adequate. Winter and Burris (1968) reported that precipitation with acetone or alcohol yielded a mixture of the Mo-Fe and Fe proteins with a specific activity of about 50. For more extensive purification the Mo-Fe protein and the Fe protein are separated, purified individually and then recombined.

The general purification scheme employed for separation from *C. pasteurianum* of the two proteins that are involved in N_2 fixation was developed primarily by Mortenson and associates. The purification scheme of Mortenson, Morris and Yeng (1967) is representative. All operations for separation of the active proteins must be under strictly anaerobic conditions. To reduce the viscosity and to prevent occlusion of active protein when nucleic acids were removed, the preparation was treated with ribonuclease and deoxyribonuclease for 30 min. Protamine sulfate then was added to remove residual nucleic acids (Bulen *et al.*, 1965). After removal of the precipitate, additional protamine sulfate was added to precipitate the active proteins. The proteins were released from protamine sulfate with phosphocellulose, and the soluble material was placed on an anaerobic column of Sephadex G-100 and eluted with Tris-HCl. This column separated the Mo-Fe protein from the Fe protein and also separated ferredoxin. The Mo-Fe protein was purified further on a column of Sephadex G-200. The Mo-Fe protein was designated molybdoferredoxin and the Fe-protein azoferredoxin. Later, Mortenson, Morris and Kennedy (1968) introduced an additional calcium phosphate gel fractionation.

Vandecasteele and Burris (1970) introduced a simpler procedure employing separations on DEAE cellulose and Sephadex G-100. The concentrated crude extract from *C. pasteurianum* was placed directly on an anaerobic column of DEAE cellulose which had been equilibrated with O_2-free 0.02 M Tris buffer, pH 7.4 (Munson *et al.*, 1965). The Mo-Fe protein was eluted with 0.065 M $MgCl_2$ and the Fe protein with 0.09 M $MgCl_2$ in Tris buffer (Kelly *et al.*, 1967). Ferredoxin was left on the column. As the Mo-Fe protein fraction was contaminated with hydrogenase, it was diluted and purified further by passing it through another column of DEAE cellulose and eluting with a linear gradient (0.04 to 0.14 M) of $MgCl_2$ buffered with Tris. This separates hydrogenase from the Mo-Fe protein and yields an Mo-Fe protein of specific activity 280 (in the presence of excess Fe protein but based on Mo-Fe protein only).

Immediately after the Fe-protein had been eluted from the DEAE cellulose column, it was placed on a column of Sephadex G-100 and was eluted with pH 8, 0.02 M Tris buffer containing 2 mM $Na_2S_2O_4$ and 2 mM dithiothreitol. The specific activity, based upon the protein of the Fe protein only, is about 470. This relatively simple procedure involved a minimum number of manipulations and yielded proteins of high specific activity. Neither the Mo-Fe protein nor the Fe protein had activity by itself, but enzymatic activity was restored by mixing the two proteins without further protein additions.

This method employs $MgCl_2$ as the eluting salt as described by Kelly *et al.* (1967), whereas many workers employ NaCl. Each salt is satisfactory. Lower molar concentrations of $MgCl_2$ are required, because its ionic strength is greater than that of NaCl. Mg^{2+} is required in the reaction mixture, and judicious dilution often is sufficient without dialysis to permit further column chromatography. There is evidence, that the Fe protein is less stable and forms an inactive precipitate more rapidly in the presence of $MgCl_2$ than of NaCl.

The method is rapid and simple, however, some preparations of dried cells give anomalous results. Purification is less successful with extracts which have low initial activity. Variation between batches of *C. pasteurianum* appears to depend primarily upon the drying procedure. Cells should be dried rapidly without freezing and should be evacuated until thoroughly dry. The pooled dry cells then can be ground in a mortar and pestle, placed in test tubes, evacuated thoroughly with a high vacuum pump and sealed for storage at −20°. When properly prepared, such cells retain their activity for years; reduction in activity within a week is observed with improperly prepared cells.

These separations are simple in principle, but in practice the investigator must develop skill in anaerobic operations before achieving consistent success in the preparation of active materials of high purity. Assigning a percentage purity to these preparations is somewhat hazardous, because examination by polyacrylamide gel electrophoresis always reveals some impurity in the material. However, it seems likely that the best of these preparations from the clostridia are at least 90% pure.

Azotobacter   Nicholas and Fisher (1960) reported N_2 fixation in extracts from *A. vinelandii*. The fixation was minimal and no higher than that often obtained earlier by others. Nicholas, Silvester and Fowler (1961) also reported fixation of radioactive $^{13}N_2$ by extracts. These preparations, like the ones reported earlier, were not useful for studying the biochemistry of N_2 fixation, because their activity was low and erratic. Other investigators found no evidence that the supernatant medium played any role in the fixation process as reported by Nicholas and coworkers.

The first experimentally useful preparations from the azotobacter were reported by Bulen, Burns and LeComte (1964) and procedures subsequently reported depend heavily on their methods. An extract from *A. vinelandii* was supplemented with an ATP-generating system and an electron donor system consisting of hydrogenase and ferredoxin preparations from *C. pasteurianum*. This preparation soon was replaced by a preparation which employed hydrosulfite rather than ferredoxin as the electron donor (Bulen, Burns and LeComte, 1965). Since then, most investigators have used hydrosulfite as the electron donor in all systems, as it functions directly as an electron donor without requiring added ferredoxin.

Bulen and LeComte (1966) established that the nitrogenase of the azotobacter consisted of two proteins working in conjunction. They disrupted the azotobacter (the paste of *A. vinelandii* cells recovered by centrifugation of the culture can be stored at $-20°$ for months with little loss in activity) with a French pressure cell, and centrifugation at 144 000 xg for 30 min left 90-95% of the activity in the supernatant. The supernatant was treated with protamine sulfate to remove nucleic acids, and then additional protamine sulfate was added to precipitate the active material at $0°$. The precipitate in turn was solubilized by binding the protamine sulfate to cellulose phosphate. The soluble material was fractionated first on a column of Bio-Gel P-200, which removed cytochromes and other low molecular weight compounds, followed by a column of DEAE cellulose. The column was washed with Tris buffer at pH 7.2 and then was eluted stepwise with 0.15, 0.25, and 0.35 M NaCl. Although crude preparations from the azotobacter are not sensitive to oxygen in the first centrifugation steps, they are highly sensitive to oxygen when they are being separated chromatographically, and all operations must be anaerobic. The DEAE cellulose column separated the material into the Mo-Fe protein and the Fe protein; some Mo-Fe protein contaminated the Fe protein. Bulen and LeComte (1966) suggested that the two fractions were 70% pure. Only when they were mixed did they exhibit substantial enzymatic activity.

Kelly, Klucas and Burris (1967) employed a simplified method for purifying the Mo-Fe and the Fe proteins from *A. vinelandii*. Initial steps were those described by Bulen and LeComte, and the material regenerated from protamine sulfate precipitation was placed on a

DEAE cellulose column and eluted with solutions of NaCl followed by $MgCl_2$. 0.15 M NaCl eluted a cytochrome fraction, and 0.035 M $MgCl_2$ eluted the Mo-Fe protein. 0.060 M $MgCl_2$ removed an intermediate fraction which was discarded, and following this 0.090 M $MgCl_2$ removed the Fe protein. The fractions had high specific activity and proved stable to storage in liquid nitrogen.

Hwang (1968) employed a variety of preparative procedures with *A. vinelandii,* and adopted a rapid procedure which gave a preparation of reasonably high activity. The crude extract, prepared by disruption of cells with a French pressure cell, was centrifuged, and the supernatant material was heated anaerobically at 60° for 10 min. Centrifugation at 37 000 x *g* for 15 min sedimented an inactive protein fraction. The heated supernatant then was treated with streptomycin-SO_4 to remove nucleic acids. Streptomycin was chosen rather than protamine, because with it the results are more reproducible. The inactive precipitate was discarded, and the supernatant then was centrifuged at 144 000 x *g* for 60 min. The supernatant was discarded and the sedimented material was retained. The sediment was resuspended and recentrifuged at 37 000 x *g* for 15 min, and the supernatant material was retained as a preparation with activity equal to those of Bulen and LeComte (1966). Chen (unpublished observations) achieved further purification of Hwang's preparations by passing them through Sephadex columns.

Burns, Holsten and Hardy (1970) have reported the crystallization of the Mo-Fe protein from *A. vinelandii.* They observed that their DEAE-column fractions became turbid upon dilution, and that crystals were responsible for the turbidity. The Mo-Fe protein was solubilized by salts, and upon dilution it crystallized from the medium. The procedure for crystallization involved the following steps: The extract from cells broken in a French press was centrifuged for 16 h at 10 000 rpm. Nucleic acids were removed from the supernatant with protamine sulfate, and the extract was heated at 60° for 10 min. The nitrogenase remaining in solution was precipitated with protamine sulfate and resolubilized with cellulose phosphate. After treatment with $Na_2S_2O_4$, the material was fractionated on a column of DEAE cellulose into the two protein fractions, the Mo-Fe protein being eluted with 0.25 M NaCl and the Fe protein with 0.35 M NaCl. The Mo-Fe protein was concentrated by ultrafiltration, and when it was diluted approximately 10-fold, a substantial amount of material came out of solution. This material was recovered by centrifugation and the pellet was washed and then redissolved in NaCl solution; needles appeared upon dilution with buffer. These crystals were approximately 3 x 50 microns and could be collected by centrifugation; though they formed a brown solution they were themselves white. Ultracentrifugation to demonstrate homogeneity was reported for a single pH, and the absorption spectrum of the reduced compound suggests possible contamination with cytochrome. The yield of crystals has been somewhat variable, and others have experienced difficulty in obtaining active crystalline preparations. However, Burns *et al.* (1970) found that their crystals of Mo-Fe protein retained their specific activity and amino acid compositions constant through three recrystallizations. These crystals (plus the Fe protein) were active in the reduction of C_2H_2 and other substrates and were functional in the ATP-dependent H_2-evolution reaction.

Kelly and coworkers have employed *A. chroococcum* as a test organism. Its properties are very similar to those of *A. vinelandii.* Kelly (1969a) separated the Mo-Fe protein and the Fe protein from *A. chroococcum* by chromatography on DEAE cellulose and elution with $MgCl_2$ solution much as described for *A. vinelandii* (Kelly *et al.,* 1967).

Other N_2-fixing organisms. Considerably less attention has been paid to purification of nitrogenase components from organisms other than *C. pasteurianum* and the azotobacter. Schneider *et al.* (1960) reported fixation by extracts from *Rhodospirillum rubrum* prepared much like the preparation of clostridia described by Castle *et al.* (1960b). Although the data appeared convincing, later workers in the same and in other laboratories were unable to verify the results. Burns and Bulen (1966) disrupted fresh cells by sonic oscillation while maintaining the pH reasonably constant by addition of base. This preparation was reported as consistently active, but Munson and Burris (1969) had variable success with it, and Burns (personal communication) later also had difficulty in obtaining consistent N_2 fixation. The key to uniform preparations in Munson's hands was continuous culture of *R. rubrum.* Cells recovered from continuous culture regularly produced active cell-free preparations whether disrupted immediately after harvesting or after storage of the intact cells in liquid N_2. There has been little effort to purify the Fe protein and the Mo-Fe protein from *R. rubrum,* but there are no indications that these proteins differ substantially from those of other N_2-fixing agents.

Recently Winter and Arnon (1970) have prepared consistently active extracts from *Chromatium* strain D; earlier preparations from this organism had scarcely detectable activity (Arnon *et al.,* 1961).

Extracts were prepared by sonication or by disruption of the cells with a Ribi cell fractionator. No separation into the individual protein components or their purification was reported.

Witz, Detroy and Wilson (1967) compared N_2 fixation in extracts from *C. pasteurianum, Bacillus polymyxa* and *Bacillus macerans*. Witz and Wilson (1967) fractionated nitrogenase from *B. polymyxa,* and Mahl and Wilson (1968) obtained active cell-free preparations from *Klebsiella pneumoniae*. In conjunction with their studies on reconstruction of active nitrogenase from individual proteins isolated from different N_2-fixing organisms, Dahlen, Parejko and Wilson (1969) have purified the Mo-Fe and Fe proteins from *B. polymyxa* and *K. pneumoniae*. The proteins from these sources can be purified much as described by Kelly *et al.* (1967).

Although certain strains of the sulfate-reducing bacteria fix N_2 (Riederer-Henderson and Wilson, 1970), we are not aware of the preparation of any consistently active N_2-fixing extracts from them.

Biggins and Postgate (1969) disrupted cells of *Mycobacterium flavum* by sonic oscillation and recovered a cell-free extract capable of reducing N_2, protons, acetylene, KCN and CH_3NC. An ATP-generating system and $Na_2S_2O_4$ were supplied. The enzyme complex was sedimented in 3.5 h at 145 000 x g; in this respect it resembled the system from *A. vinelandii* rather than *C. pasteurianum* but, unlike azotobacter particles, it was sensitive to oxygen.

Schneider *et al.* (1960) obtained N_2 fixation with extracts from blue-green algae. Cox and Fay (1967) observed that N_2 fixation by extracts from *Anabaena cylindrica* was enhanced by addition of pyruvate.

One of the most interesting advances in recent years has been the demonstration of cell-free N_2 fixation with extracts from nodules of leguminous plants. The subject is discussed in detail in Chapter 6. The components of the symbiotic system had defied separation for years, and Bergersen (1960) had contended that the site of N_2 fixation in leguminous root nodules was a membrane surrounding the bacteroids. However, Bergersen (1966a) demonstrated that nodules crushed anaerobically and exposed subsequently at a low partial pressure of O_2 fixed N_2. Bergersen (1966b) presented evidence suggesting that the O_2 required was for the generation of ATP. Work from the laboratories of Evans and of Bergersen established that the seat of N_2 fixation was the bacteroids (Bergersen and Turner, 1967). Koch, Evans and Russell (1967) prepared active extracts from soybean nodule bacteroids, and Klucas *et al.* (1968) and Koch *et al.* (1970) described fractionation of the extract from root nodule bacteroids into the Mo-Fe protein and the Fe protein. Fractionation methods can be applied similar to those used with extracts from the free-living bacteria, as the properties of the isolated proteins are similar. Bergersen and Turner (1970) have reported their separation on Sephadex G-200 of the Mo-Fe protein and the Fe protein from soybean bacteroids. They assigned approximate molecular weights of 180 000 and 50 000, respectively, to these proteins.

4.2.4   *Properties of the Mo-Fe protein and Fe protein*
Characteristic fractionation schemes rely upon the separation of the Mo-Fe protein from the Fe protein by protamine sulfate precipitation or on columns of DEAE cellulose or Sephadex. The Mo-Fe protein is eluted first from the column, and with proper operation is separated rather cleanly from the Fe protein. Both proteins form brown bands on the columns which elute as brown solutions.

Information on the molecular weight of the clostridial proteins has come primarily from Mortenson's laboratory. Mortenson, Morris and Jeng (1967) reported that the Mo-Fe protein was 78% pure and had a molecular weight of 100 000. Later, Mortenson, Morris and Kennedy (1968) stated that the protein had a molecular weight of 125 000 to 150 000, and more recently Dalton and Mortenson (1970) reported a molecular weight of 200 000. In unpublished work from our laboratory based upon Sephadex gel filtration, and glycerol-gradient centrifugation, Chen has found a molecular weight of 212 000 for the Mo-Fe protein from *A. vinelandii,* and Tso has observed a molecular weight of 210 000 for the Mo-Fe protein from *C. pasteurianum.* The crystalline Mo-Fe protein isolated by Burns *et al.* (1970) from *A. vinelandii* had a molecular weight of 270 000 to 300 000 as determined by ultracentrifugation.

Although the proteins from *C. pasteurianum* and *A. vinelandii* do not necessarily have the same molecular weight, the high molecular weight of the *A. vinelandii* protein suggests that the more recent value of 200 000 for *C. pasteurianum* probably is approaching the proper value.

The Mo-Fe protein quite possibly consists of subunits. Chen's unpublished data for *A. vinelandii* suggested four subunits on the basis of dissociation with sodium dodecylsulfate. As there are only two atoms of Mo per aggregate, it is evident that not all of them can carry Mo if there are more than two subunits. No active subunit has been recovered, and the experimental data suggest that the aggregate of subunits must be present for enzymatic activity. It must be established whether some subunits of the molecule carry Fe but not

Mo whereas other subunits carry both Fe and Mo. Burns *et al.* (1970) list two Mo atoms for each protein unit of 270 000 to 300 000 and indicate the ratio of Mo:Fe:CySH:labile S as 1:20:20:15; doubling these numbers gives the content in atoms per molecule.

The Fe and Mo content of the Mo-Fe protein from *C. pasteurianum* is not entirely clear. Mortenson *et al.* (1967) reported one atom of Mo per molecule of the Mo-Fe protein of 100 000 molecular weight, 12 atoms of Fe and three atoms of sulfide. Mortenson *et al.* (1968) reported that a unit of molecular weight of 125 000 to 150 000 carried two Mo and eight Fe per molecule. Dalton and Mortenson (1970) stated that each unit of 200 000 molecular weight contain two Mo, 15 Fe, 15 acid labile sulfide and 95 acid-stable sulfur atoms. The analysis of Vandecasteele and Burris (1970) indicating one atom of Mo and 15 atoms of Fe per unit of 160 000 molecular weight is closer to the value observed by Burns *et al.* (1970) for the crystalline Mo-Fe protein from *A. vinelandii.* As the composition of the crystalline Mo-Fe protein from *A. vinelandii* is at variance with all reported values from *C. pasteurianum,* it must be resolved whether the compositions actually differ or whether incomplete purification of the protein is responsible for the differences.

Burns *et al.* (1970) reported that their crystalline Mo-Fe protein had 34 to 38 Fe and 26 to 28 labile sulfide per molecule of molecular weight 270 000. An interesting property of the crystallized protein is its low solubility in the absence of salt; it is almost insoluble at NaCl concentrations less than 0.08 molar. The spectrum of the crystalline protein shows an absorption shoulder at 410 to 420 nm. Treatment with $Na_2S_2O_4$ decreased this shoulder, and it is replaced by a peak at 420 nm plus absorption at 525 and 557 nm. Burns *et al.* (1970) allude to the measurement of Mössbauer spectra and determinations of magnetic susceptibility which suggest the predominance of high-spin Fe^{3+} and a small amount of high-spin Fe^{2+}. They also refer to EPR measurements yielding G values of 4.30, 3.67, 3.01 and 1.94 at 4°K. Matkhanov *et al.* (1969) also reported a 1.94 EPR signal from the N_2-fixing fraction from *A. vinelandii.*

Although the Mo-Fe protein from *C. pasteurianum* has not been obtained in crystalline form, it has been recovered in a high degree of purity. As with the corresponding protein from *A. vinelandii* it can be dissociated into subunits (treatment with sodium mersalyl and Chelex 100, Dalton and Mortenson, 1970). Treatment with Cleland's reagent, EDTA, α,α'-dipyridyl, urea and at pH 10 produced as many as six dissimilar subunits. Treatment with mersalyl followed by fractionation on Sephadex G-25 removed all Mo. Unpublished measurements by Tso indicate an isoelectric point of 4.95 for the Mo-Fe protein from *C. pasteurianum.*

The molecular weight of the Fe protein has been reported only by two groups. Moustafa and Mortenson (1969) reported that the molecular weight from *C. pasteurianum* is about 40 000, and that it has two nonhaem iron atoms, two labile sulfides and eight free SH groups. Mortenson (personal communication) now has found that the molecular weight is 55 000. Bergersen and Turner (1970) found the Fe protein from soybean nodules exhibited a molecular weight of 51 000 on Sephadex G-200 and had one iron per molecule of protein. The analysis by Vandecasteele and Burris (1970) indicated approximately three iron atoms per molecule assuming the molecular weight of 40 000 to be correct. With a molecular weight of 55 000, the iron content would be revised to four iron atoms per molecule. Jeng, Devanathan, Moustafa and Mortenson (1969) found no ESR signal for the native Fe protein, but upon its exposure to air a signal appeared with g = approximately 2.0. The protein is somewhat more acidic than the Mo-Fe protein.

One of the interesting properties of the Fe protein is its cold lability. Dua and Burris (1963) reported almost complete inactivation by anaerobic storage of extracts from *C. pasteurianum* at $0°$ for 24 h. Much more activity was preserved at $20°$ than at $0°$, so it was evident that some portion of the N_2-fixing enzyme complex was inactivated by cold. The preparations could be partially reactivated by incubation under H_2, but incubation under inert gases did not effect reactivation. The special activity of H_2 suggested that the reactivation resulted from a reduction, and Dua and Burris (1965) demonstrated partial restoration of activity by incubation with reducing agents such as 2-mercaptoethanol, cysteine, thioglycolate and glutathione. Because the N_2-fixing activity in the crude extracts was lost more rapidly than the phosphoroclastic activity, it was concluded that the cold-labile enzyme was a portion of the N_2-fixing complex. After the constituents of the N_2-fixing complex had been separated, Moustafa *et al.* (1968) demonstrated that the cold labile constituent of the complex was the Fe protein. Despite speculation regarding the nature of the cold lability, no experimental basis has been reported to support any choice among the mechanisms suggested (Dua and Burris, 1965; Moustafa and Mortenson, 1969). Moustafa (1969) also has reported that the N_2-fixing enzyme system from excised leguminous nodules exhibits cold liability similar to that shown by the free-living N_2 fixers.

Although the N_2-fixing enzyme complex is sensitive at $0°$, it is not inactivated at very low temperatures. Mortenson *et al.* (1967) reported that the Fe protein lost activity under all conditions of storage attempted. This protein is much more labile than the Mo-Fe protein, and its lability creates problems in purification and in experimental use. Kelly *et al.* (1967) demonstrated that the N_2-fixing enzyme components could be stored effectively at liquid N_2 temperatures. They observed no loss of activity over a period of 14 days. Other workers have adopted this procedure.

The separated, purified proteins have been tested in several laboratories for all of the enzymatic activities which they exhibit in combination. The general experience is that when tested separately they will not reduce N_2 or other alternative substrates, nor will they hydrolyze ATP, evolve H_2 or promote H_2 exchange. Evidently the two proteins must act together to effect these various enzymatic changes.

We speak of the N_2-fixing enzyme complex, but there is little direct experimental evidence that the enzymes actually must bind together to become effective. Kelly and Lang (1970) claimed that the Mössbauer spectra of permutations of labelled and unlabelled Mo-Fe and Fe proteins from *K. pneumoniae* indicated that a change in the chemical environment of the iron in the Mo-Fe protein occurred as a result of adding ATP, hydrosulphite and Fe protein. They supported their view that complex formation was necessary for activity by the observation that CO-inhibited binding of radio-active cyanide, which they regarded as a model for N_2-binding, required both proteins as well as ATP and hydrosulfite. Such evidence does not exclude the possibility that the two proteins act sequentially, but it seems more likely that they actually bind together to create active enzymatic sites.

It is somewhat difficult to determine exactly how many molecules of the Mo-Fe protein and how many of the Fe protein act in conjunction. Kelly's (1969a) titrations of *A. chroococcum* proteins indicated that more Mo-Fe protein was optimal for cyanide reduction than for acetylene or nitrogen reduction. Burris (1969) and Vandecasteele and Burris (1970) have presented evidence suggesting that two molecules of the Fe protein combine with one molecule of the Mo-Fe protein to produce the most active combination.

4.2.5   *The problem of a third protein*
The discussion of purification of the components of nitrogenase has

centered around the Fe protein and the Mo-Fe protein. These two highly purified proteins plus an ATP-generating system and $Na_2S_2O_4$ are sufficient to generate an active N_2-fixing enzyme system in the hands of most investigators. However, there have been statements in the literature that a third protein is required.

Mortenson (1966) and Mortenson, Morris and Jeng (1967) discussed the purification of the Mo-Fe and Fe proteins and suggested that an additional protein might be needed. This viewpoint was rejected by Kennedy, Morris and Mortenson (1968). Taylor (1969) again raised the issue and contended that N_2 fixation required a third protein which was not required for acetylene reduction. Kajiyama et al. (1969) also reported the necessity for three components from A. vinelandii. Jeng, Devanathan and Mortenson (1969) vigorously attacked the position of Taylor (1969), and presented convincing evidence that no third protein is required for N_2 fixation. The consensus clearly favors the adequacy of two proteins for the formation of nitrogenase.

4.2.6   Assays for fixation of N_2

Methods for estimating N_2 fixation will be discussed very briefly; for a more detailed description consult the section by Burris on 'Nitrogen Fixation—Assay Methods and Techniques' in Methods in Enzymology, edited by Colowick and Kaplan (specific volume edited by San Pietro). A survey of the acetylene assay appears as Appendix 2 of the present volume.

Fixation of $^{15}N_2$ can serve as a primary reference standard for N_2 fixation by enzyme preparations. Reactants minus one crucial component, e.g. the enzyme, can be placed in 5 or 15 ml (nominal sizes) serum or vaccine bottles stoppered with rubber serum stoppers. The vessel is evacuated through a hypodermic needle piercing the stopper and then filled with 0.2 atm. $^{15}N_2$ and 0.8 atm. of argon or helium. The last reactant is introduced at time zero by injection with hypodermic needle and syringe, and the reactants are incubated with shaking for an hour (it is advisable to inactivate replicates at intervals to establish the time course of the reaction) before inactivating with 0.1 volume of 5 N H_2SO_4. The entire contents of the vessel can be added to a small Kjeldahl flask for digestion, distillation and conversion to $^{15}N_2$ for analysis. Alternatively, NH_3 formed can be recovered by microdiffusion, and if sufficient is available it can be analyzed for ^{15}N. Without knowledge of the total nitrogen present, it is difficult to translate results into absolute terms.

A similar method can be employed to follow NH_3 formation without the use of $^{15}N_2$. Replicate reaction mixtures are inactivated at intervals by the addition of saturated K_2CO_3; a stopper carrying a glass rod dipped in H_2SO_4 is inserted into the vessel and NH_3 diffuses to the rod. The NH_3 is analyzed with the Nessler, ninhydrin or indophenol method. Some confusion may be introduced by production of ammonia from adenine and its derivatives (Hwang, 1968; Kirsteine et al., 1968; Lyubimov et al., 1969).

As the N_2-fixing system takes up N_2 and H_2 (when H_2 serves as the electron donor), the change in gas pressure can be measured as an indication of N_2 fixation as described by Mortenson (1964a). The μmoles of NH_3 formed are half the total μmoles of $N_2 + H_2$ that disappear.

Reduction of acetylene to ethylene provides a simple and highly sensitive test for the N_2-fixing enzyme system. About 0.2 atm. acetylene is injected into a reaction bottle as described for tests of $^{15}N_2$ reduction. After a short time, the system is inactivated with 0.1 volume of 5 N H_2SO_4 and a gas sample is withdrawn with a hypodermic syringe for injection into a gas chromatographic column. Measurement of the ethylene formed is an index of nitrogenase activity. The assay can be used as an indirect means for measuring available phosphorus (Stewart, Fitzgerald and Burris, 1970).

4.3   COFACTORS IN N_2 FIXATION

4.3.1   Cofactors for metabolism of pyruvate

The first consistent and vigorous cell-free N_2 fixation was supported by the metabolism of pyruvate (Carnahan et al., 1960b). The pyruvate was supplied in high concentration, because it was utilized rapidly by the phosphoroclastic reaction which produces acetyl phosphate, CO_2 and H_2 (Koepsell and Johnson, 1942). The optimal pyruvate concentration varied with the batch of C. pasteurianum, but was generally near 0.15 M. Although the point was not immediately evident, the metabolism of pyruvate furnished ATP as well as an electron donor in support of N_2 fixation.

Conversion of pyruvate to acetyl phosphate requires the mediation of ATP, and this in turn implies a need for a kinase and Mg^{2+}. Munson et al. (1965) described a method for producing anaerobic columns suitable for separation of constituents of the N_2-fixing complex, and with anaerobic columns of Sephadex G-25 they were able to remove low molecular weight cofactors from extracts of C. pasteurianum. The treated extracts had negligible activity for N_2

fixation, but activity greater than that of the original crude extract was restored with optimal concentrations of Mg^{2+}, ATP, ortho-phosphate, thiamine pyrophosphate and coenzyme A (ferredoxin was not removed in these experiments). Mortenson (1964a) had demonstrated the need for ferredoxin as an electron donor for the reduction of N_2. It was clear that cofactor concentrations were not optimal in the crude extracts. Among the cofactors required, ATP and Mg^{2+} were involved directly in N_2 fixation, and the ortho-phosphate and thiamine pyrophosphate were concerned with ATP regeneration and pyruvate metabolism. Ferredoxin served as the physiological electron donor for N_2 fixation.

4.3.2 Cofactors for the reduction of N_2

The introduction of $Na_2S_2O_4$ as an electron donor for N_2 fixation (Bulen, Burns and LeComte, 1965) and the demonstration that substrate quantities of ATP will support various reactions by nitrogenase (Dilworth *et al.*, 1965; Hardy and Knight, 1966b; Moustafa and Mortenson, 1967) provided a simplified enzymatic system. In such a system $Na_2S_2O_4$ serves as the electron donor without the necessity for ferredoxin, and ATP + Mg^{2+} function as the only cofactors required for N_2 fixation. In the clostridial cell, ATP + Mg^{2+} and ferredoxin are considered as cofactors functioning directly in N_2 fixation, and all other cofactors are considered to have an indirect role by regenerating ATP and reduced ferredoxin.

4.3.3 Ferredoxin and its properties

Mortenson, Valentine and Carnahan (1962) described a low potential electron transport factor from *C. pasteurianum,* and Mortenson (1964a, b) clearly showed that it could serve as an electron donor in N_2 fixation. It became evident that the substance, named ferredoxin, was closely related to the methaemoglobin-reducing factor described by Davenport, Hill and Whatley (1952), Davenport and Hill (1960) and crystallized by Hill and Bendall (1960), and to the photo-synthetic pyridine nucleotide reductase studied by San Pietro and Lang (1958). The ferredoxins have a remarkably low oxidation-reduction potential and serve as electron donors in photosynthesis, N_2 fixation and a variety of other biological reactions. The properties of ferredoxins and other nonhaem iron compounds have been reviewed by Valentine (1964), Beinert and Palmer (1965), Malkin and Rabinowitz (1967a), Hall and Evans (1969) and Buchanan and Arnon (1970).

Ferredoxin was first shown to function in N_2 fixation in *C. pasteurianum*. The evidence strongly supports its role as an electron donor for N_2 reduction in this organism, but evidence for functional ferredoxins in other N_2-fixing organisms has accumulated slowly. Although ferredoxin is present in blue-green algae and is presumed to have a role there in photosynthesis, its functioning in N_2 fixation has not been demonstrated. There has been more interest in its function in the photosynthetic bacteria. Tagawa and Arnon (1962) demonstrated that ferredoxin from *C. pasteurianum* could replace spinach ferredoxin in photochemical and dark reduction of NADP, as could ferredoxins from *Chromatium* sp. and *R. rubrum*. The systems could be coupled with hydrogenase to release H_2. Evans and Buchanan (1965) reported that all photosynthetic bacteria examined yielded ferredoxin. They also reported the photochemical generation of reduced ferredoxin in a strictly bacterial system; particles of *Chlorobium thiosulfatophilum* reduced ferredoxin from this organism and from *Chromatium* sp. in the presence of an electron donor such as sodium sulfide or mercaptoethanol. The reduced ferredoxin in turn functioned in reductive carboxylation to form pyruvate.

Ferredoxins have been reported from about 20 higher plants and bacteria. Although their occurrence appears to be universal in green plants, it is irregular in bacteria and questionable in animal tissues. The ferredoxins assume various crystalline forms and vary in molecular weight from 6000 to 18 000. The oxidation-reduction potentials which have been measured are in the range -395 to -490 mV for the true ferredoxins. The definition of a ferredoxin still has not been clarified officially, so some authors interpret the term more broadly than others. Most would prefer to restrict the term to the low potential, relatively low molecular weight nonhaem iron proteins containing labile sulfide.

Little disagreement remains on the number of electrons transferred by various species of ferredoxin. It is generally accepted that spinach ferredoxin transfers one electron per molecule; although Tagawa and Arnon (1968) have contended that this also is true of ferredoxin from *C. pasteurianum,* the preponderance of evidence shows that ferredoxins from *C. pasteurianum* and *Chromatium* sp. transfer two electrons.

The Fe content of ferredoxins ranges from two atoms per molecule in spinach ferredoxin to eight atoms per molecule of *C. pasteurianum* ferredoxin. The labile sulfur is released as inorganic sulfide upon acidification of ferredoxin (Fry and San Pietro, 1962).

Various explanations have been advanced to describe a structure of clostridial ferredoxin which will incorporate 8 Fe and 8 S. Blomstrom *et al.* (1964) reported that the Fe atoms were not equivalent, and they proposed that the Fe atoms were arranged linearly and were bound together with sulfur bridges furnished by cysteine residues and inorganic sulfide atoms. As Gersonde and Druskeit (1968) point out, there remains an uncertainty about the origin and nature of the labile S; they found that ferredoxin reacted the same with *p*-chloromercuribenzoate whether denatured or not. Malkin and Rabinowitz (1966) prepared apoferredoxin by treatment with Na mersalyl and α,α'-dipyridyl; the apoferredoxin was free of Fe and labile S and inactive. It could not be reactivated. If the H_2S originated from cysteine residues, dehydroalanine should have been formed in the process, but dehydroalanine was not detected in the apoferredoxin; apparently the labile S does not arise from cysteine. Jeng and Mortenson (1968) agreed with this conclusion and demonstrated that ^{35}S sulfide exchanged with the labile S of ferredoxin.

Malkin and Rabinowitz (1967b) found that ferredoxin from *Clostridium acidiurici* reacts slowly with *o*-phenanthroline anaerobically, but after reduction of ferredoxin one of its Fe atoms reacts rapidly, and the others are unreactive. In high concentrations of urea or guanidine · HCl and under aerobic conditions, all the Fe reacts with *o*-phenanthroline. Native ferredoxin did not react with 5,5′-dithiobis(2-nitrobenzoic acid) (DTNB), an SH reagent; however, in 4 M guanidine · HCl aerobically or anaerobically 14 moles of DTNB reacted/mole of ferredoxin. This reaction involves only the inorganic sulfide of ferredoxin. If EDTA is added to the mixture, cysteine residues as well as inorganic sulfide react anaerobically with DTNB. After the inorganic sulfide has reacted with DTNB the ferredoxin no longer is oxygen sensitive, so Malkin and Rabinowitz concluded that the site of oxygen sensitivity is the inorganic sulfide portion of ferredoxin.

The properties of ferredoxin also have been examined by physical methods. Sieker and Jensen (1965) studied the X-ray pattern of bacterial ferredoxin and produced a Patterson map at 5 Å resolution. Prominent peaks were provided from the Fe-Fe and Fe-S interactions. Although a detailed structure could not be derived, as ferredoxin presents substantial difficulties in X-ray analysis, they were of the opinion that the iron was not arrayed in linear fashion as had been suggested by others. Palmer, Sands and Mortenson (1966) studied the EPR spectrum of ferredoxin from *C. pasteurianum*. The *g*

= 2.0 signal typical of high spin ferric iron was present, and after reduction with dithionite, signals appeared at 1.89 and 1.96. It was suggested that the 1.94 class of signal (1.96 observed) arose from low spin ferric iron in tetrahedral symmetry. Hollocher and Luechauer (1968) studied the combined nonhaem iron proteins of *A. vinelandii* and found that the EPR spectrum from cells grown on ^{15}N was very little different from those grown on ^{14}N. The compounds were not isolated, but it was concluded that the nitrogen was not a ligand of the iron in the nonhaem iron proteins of the organism. Poe *et al.* (1970) also measured the proton magnetic resonance of ferredoxin from *C. pasteurianum* and presented a model of the compound. Druskeit, Gersonde and Netter (1967) on the basis of spectral and magnetic properties of ferredoxin from *C. pasteurianum* concluded that iron occurs in two oxidation states in the oxidized form of ferredoxin.

Tanaka *et al.* (1966) established the sequence of the 55 amino-acid residues of *C. pasteurianum* ferredoxin. Benson, Mower and Yasunobu (1967) also determined the sequence in ferredoxin from *C. butyricum*. The sequence differed in nine positions from that of the ferredoxin from *C. pasteurianum,* but positions of the cysteine and proline residues did not vary between the two ferredoxins.

Although physiological studies of the role of ferredoxin in N_2 fixation have centered on *C. pasteurianum,* a role for ferredoxin in other organisms also has been sought. The electron transport pathway for N_2 fixation in *A. vinelandii* is more complex than in the anaerobic clostridia. Shethna and his colleagues reported the isolation of three nonhaem iron proteins and a flavoprotein from *A. vinelandii* (Shethna *et al.,* 1964, 1966, 1968; Shethna, 1970a). Only nonhaem Fe protein III has the properties of a typical ferredoxin. Shethna (1970a) reported that it had six to seven iron atoms per molecule, an equivalent number of labile sulfides and a molecular weight of approximately 13 000. Its properties are quite similar to those of clostridial ferredoxin. Recently Benemann *et al.* (1969) and Yoch *et al.* (1969) have utilized these electron carriers from the azotobacter in reconstructing artificial systems for N_2 fixation. The flavoprotein reported by Shethna, Wilson and Beinert (1966) and crystallized and characterized by Hinkson and Bulen (1967) has been named azotoflavin by Benemann *et al.* (1969). They mixed an extract of *A. vinelandii* cells with spinach chloroplasts and demonstrated that the added flavin from azotobacter could be photoreduced. Electrons donated by the flavoprotein were effective in reducing acetylene to ethylene and N_2 to ammonia. At the Airlie

House Conference in 1968, Shethna reported the isolation of a nonhaem iron compound from *A. vinelandii* with properties similar to bacterial ferredoxins; Shethna (1970a) described the properties of the ferredoxin in detail. Yoch *et al.* (1969) utilized this ferredoxin in a reconstructed system similar to that employed with the flavoprotein. A chloroplast system in the light reduced the ferredoxin from azotobacter, as it did its native chloroplast ferredoxin, and the electrons from the ferredoxin reduced NADP and also acetylene to ethylene. The ferredoxin from azotobacter was considerably more effective as an electron donor than the flavoprotein from the same organism. Since these agents probably have a physiological role in electron transfer for N_2 reduction in azotobacter, it should be determined what system in *A. vinelandii* functions in lieu of the strong reducing power of the chloroplast in light.

Yates and Daniel (1970) prepared particles from *A. chroococcum* which reduced acetylene to ethylene with electrons from NADH, glucose 6-phosphate, and other physiological donors. NADH and glucose 6-phosphate were the best donors, but they were less than 10% as effective as $Na_2S_2O_4$ in he presence of benzyl viologen and less than 1% as effective in its absence. As no electron carriers were isolated from the organism, there was no indication that ferredoxin or flavoproteins were involved.

Apparently *B. polymyxa* is the only other organism whose electron transfer in N_2 fixation has been studied in any detail. Shethna (1970b) reported the isolation of ferredoxin from *B. polymyxa* strain Hino. It has an EPR signal and an optical absorption spectrum comparable to other bacterial ferredoxins, and can be reduced by H_2 plus hydrogenase from *C. pasteurianum* or *B. polymyxa*. Shethna, Stombaugh and Burris (1971) have purified the ferredoxin from *B. polymyxa* to homogeneity as measured by polyacrylamide gel electrophoresis. When ferredoxin was removed from a cell-free extract of *B. polymyxa* its capacity for acetylene reduction with pyruvate as substrate was greatly reduced. The addition of pure ferredoxin from the organism restored activity. A 20-fold increase in activity was attained in some experiments, although restoration of this magnitude is irregular. In *B. polymyxa,* ferredoxin clearly can serve as the electron donor for acetylene reduction and by implication for N_2 fixation.

Ferredoxin has been biosynthesized with cell-free preparations. Trakatellis and Schwartz (1969) used a cell-free extract from *C. pasteurianum* and demonstrated the incorporation of ^{14}C-labeled L-alanine into the ferredoxin which was isolated in crystalline form.

Studies with radioactive iron also suggested that the iron may be combined with apoprotein after the latter is almost completely synthesized on polysomes. Nepokroeff and Aronson (1970) also observed incorporation of amino acids into ferredoxin formed by a cell-free preparation from *C. pasteurianum* protoplasts. They suggested that incorporation of amino acids was for the completion of pre-existing peptide chains rather than for a complete synthesis of the protein.

4.3.4 *Fe, Mo and other elements*

It has been recognized for many years that N_2-fixing organisms require Fe and Mo for growth and N_2 fixation, and that a number of additional elements are required for growth. Separation of the components of nitrogenase now makes it evident that Fe and Mo are integral parts of the nitrogenase enzyme complex. The role of other elements, e.g. Co, is indirect; they influence growth and general metabolism in contrast to the direct participation of Fe and Mo in N_2 fixation.

4.4   REACTIONS CATALYZED BY NITROGENASE

4.4.1 *Complementary functioning of Mo-Fe and Fe proteins*

Demonstration that nitrogenase involves the joint functioning of an Mo-Fe protein and an Fe protein raised the question whether the proteins can be interchanged from various sources to generate active N_2-fixing complexes. Detroy *et al.* (1968) demonstrated that, in fact, active preparations can be made with the Mo-Fe component from one organism plus the Fe protein from another organism. Their investigation involved preparation of separated Fe protein and Mo-Fe protein from *C. pasteurianum, A. vinelandii, K. pneumoniae* and *B. polymyxa*. Of the 16 possible crosses between the two components from these organisms four are homologous crosses involving constituents from the same organism. Among the twelve remaining crosses, six generated active preparations and six were inactive. The organisms with extreme properties, such as the strictly aerobic *A. vinelandii* and the strictly anaerobic *C. pasteurianum*, did not generate active crosses. In contrast, more closely allied organisms such as *K. pneumoniae* and *B. polymyxa* gave activity from all crosses tested. *C. pasteurianum* generated only one active non-homologous cross, the cross between the Mo-Fe protein of *C. pasteurianum* and the Fe protein of *B. polymyxa*.

Dahlen, Parejko and Wilson (1969) examined the activity of mixtures of Mo-Fe and Fe proteins for reducing substrates other than N_2; they determined the activity on acetylene, azide and cyanide and showed the same complementarity that had been observed for N_2 fixation. In addition to demonstrating the complementarity between specific organisms, these results also supported the concept that capacity for N_2 fixation is reflected in capacity for reduction of alternate substrates; evidently the same enzyme system catalyzes all these reactions. Kelly (1969b) also has compared the cross reactions between *K. pneumoniae, A. chroococcum* and *B. polymyxa*. His extracts from *C. pasteurianum* had unusually low activity, and he did not test this organism for cross reactivity. The results were the same as reported by Detroy *et al.* (1968) with the exception that Kelly observed activity in the cross between the Mo-Fe protein of *A. chroococcum* and the Fe protein of *B. polymyxa*. Kelly used *A. chroococcum*, whereas Detroy *et al.* used *A. vinelandii*, and this may explain the discrepancy. Table 1 is a composite table of cross reactions so far known to the writer and includes unpublished data of Biggins, Kelly and Postgate (personal communication).

Studies on complementarity between the Mo-Fe protein and the Fe protein from N_2-fixing organisms emphasized the close similarity among organisms of diverse origins. These studies should be extended

TABLE 4.1. Summary of cross reactions between nitrogenase components from N_2-fixing bacteria.

	Mo-Fe protein						
	Azot.v	Azot.c	Kleb.	Bac.	Clos.	Rhod.	Myc.
Fe protein							
Azot.v	+	...	+	+	−	...	...
Azot.c	...	+	+	±	...	tr	±
Kleb.	+	+	+	+	−	+	+
Bac.	−	tr	+	+	+	±	±
Clos.	−	...	−	+	+	−	−
Rhod.	...	...	...	...	...	+	...
Myc.	...	+	+	+	...	+	+

+ means 80 to 100% of activity of homologous cross;
± = about 50%; tr = about 10%; − = no activity; ... = no test recorded

Azot.v = *Azotobacter vinelandii*; Azot.c = *A. chroococcum*; Kleb. = *Klebsiella pneumoniae*; Bac. = *Bacillus polymyxa*; Clos. = *Clostridium pasteurianum*; Rhod. = *Rhodospirillum rubrum*; Myc. = *Mycobacterium flavum* 301.

to the photosynthetic bacteria, the blue-green algae, and preparations from the bacteroids of leguminous root nodules. The evolutionary development of the N_2-fixing enzyme system has not involved great variability as evidenced by the frequent cross reactions among the proteins. However, organisms widely disparate in their properties do not regenerate an active complex when their two proteins are crossed.

The studies of the assembly of nitrate reductase from protein subunits of a *Neurospora* mutant and subunits from other organisms has interesting implications for nitrogenase (Ketchum *et al.*, 1970) and is discussed further in Chapter 6.

4.4.2   Mutant organisms

The fact that nitrogenase can be resolved into an Mo-Fe protein and an Fe protein invited experiments to determine whether mutants deficient in each or both proteins could be produced. Fisher and Brill (1969) produced mutants of *A. vinelandii* by treatment with nitrosoguanidine. Two mutants selected for study grew on combined nitrogen but not on N_2. Extracts made from the mutants did not contain an ATP-dependent H_2 evolving system nor could they fix N_2, reduce acetylene, cyanide or azide. Tests with purified components from the wild-type organism established that one of the mutants was deficient in the Mo-Fe protein and the other was deficient in both the Mo-Fe and the Fe protein. Subsequently, mutants deficient in the Fe protein only were produced. These organisms should prove very helpful in studies of the mechanism of N_2 fixation. Sorger and Trofimenkoff (1970) also produced mutants from the azotobacter which grew on ammonia but not on N_2; they designated the two types of mutants as carrying enhancement and enhanceable factors. Their studies suggested that the enhancement factor was equivalent to the Fe protein and the enhanceable factor to the Mo-Fe protein.

4.4.3   Reductants

As indicated earlier, a low potential reductant is required to effect the reduction of N_2 and other substrates; metabolites from pyruvate reduced N_2 in *C. pasteurianum*. Crude extracts metabolized pyruvate and yielded reduced ferredoxin to serve as the physiological electron donor. The only other compound which supported N_2 fixation was α-ketobutyrate. Hwang (1968) demonstrated that formate could serve as an electron donor for extracts from *C. pasteurianum*.

Electron transport pathways for the various N_2-fixing organisms may often constitute the means for generating ATP as well as the reductants necessary for N_2 fixation; therefore there has been considerable interest in these pathways. In most N_2-fixing organisms it is not clear what steps in electron transport are involved directly in generation of functional reductants. Intensive study has centered on the azotobacter because of their unusually vigorous respiration through the cytochrome chain and their convenience for study. Cytochrome absorption bands are evident in intact suspensions of the azotobacter; Tissiéres (1956) purified two c-type cytochromes from *A. vinelandii* and designated them cytochromes c_4 and c_5. Neumann and Burris (1959) crystallized these two cytochromes and described their physical characteristics. Swank and Burris (1969b) observed a third cytochrome very similar in its properties to cytochrome c_4 but with half its molecular weight. In addition to cytochromes c_4 and c_5, Jones and Redfearn (1967a) demonstrated by spectral observations the presence of cytochromes b_1, a_1, a_2 and o; the last three were considered to be terminal oxidases. Observations by Jones and Redfearn with specific inhibitors suggested that the cytochrome system is branched and passes through two terminal oxidase systems, a matter discussed further in Chapter 5.

Ivanov and Gogoleva (1968) reported that inhibitors of the cytochrome system also blocked N_2 fixation. Ivanov (1967) suggested that N_2 may be reduced by cytochrome, but the validity of the experimental support has been questioned (see Appendix 1), and the potential of measured cytochrome systems is too positive for this reduction. Gvozdev *et al.* (1968) and others (see Chapter 5) have questioned Ivanov's conclusions, because they can observe no quantitative nor qualitative differences in the cytochromes of azotobacter capable of fixing N_2 and azotobacter grown on combined nitrogen and incapable of N_2 fixation. It is highly improbable that the cytochromes operate directly in N_2 fixation, and any function probably is indirect, in ATP generation coupled to electron transfer. Jurtshuk *et al.* (1969) have studied a variety of carriers of the azotobacter chain, but their relation to N_2 fixation has not been established.

The flavoprotein from *A. vinelandii* is reduced to the semiquinone form with little further reduction (Shethna *et al.*, 1966). Hinkson (1968) dissociated the pure compound (Hinkson and Bulen, 1967) into FMN and the apoprotein and reconstituted the active holoprotein. Ivleva, Sadkov and Yakovlev (1969) presented evidence that electrons were transferred to N_2 at the level of the flavin components of the respiratory chain. As indicated earlier, Benemann *et al.* (1969) captured light energy with spinach chloroplasts and coupled this energy through azotobacter flavoprotein for the reduction of acetylene and N_2.

Ubiquinone also can function in the electron transfer chain of *A. vinelandii* (Temperli and Wilson, 1962). Swank and Burris (1969a) demonstrated that NADH oxidase activity could be greatly reduced in particles from *A. vinelandii* by extraction with pentane. About 80% of the activity could be restored by the addition of ubiquinone. Ubiquinone 8, the form found in *A. vinelandii* (Lester and Crane, 1959), is abundant both in the red and green electron transport particles from the azotobacter (Jones and Redfearn, 1967b). Knowles and Redfearn (1969) observed that the content of ubiquinone 8 was greatly reduced in organisms grown on fixed nitrogen.

D'Eustachio and Hardy (1964) demonstrated that, in addition to pyruvate, H_2, KBH_4 and NADH can furnish electrons for N_2 fixation in extracts from the clostridia. Electrons could be transferred from these agents by way of ferredoxin or methyl viologen. Naik and Nicholas (1969) reported NADH oxidase activity in small particles from *A. vinelandii* capable of fixing N_2. These particles had a transhydrogenase (for transferring hydrogen between NAD and NADP), an enzyme studied in detail by Chung (1970). Yates and Daniel (1970) investigated the role of physiological electron donors serving instead of $Na_2S_2O_4$ in extracts from *A. chroococcum;* they obtained membrane-associated preparations which reduced acetylene with NADH or with NADH-generating substrates but which required no artificial carrier.

In *C. pasteurianum* the electron transfer is an anaerobic process and does not involve the multiplicity of carriers characteristic of the azotobacter. Apparently ATP and reduced electron donors normally are generated by metabolism of pyruvate, a common metabolite for the clostridia. The phosphoroclastic breakdown of pyruvate and accompanying reactions (Mortenson, Valentine and Carnahan, 1963) yield both ATP and reduced ferredoxin (Mortenson, 1963) which can be utilized to reduce N_2 or other substrates or to liberate H_2. Mortenson (1964a) first clearly showed the requirement of ferredoxin for N_2 fixation by extracts from *C. pasteurianum*. Removal of ferredoxin drastically reduced the N_2 fixing activity, and its readdition to the preparations restored their capacity for N_2 fixation. Jungermann *et al.* (1969) also demonstrated the requirement for ferredoxin in H_2 evolution by *Clostridium kluyveri;* transfer of electrons was most effective from NADPH.

Rubredoxin is another interesting electron transfer agent isolated from *C. pasteurianum* (Lovenberg and Sobel, 1965; Lovenberg, Williams and Bruckwick, 1969). To date there is no evidence for a function of rubredoxin in electron transfer in the process of N_2 fixation in the clostridia.

Flavodoxin is a low molecular weight flavoprotein which will substitute for ferredoxin in N_2 fixation, H_2 release and pyruvate metabolism. It is somewhat less effective than ferredoxin but functions well in all of these reactions. It appears in *C. pasteurianum* which is grown on an iron-deficient medium. Under these growth conditions a very small amount of ferredoxin is formed, and the continuing growth and N_2 fixation by the organisms appears to be mediated by flavodoxin (Knight and Hardy, 1966).

Desulfovibrio desulfuricans represents another strictly anaerobic N_2-fixing organism. Little is known of the electron transfer pathway for N_2 fixation in the organism, but Newman and Postgate (1968) have isolated rubredoxin from the organism, and it contains a low potential cytochrome c_3.

The facultative organisms such as *K. pneumoniae* and the microaerophilic bacilli such as *B. polymyxa* fix N_2 only under anaerobic conditions, and hence one would anticipate that their electron transfer mechanisms might be similar to those demonstrated in the clostridia. To date, the only clear evidence of electron donation for N_2 fixation in these organisms is the observation by Shethna, Stombaugh and Burris (1970) that a ferredoxin donates electrons for N_2 fixation in *B. polymyxa*.

Arnon and coworkers studied the electron transfer supporting N_2 fixation by the photosynthetic bacteria. They demonstrated a low level of N_2 fixation by *Chromatium* sp. supported by H_2 (Arnon *et al.*, 1961). In 1966 Bachofen and Arnon crystallized ferredoxin from chromatium, and Buchanan and Bachofen (1968) demonstrated that the reduced ferredoxin could reduce NAD. H_2 could reduce ferredoxin in the preparations from this organism. Yoch, Mortenson and Lindstrom (1968) demonstrated that thiosulfate could serve as electron donor to support N_2 fixation in *Rhodopseudomonas palustris*. Yoch and Arnon (1970) reconstructed a system capable of fixing N_2 or reducing acetylene with particles from the chromatium able to generate ATP by cyclic photophosphorylation. This preparation was combined with a preparation of spinach chloroplasts to generate reduced ferredoxin. The ferredoxin either from chromatium or *C. pasteurianum* was functional, but the ferredoxin from spinach chloroplasts was inactive.

Investigation of electron transfer in root nodules poses special problems because of the necessity for isolation of the bacterial system from the plant system. Matters such as the role of haemoglobin in root nodules, which now seems to have no direct function in N_2 fixation but probably serves only indirectly in control of the O_2 at a level sufficient to support ATP formation but insufficient to inactivate the N_2-fixing system, are discussed by Professor Evans in Chapter 6. Yoch *et al.* (1970) utilized the N_2-fixing preparation developed by Evans and coworkers to demonstrate in a reconstructed system that an electron carrier, apparently similar to the iron-sulfur protein III of *A. vinelandii*, functioned in the reduction of acetylene to ethylene. There was no demonstration of reduction of N_2. The carrier was utilized as a crude, heat-treated extract of bacteroids or as a preparation which had been partially purified by adsorption and elution from DEAE cellulose. A broad shoulder of absorption between 360 and 416 nm and the formation of a brown band on DEAE cellulose constitute the only evidence that the compound is similar to the iron-sulfur protein III of Shethna which they refer to as azotobacter ferredoxin.

4.4.4   Utilization of ATP

The first experimental evidence that ATP is required in support of N_2 fixation was supplied by McNary and Burris (1962). They demonstrated that arsenate or glucose, which served with hexokinase to trap ATP, effected a strong inhibition of N_2 fixation. Although the evidence supported a role for ATP in N_2 fixation, the report was initially greeted with skepticism and others suggested alternative explanations for the observed effects of arsenate and glucose. However, when other investigators checked the results experimentally, they found that in fact ATP is required for N_2 fixation. For example, Hardy and D'Eustachio (1964) demonstrated that ATP was required for N_2 fixation by extracts of *C. pasteurianum* and that it could be generated from acetyl phosphate; in the presence of an electron donor such as KBH_4, N_2 was reduced. Mortenson (1964) also observed that ATP was required by extracts from *C. pasteurianum* in addition to reduced ferredoxin to reduce N_2 to ammonia. Employing creatine phosphate plus creatine phosphokinase for ATP generation, Bulen, Burns and LeComte (1964) demonstrated the necessity for ATP in extracts from *A. vinelandii*. Dilworth *et al.* (1965) removed ATP with activated carbon from extracts of *C. pasteurianum* and restored activity both with an ATP-generating system and by direct addition of small amounts of ATP. Munson, Dilworth and Burris (1965) reduced N_2 fixation virtually to zero by removing ATP from extracts of *C. pasteurianum* and restored activity by the addition of optimum concentrations of cofactors to a level greater than that initially observed with crude extracts. Bergersen and Turner (1968) demonstrated that preparations from bacteroids of soybean nodules require ATP and a strong reductant exactly as do other N_2-fixing systems. These demonstrations with various organisms established the absolute requirement of biological N_2 fixation for ATP. The requirement is highly specific and ATP analogues are not functional in the reaction. The need for ATP is general, and all organisms tested to date require it for N_2 fixation and the reduction of alternative substrates.

The question arises how much ATP is needed to support N_2 fixation. There has been disagreement on this point, but results generally have supported a requirement of 4-5 ATPs per pair of electrons transferred. Thus, 12-15 ATP molecules are required for each N_2 reduced to 2 NH_3. Although Hardy and Knight (1966b) reported a requirement of only two ATPs per pair of electrons transferred in extracts from *C. pasteurianum*, Mortenson (1966) originally found a requirement of four but more recently reported two in the H_2-evolving reaction when corrected for ATP hydrolysis in the absence of reductant (Jeng, Morris and Mortenson, 1970). The experiments of Winter and Burris (1968) indicated a requirement between 4 and 4.6 dependent upon the pH of the reaction medium with *C. pasteurianum,* and those of Bulen and LeComte (1968) and Hadfield and Bulen (1969) 5 ATPs by *A. vinelandii* preparations per pair of electrons transferred for N_2 or acetylene reduction under various conditions. The number of ATP molecules needed decreased as the temperature was lowered. ATP/N_2 ratios in living cells are discussed in Chapter 5.

The generation of ATP is dependent upon the electron transfer process in the N_2-fixing organisms. Although high P/O ratios are obtained routinely for mitochondria isolated from plant and animal sources, it has proved considerably more difficult to get high efficiency for ATP formation in cell-free preparations from bacteria. Preparations from *A. vinelandii* generally have had P/O ratios near 0.2. Ackrell and Jones (1970) reported a membrane preparation from *A. vinelandii* with a P/O ratio of 1.1 for NADH. Knowles and Smith (1970) found that growing cells of *A. vinelandii* have a P/O ratio of 2.0 with β-hydroxybutyrate as substrate. Not only may ATP formation be more efficient in intact organisms, but the number of ATPs required per pair of electrons transferred may be less for intact organisms than for extracts.

The function of ATP in biological N_2 fixation remains unsolved. Suggestions have been made without experimental support that ATP has an allosteric effect or that it activates some portion of nitrogenase. An analogy has been drawn with the ATP requirement for reverse electron transfer (e.g. formation of reduced pyridine nucleotides from succinate oxidation in the presence of ATP). Biological N_2 fixation is not particularly demanding from an energetic standpoint, but ATP clearly is required to drive the process. The definitive experiments to establish the role of ATP remain to be performed.

Hardy and Knight (1966b) inferred that the products of ATP hydrolysis are ADP and orthophosphate, and Kennedy, Morris and Mortenson (1968) clearly established this point. This also is implicit in the use of creatine kinase for the regeneration of ATP from ADP. There is no evidence for the formation of AMP and pyrophosphate in the N_2-fixation reaction. Biggins and Dilworth (1968) presented evidence that ADP influences the concentration of acetyl phosphate and serves as a control agent in the phosphoroclastic reaction. Addition of arsenate or ATP + ATPase enhanced pyruvate breakdown by preventing the accumulation of acetyl phosphate suggesting that acetyl phosphate may serve as a coarse control and ADP as a fine control on pyruvate metabolism in *C. pasteurianum*.

Hardy and Knight (1966b) described an ATPase requiring the presence of $Na_2S_2O_4$ as a reductant in extracts from *A. vinelandii* and found that this ATPase activity was purified in parallel with the N_2-fixing activity. The purified Mo-Fe protein plus the Fe protein are necessary to support its activity (Bulen and LeComte, 1966; Kennedy, Morris and Mortenson, 1968). Apparently the nitrogenase *per se* functions as an ATPase during concomitant electron transfer; neither component by itself supports hydrolysis of ATP to ADP plus P_i. Bui and Mortenson (1969) verified their earlier report that ATP hydrolysis does not require the presence of an electron donor such as dithionite at low pH. They suggested that ATP may react with nitrogenase before the nitrogenase is reduced. They also proposed that the ATP-hydrolyzing complex may not require the same balance of Mo-Fe protein and Fe protein molecules as does the N_2-reducing system.

4.4.5   Metabolism of H_2

Evolution of H_2 from $Na_2S_2O_4$ by nitrogenase, like N_2 fixation, requires ATP (Bulen *et al.*, 1965; Burns, 1965; Burns and Bulen, 1965; Hardy, Knight and D'Eustachio, 1965). The metabolism of H_2

is complicated in N_2-fixing organisms, because H_2 may be evolved in two independent pathways. One route, mediated by the N_2-fixing enzyme complex, has an absolute requirement for ATP and is not sensitive to CO, whereas H_2 evolution by hydrogenase from anaerobic organisms is inhibited by CO and has no requirement for ATP. The picture is complicated further because H_2 is a competitive and specific inhibitor of N_2 fixation. Not only is H_2 an inhibitor, but it also can function as an electron donor in support of N_2 fixation by anaerobic organisms. It is not surprising that this complex series of reactions with H_2 was confused in studies of intact N_2-fixing organisms.

Hydrogenase is a metalloprotein which is inhibited by CO. Hydrogenase from *C. pasteurianum* can reduce ferredoxin with H_2, and ferredoxin in turn can transfer electrons for the reduction of N_2. Bulen, Burns and LeComte (1964) reconstructed an N_2-fixing system from *A. vinelandii* by using clostridial hydrogenase and clostridial ferredoxin to supply electrons. Such a system could not be reconstructed with similar components from *A. vinelandii*, an organism which evolves H_2 only via nitrogenase and not via hydrogenase.

Shug, Hamilton and Wilson (1956) purified hydrogenase from *C. pasteurianum* and reported that it contained molybdenum. Westlake, Shug and Wilson (1961) believed that Mo normally did not function in hydrogenase but was required to act with one electron acceptors. Hydrogenases earlier had been reported to be iron proteins, and the presence of molybdenum in the enzyme was questioned by a number of investigators. For example, Yakovlev and Gvozdev (1967) indicated the Mo participates in activation of N_2 but is not associated with hydrogenase function. Kleiner and Burris (1970) have verified the presence of Mo and have demonstrated that it has a catalytic role in hydrogenase. If part of the Mo is removed from the enzyme, its activity is decreased; addition of Mo as permolybdic acid partially restores enzymatic activity for methyl viologen reduction, H_2 evolution and H_2O-D_2 exchange. Molybdate does not restore activity and high concentrations of Mo are inhibitory.

Hydrogenase has only an indirect relationship to N_2 fixation; H_2 evolution by hydrogenase competes for electrons with N_2 fixation. To avoid confusion between H_2-evolving systems when one is studying the stoichiometry of ATP utilization, the hydrogenase can be blocked by CO without inhibiting the ATP-dependent H_2-evolving system. Burns (1965) showed clearly that *C. pasteurianum* not only carries hydrogenase but in addition evolves H_2 from $Na_2S_2O_4$

through a CO-insensitive ATP-requiring reaction. The optimum pH for H_2 evolution was almost the same as for N_2 fixation, and H_2 evolution was dependent upon ATP and Mg^{2+} concentrations. Capability for N_2 fixation and for ATP-dependent H_2 evolution decreased in parallel during enzyme storage. Bulen, Burns and LeComte (1965) first observed ATP-dependent H_2 evolution in extracts from *A. vinelandii*, and Hardy and Knight (1966b) verified the observations. Burns and Bulen (1966) also reported the reaction in extracts from *R. rubrum*. Four ATPs are required per pair of electrons transferred for H_2 evolution as for N_2 fixation according to Winter and Burris (1968) and other investigators; the recent report of Jeng, Morris and Mortenson (1970) assigns a requirement of 2 ATPs per H_2 evolved when corrected for reductant-independent ATP hydrolysis.

N_2-fixing organisms catalyze H_2 exchange. This can be measured in a variety of ways, e.g. by the formation of HD from H_2 and D_2 or the formation of HD from D_2 and H_2O, and isolated hydrogenase will catalyze all of these reactions. Although it has been assumed that the nitrogenase complex also catalyzes H_2 exchange, it is not clear that the preparations used to date have been entirely devoid of hydrogenase. Vandecasteele (1968) found that highly purified Mo-Fe and Fe proteins alone or in combination supported a minimal exchange reaction. Vandecasteele and Burris (1970) reported that little HD production from D_2 was found either under an atmosphere of N_2 or argon in the presence of dithionite, the ATP generating system and the Mo-Fe and Fe proteins.

The H_2-exchange reaction appears to be influenced by the presence of N_2 (Hoch, Schneider and Burris, 1960); in the absence of N_2 the formation of HD was decreased. It was suggested that this implied that N_2 initially was bound to the enzyme surface, and that subsequently H_2 was bound to the N_2 before H_2 dissociation and exchange occurred.

Bergersen (1963) also observed the H_2 exchange reaction in nodules from soybeans. HD formation, H_2 evolution and N_2 fixation were inhibited by CO. Jackson and Hardy (1967) reported that the nitrogenase in extracts from *A. vinelandii* also catalyzed the H_2-exchange reaction, and that requirements for the reaction were the same as those for N_2 fixation, namely, a reductant and ATP. As had been observed earlier by Hoch, Schneider and Burris (1960), N_2 enhanced the exchange reaction and formation of HD; alternative substrates did not effect such an enhancement. Kelly (1968) also found that N_2 enhanced the exchange of hydrogen in lupine nodules but, contrary to the earlier report, his preparations from *A. vinelandii* or *A. chroococcum* evolved less H_2 and formed less HD under N_2 than under argon. Exchange in these organisms required the presence of an ATP generating system and $Na_2S_2O_4$. Kelly suggested that it was unwise to conclude that a compound such as diimide was formed as a precursor for the exchange reaction. Turner and Bergersen (1969) extended their studies to cell-free extracts from soybean bacteroids and showed that the partially purified extracts exchanged H_2 and the rate was enhanced by N_2. Alternative substrates, such as acetylene and methylisocyanide, did not increase the rate of exchange.

Hydrogenase is of general occurrence in N_2-fixing organisms. It is present in root nodules, the clostridia and azotobacter, blue-green algae (Fujita and Myers, 1965), photosynthetic bacteria (Ormerod, Ormerod and Gest, 1961), and in a number of other organisms (Gray and Gest, 1965). In addition to catalyzing H_2 uptake, evolution and exchange, hydrogenase in pea nodule bacteroids can catalyze the oxyhydrogen reaction which yields water from H_2 and O_2 (Dixon, 1968).

Hydrogenase takes various forms, and isoenzymes of the hydrogenase from *C. pasteurianum* have been reported (Ackrell, Asato and Mower, 1966; Kidman, Ackrell and Asato, 1968). The three isoenzymes all have molecular weights in the 50 000 to 60 000 range. The activation energy is 14.6 kcal/mole, as observed by Kleiner and Burris (1970) for hydrogenase from *C. pasteurianum*.

Desulfovibrio desulfuricans has a highly active hydrogenase which has been purified (Krasna, Riklis and Rittenberg, 1960; Yagi, Honya and Tamiya, 1968). It is reported to evolve H_2 from $Na_2S_2O_4$ in the presence of cytochrome c_3 or methyl viologen but not in the presence of ferredoxin.

Although exchange of H_2 is a clear-cut phenomenon in the N_2-fixing organisms it never has been possible to demonstrate an exchange of N_2. Burris and Miller (1941) reported negative results in tests for such an exchange. Klucas (1964) used a variety of tests and ran experiments with cell-free extracts in the presence of several inhibitors of N_2 fixation, but under no circumstances could he demonstrate any exchange reaction of N_2.

4.4.6 Reduction of N_2

Biological N_2 fixation is an anaerobic process. Although fixation occurs in certain aerobic organisms, these organisms have adapted their systems to maintain a low oxidation-reduction potential. The azotobacter have the most vigorous respiration of any known organism, and this aids in keeping a proper potential for N_2 fixation. When extracts are purified from the azotobacter, they operate under strictly anaerobic conditions and are inactivated in air. It has been suggested that the blue-green algae (Stewart, Haystead and Pearson, 1969) fix N_2 in their heterocysts, which are devoid of chlorophyll b and hence do not liberate O_2 photosynthetically, and that substrates are fed to heterocysts from the photosynthesizing cells. It now appears that certain non-heterocystous blue-green algae can fix N_2 in their vegetative cells under anaerobic conditions (Stewart and Lex, 1970). N_2-fixing facultative bacteria fix N_2 only under anaerobic conditions, although they can grow on combined nitrogen either aerobically or anaerobically. The necessity for anaerobic conditions is hardly surprising, because N_2 fixation is a reductive process. As has been noted, extracts require a strong reductant such as reduced ferredoxin or $Na_2S_2O_4$ to support N_2 fixation.

N_2 is bound to the enzyme surface preliminary to its reduction. There is no convincing evidence establishing whether the position of the active site on the enzyme is on the Mo-Fe protein or on the Fe protein of the complex. Matkhanov et $al.$ (1969) have reported that the EPR signal of nonhaem iron from $A.$ $vinelandii$ at $G = 1.94$ disappears in an atmosphere of N_2, the implication being that the N_2 is bound to the component giving this signal. Bui and Mortenson (1968) presented evidence based on co-chromatography with labeled ADP that the clostridial Fe protein does not bind N_2 but binds ATP; on the basis of the binding of the alternative substrate cyanide, they concluded that N_2 is bound to the Mo-Fe protein. Their work was criticized by Biggins and Kelly (1970) on the grounds that the apparent specificity was a matter of the molecular weight of the proteins and that comparable binding reactions could be mimicked with albumin or cytochrome. As mentioned earlier, Kelly and Lang (1970) claimed, from equilibrium dialysis experiments with radioactive cyanide, that both proteins, as well as ATP and hydrosulfite, are all needed for substrate binding.

The binding of N_2 is blocked at the active site by certain competitive inhibitors of N_2 fixation. H_2 and N_2O are clearly competitive inhibitors of N_2 fixation and are specific in the sense that they do not block the growth of N_2 fixing organisms supplied ammonia. The nature of inhibition by other agents will be discussed later. It is evident that the active site is not particularly specific for

N_2 as it will bind a variety of alternative substrates as well as inhibitors.

Studies of the apparent Michaelis constant for N_2 fixation in intact organisms showed requirements ranging from 0.01 to 0.095 atm. N_2 for half maximum N_2 fixation. Surprisingly, when consistently active cell-free preparations became available, it was evident that the Michaelis constant was substantially higher than had been observed in intact organisms (range 0.056 to 0.165). Strandberg and Wilson (1967) reported that the Michaelis constant for N_2 fixation by extracts from *A. vinelandii* was 0.16 atm., whereas observations with intact cells indicated a constant between 0.01 and 0.066 atm. The basis for the discrepancy between intact organisms and cell-free extracts is not clear; perhaps the cell can concentrate N_2 at the active site, so that effectively at that site there is a higher concentration of N_2 than in the surrounding medium.

Hardy and Burns (1968) have suggested that the N_2 is bound end-on to the active site. Information on this topic has been derived primarily from an examination of model systems, and is the subject of discussion by Chatt and Richards in Chapter 3.

The reduction of N_2 is accomplished by an electron transfer agent such as reduced ferredoxin, but this must be aided by the energy from ATP. Ferredoxin and $Na_2S_2O_4$ have very negative potentials, and together with ATP readily effect reduction. In reconstructed systems from the azotobacter employing iron sulfur protein III (Yoch *et al.*, 1969) or flavoprotein (Benemann *et al.*, 1969) it has been necessary to use light and chloroplasts to maintain suitable reducing conditions for N_2 fixation. Burns (1969) reported that the activation energy for N_2 fixation by extracts from *A. vinelandii* was 14.6 kcal/mole above $21°$; below this temperature the activation energy was higher.

It has been suggested that N_2 fixation progresses by two electron transfer steps (Hoch *et al.*, 1960). This concept is supported by observations that a variety of molecules other than N_2 can be reduced by nitrogenase. The reduction of acetylene to ethylene terminates after a two electron transfer. Reduction of hydroxylamine to NH_3 also requires two electrons, and hydroxylamine reductase has been observed in a number of N_2-fixing organisms (Kretovich, Bundel and Borovikova, 1967; Suzuki and Suzuki, 1954).

A number of logical intermediates in the reduction of N_2 are suggested by assuming a two electron transfer system. The evidence in support of ammonia as the end product of N_2 fixation is extensive and unequivocal (Burris, 1966). Bergersen (1965, 1969) has verified that ammonia is the main product of N_2 fixation both in intact soybean nodules and in extracts from bacteroids. Using 'pulse' experiments with $^{15}N_2$, Kennedy (1966a, b) followed N_2 fixation in serradella nodules; the results supported the concept that all of the N_2 fixed is assimilated by way of ammonia.

Garcia-Rivera and Burris (1967) examined vigorously-fixing cell-free extracts from *C. pasteurianum* for intermediates other than NH_3. Although the preparations rapidly converted $^{15}N_2$ to ammonia, no evidence was obtained of hydrazine or hydroxylamine as free intermediates. Likewise, Burris *et al.* (1965) found no experimental evidence for diimide or carbamyl phosphate as intermediates in N_2 fixation.

Although no demonstrable level of free intermediates of N_2 fixation other than NH_3 appear in the medium, demonstration of the reduction of certain postulated intermediates by N_2-fixing organisms is possible. The suggestion by Hanstein *et al.* (1967) that complexes between diimide and haemoglobin may be early intermediates in N_2 fixation is unlikely because the role of haemoglobin is only indirect in N_2 fixation. Harris, Diamantis and Roberts (1965) reported that the conversion of hydrazine to acetyl hydrazine in growing cultures of *A. vinelandii* was inhibited by ammonia; hydrazine decreased the uptake of ammonia by the organisms (Harris and Roberts, 1965). Saris and Virtanen (1957) reported the formation of hydroxylamine compounds in cultures of the azotobacter. Bundel, Kretovich and Borovikova (1966) found that a culture of *A. vinelandii* assimilated nitrogen from the oxime of α-ketoglutaric acid. Although the evidence remains unconvincing for the occurrence of any free intermediates between N_2 and ammonia, it is logical that the reduction should proceed by two electron steps, and that bound intermediates corresponding in reduction level to diimide, and hydrazine or hydroxylamine are held by the enzyme, as concluded by Garcia-Rivera and Burris (1967).

Ammonia can be considered the end product of N_2 fixation, and metabolism beyond that point can be interpreted as nitrogen assimilation. Assimilation requires that the ammonia be incorporated into an organic compound. Evidence with a wide variety of N_2-fixing organisms has indicated that the main pathway of assimilation is by reductive amination of α-ketoglutarate to yield glutamic acid. The glutamic acid in turn is transaminated to form the other amino acids necessary for building protein (Wilson and Burris, 1953). This means that the organisms must provide α-ketoglutaric acid as the ammonia acceptor and a variety of other ketoacids as acceptors in the transamination process. Many bacterial species, unlike animals, are able to form all of these necessary acceptors, and thus can live on simple inorganic nitrogen compounds as a sole source of nitrogen. Geiko, Lyubimov and Kretovich (1965) have demonstrated a variety of α-ketoacids in *A. vinelandii*.

Although Pengra and Wilson (1958) observed diauxic growth of *K. pneumoniae* (then known as *A. aerogenes*) when grown on limiting quantities of ammonia in an atmosphere of N_2, serious interest in establishing the control of N_2 fixation has developed only recently. The important observation of Lindsay (1963) confirmed by Yoch and Pengra (1966) that addition of amino acids greatly reduced the lag period of several hours required by this organism to shift from growth on NH_4^+ to fixation of N_2, and the availability of cell-free extracts capable of fixation provided the techniques necessary for detailed investigation. Definite but much shorter periods of lag before nitrogenase was detectable in cells grown on limiting quantities of ammonia in an N_2 atmosphere were established for *A. vinelandii* (Strandberg and Wilson, 1968; Oppenheim and Marcus, 1970) and for *B. polymyxa* and *B. macerans* (Witz *et al.*, 1967). Chloramphenicol prevented the formation of nitrogenase in the azotobacter but did not affect the N_2-fixing activity of cell-free extracts. Ammonia also blocked formation of nitrogenase in *C. pasteurianum* (Detroy and Wilson, 1967).

It has been difficult to determine unequivocally whether the control of nitrogenase is effected through induction by N_2 or repression by NH_4^+ because of the great difficulty of excluding N_2 completely during the experiment. Although Mahl and Wilson (1968) tentatively concluded regulation of nitrogenase synthesis by *K. pneumoniae* was through induction by N_2, later experiments by Parejko and Wilson (1970) in which great care was taken to control the gas environment lead to the conclusion that simple repression by ammonia is the controlling mechanism, a view advanced by Munson and Burris (1969) for control in *Rhodospirillum rubrum* and by Daesch and Mortenson (1968a) for *C. pasteurianum*.

4:4.7 Reduction of substrates other than N_2
The observations of Wilson and Umbreit (1937) that H_2 specifically and competitively inhibits N_2 fixation first indicated that the active site of nitrogenase is not highly specific. The competitive nature of the inhibition suggested that H_2 and N_2 are bound at the same

enzymatic site. The first substrate other than N_2 shown to be reduced by nitrogenase was N_2O (Mozen and Burris, 1954; Hoch, Schneider and Burris, 1960). ^{15}N from N_2O, which also is a specific competitive inhibitor of N_2 fixation, was incorporated by *A. vinelandii* and by soybean nodules. Lockshin and Burris (1965) observed that cell-free extracts from *C. pasteurianum* also reduced N_2O, as did Hardy and Knight (1966a) using extracts from both *C. pasteurianum* and *A. vinelandii*. Schöllhorn and Burris (1966) reported the reduction of acetylene and azide by the N_2-fixing system from *C. pasteurianum*; these compounds are inhibitors of N_2 fixation as well as substrates for nitrogenase. Independently, Dilworth (1966) observed the reduction of acetylene and demonstrated clearly that the product of the reduction, ethylene, was not reduced further. The reducing system was stereochemically specific, as all of the ethylene produced was of the *cis* form; it is apparently not completely stereospecific in extracts of *A. chroococcum* (Kelly, 1969b). The reduction required both ATP and ferredoxin; extracts of *C. pasteurianum* grown on ammonia did not reduce acetylene. Acetylene reduction and N_2 fixation appear to be parallel processes, and, as has been mentioned, the fact that ethylene is the terminal product of acetylene reduction (two electron reduction) suggests that N_2 reduction also occurs by transfer of two electrons at a step.

Azide has N—N bonds, so its reduction by an N_2-fixing system is not particularly surprising. It is reduced to N_2 and NH_3, and the reaction requires an electron donor and ATP. Preparations which reduced azide did not reduce hydrazine, and cells of *C. pasteurianum* grown on ammonia are incapable of reducing azide (Schöllhorn and Burris, 1967a). Hardy and Knight (1967) reported a Michaelis constant for azide of 1 mM.

Hardy and Knight (1967) first demonstrated the reduction of cyanide by nitrogenase; the major products of reduction are methane and ammonia, although some methylamine also is formed. The reported K_m varied from 0.4 to 4 mM. Hardy and Jackson (1967) indicated a number of analogs of cyanide and acetylene which are reduced by nitrogenase. Kelly, Postgate and Richards (1967) observed and studied the reduction of methylisocyanide. The reduction of methylisocyanide was about 10 times as rapid as the reduction of ethylisocyanide. Methane, ethane, ethylene and methylamine are principal products of the reaction; these products are characteristic of the reduction of metal-bound isocyanide complexes and provide circumstantial evidence that the substrates become bound to a metal in the enzyme before reduction. Simple homologs of acetylene, cyanide and methylisocyanide are reduced by nitrogenase with various degrees of efficiency and some research groups have proposed 1- or 2-site theories of N_2 fixation. Hardy and Burns (1968) discussed the bearing of homolog studies on a 2-site theory in considerable detail.

The reduction of acetylene furnishes a very useful qualitative and quantitative test for N_2 fixation in cell-free preparations and in intact cells. Small quantities of ethylene can be measured by gas chromatography, and there is every indication that acetylene reduction is dependent on the same enzyme system responsible for the reduction of N_2. Although Schöllhorn and Burris (1966) and Dilworth (1966) were independently responsible for the observation that acetylene is reduced by the N_2-fixing enzyme system, a number of people contributed to its application as a tool for measuring nitrogenase activity. R. W. F. Hardy had discussed with H. Evans and W. Silver the possibility of using the reduction of cyanide as a measure of N_2 fixation. Burris reported the reduction of acetylene informally at the 'Sage Hen Conference' in late 1965. Hardy advised Evans and Silver concerning suitable gas chromatographic techniques for measuring ethylene. Priority of publication on acetylene reduction as a measure of N_2 fixation goes to Koch and Evans (1966), but other reports of application of the method followed quickly (Sloger and Silver, 1967; Stewart, Fitzgerald and Burris, 1967; Hardy *et al.*, 1968). The method now is widely applied in laboratory studies and in field experiments. Although it is extremely sensitive and highly useful, it should be employed with caution, because it is an indirect method. It should be checked periodically against measurements of N_2 fixation *per se.*

Not only has the acetylene reduction method been used for quantitative field studies, but it also has aided greatly in the measurement of the potential for N_2 fixation by preparations of relatively low activity. In a search for N_2 fixation by extracts from leguminous root nodules, Koch, Evans and Russell (1967a, b) used this method routinely for following nitrogenase activity. In recent studies, potential for nitrogenase activity has been demonstrated with systems reconstructed from the azotobacter and chloroplasts by measuring acetylene reduction (Yoch *et al.,* 1969).

4.5   INHIBITORS OF N_2 FIXATION

4.5.1   *Specific, competitive inhibitors*
As described by Wilson in Chapter 1, H_2 was being employed as an

inert diluent gas in tests to establish the Michaelis constant for N_2 fixation, when it was discovered that H_2 in fact was an inhibitor for N_2 fixation in red clover. Investigation of the inhibition revealed that H_2 was specific in the sense that it did not inhibit the growth of the leguminous plants on combined nitrogen but only their growth on N_2. The response was clearly competitive suggesting that the N_2 and H_2 were bound at the same site. Inhibition of N_2 fixation by H_2 has been demonstrated in a wide variety of N_2-fixing agents, and all agents carefully examined have exhibited H_2 sensitivity.

Although Rosenblum and Wilson (1950) experienced difficulty in showing H_2 inhibition in growing cultures of *C. pasteurianum*, because the organism produced large amounts of H_2 as a metabolic product, Westlake and Wilson (1959) demonstrated clearly that H_2 is a specific competitive inhibitor of N_2 fixation in *C. pasteurianum*. They found a Michaelis constant of 0.03 atm. for N_2 and an inhibitor constant for H_2 of 0.5 ± 0.05 atm.

Hiai *et al.* (1957) isolated an organism from soil, which they identified initially as a clostridial species, and showed that its N_2 fixation was competitively inhibited by H_2. The Michaelis constant was 0.03 atm. N_2, and no inhibition by H_2 was observed when ammonia was supplied to the organism. Hino and Wilson (1958) found that the organism studied was a strain of *B. polymyxa* rather than *Clostridium* sp., and that it fixed N_2 only under anaerobic conditions. They reported that N_2 initially inhibited H_2 evolution from glucose but later markedly enhanced H_2 evolution. Bond (1960) demonstrated that N_2 fixation by non-leguminous root nodules was inhibited by H_2.

Although many investigators have studied H_2 inhibition and have found that the inhibition is of the classical competitive type, Parker and Dilworth (1963) reported that H_2 inhibition of N_2 fixation in *A. vinelandii* is non-competitive. Using cell-free extracts of this organism Strandberg and Wilson (1967) and Hwang (1968) re-examined the question and again observed clear-cut competitive inhibition. Lockshin and Burris (1965) employed extracts from *C. pasteurianum* and found competitive inhibition, but with abnormally high levels of phosphate, the inhibition became uncompetitive. Conceivably this response can explain the aberrant results of Parker and Dilworth, but is is not clear that excessive phosphate was responsible for their results.

Lind and Wilson (1941) first described CO as a specific inhibitor for N_2 fixation. In red clover and in the azotobacter, the inhibition appeared to be mixed competitive and non-competitive whereas, in the blue-green alga *Nostoc muscorum* the inhibition was described as competitive (Burris and Wilson, 1946). When active cell-free preparations became available, Lockshin and Burris (1965) re-examined the problem with extracts from *C. pasteurianum* and observed competitive inhibition by CO. Later Hwang (1968) examined preparations from *C. pasteurianum* and *A. vinelandii* and found that although both were competitively inhibited by H_2, N_2 fixation by *A. vinelandii* extracts was non-competitively inhibited by CO. At present the nature of inhibition by CO remains confused.

Observations by Krasna and Rittenberg (1954) that NO inhibits hydrogenase suggested that it should be tested as an inhibitor of biological N_2 fixation. In limited experiments Mozen (1955) found that 0.1% NO in N_2 completely blocked N_2 fixation by *C. pasteurianum* cells. NO inhibits at a somewhat lower partial pressure than CO, but based on molar concentrations of dissolved gases the inhibitory capacities of NO and CO are similar. NO has been studied very little because of its reactivity and the fact that it can be employed only under strictly anaerobic conditions. Lockshin and Burris (1965) studied NO as an inhibitor of cell-free extracts from *C. pasteurianum*. Although the very low concentrations of NO employed made its experimental use somewhat difficult, the evidence showed that NO was a competitive inhibitor. Pyruvate metabolism was virtually unaffected by 0.02% NO, whereas N_2 fixation was inhibited 90%.

4.5.2 Inhibitors which also serve as substrates

As noted, the N_2-fixing enzyme system is versatile and will reduce a variety of substrates. It is hardly surprising that these substrates while being activated at the N_2-fixing site will inhibit N_2 reduction. Molner, Burris and Wilson (1948) observed that N_2O functioned as a specific inhibitor of N_2 fixation by *A. vinelandii*. Repaske and Wilson (1952) established that the inhibition was competitive, and with *A. vinelandii* the K_i for N_2O was 0.08 atm. In cell-free extracts of *C. pasteurianum*, supported by pyruvate metabolism, inhibition by N_2O again proved to be competitive (Lockshin and Burris, 1965). The report by Virtanen and Lundbom (1953) that N_2O inhibits the utilization of nitrate nitrogen as well as utilization of N_2 is not compatible with the observations of other workers including Mozen *et al.* (1955).

Azide has an N—N bond, is reduced to ammonia and N_2 by nitrogenase, and inhibits N_2 reduction. Schöllhorn and Burris (1967a) did not study the details of the inhibition of N_2 fixation by azide but observed that azide reduction was inhibited by CO, NO, acetylene and nitroprusside. Hwang (1968) investigated the inter-actions of azide with other substrates for nitrogenase and observed that azide was non-competitive with acetylene and acetylene was non-competitive with azide. Rakestraw and Roberts (1957a, b) had reported that azide was a specific and reversible inhibitor for N_2 fixation by intact organisms, and that it might be competitive. They also found that cyanate weakly inhibited respiration and N_2 fixation in *A. vinelandii*.

Schöllhorn and Burris (1967b) reported that acetylene is a competitive inhibitor of N_2 fixation by preparations from *C. pasteurianum* and *A. vinelandii*, whereas, Hwang (1968) found the inhibition non-competitive. Hwang's observation that H_2 does not affect acetylene reduction, whereas it competitively inhibits N_2 reduction, suggests that the acetylene and N_2 binding sites are not entirely equivalent.

The nature of methylisocyanide inhibition of N_2 fixation has not been established. Cyanide non-competitively inhibits N_2 fixation, but both cyanide and methylisocyanide are competitive with azide (Hwang, 1968). That azide, cyanide and methylisocyanide are mutually competitive indicates that they are bound at the same site on nitrogenase. The sites for acetylene, CO and H_2 evolution seem to differ. A simplistic assumption to explain the demonstrable differ-ences as well as mutual interactions among substrates and inhibitors is that the differing substrates and inhibitors are bound at separate sites. However, it seems more probable that their presence modifies the sites somewhat to alter interactions, e.g. the normal site would bind N_2, H_2 and N_2O whereas an altered site would accommodate azide, cyanide and methylisocyanide.

4.5.3 Non-specific inhibitors

A variety of inhibitors described are not specific for N_2 fixation, because they block the growth of organisms supplied fixed nitrogen. Gogoleva and Ivanov (1969) described the effects of a variety of uncoupling agents and respiratory chain inhibitors. Bruemmer (1962) reported that fluoracetate inhibited N_2 fixation. Carnahan, Mortenson and Castle (1960) described inhibition by lipoic acid and similar compounds and indicated that the inhibition was reversed by biotin. As described earlier, ADP inhibits N_2 fixation (Bulen *et al.*, 1965; Moustafa and Mortenson, 1967). Ivleva, Likhtenshtein and Sadkov (1969) concluded that SH groups were functional at the active center of nitrogenase because activity was inhibited by *p*-chloromercuribenzoate.

There has been a great deal of interest in the effects of O_2 as an inhibitor of biological N_2 fixation (Burk and Lineweaver, 1930). Wilson and Fred (1937) described the effect of the pO_2 on N_2 fixation in leguminous plants; as the pO_2 was raised, the fixation of N_2 increased. Above 0.1 atm. the fixation rate was constant until inhibition occurred at high partial pressures of O_2. It was concluded that the effects of O_2 were not directly upon N_2 fixation but indirectly affected the respiratory activity. O_2 clearly is necessary for generation of ATP in aerobic organisms, and O_2 also is capable of inactivating the enzyme system when unprotected by vigorous respiratory activity. Bond (1961) observed maximum fixation by detached nonleguminous nodules at a pO_2 of 0.12 to 0.25 atm. He suggested the possibility of competitive inhibition by O_2 (Bond, 1964). Bergersen reported that excised soybean nodules increased their respiration and N_2 fixation with an increase in the pO_2 to 0.5 atm. He found by spectral observation that little haemoglobin in soybean nodules is oxygenated until the pO_2 exceeds 0.4-0.5 atm. (Bergersen, 1962). Haemoglobin can enhance the transport of O_2 (Hemmingsen and Scholander, 1960; Hemmingsen, 1962; Wittenberg, 1970); Yocum (1964) and others have suggested that haemoglobin serves as the O_2 carrier in N_2-fixing root nodules (see also Chapter 6).

Particulate, cell-free preparations from the blue-green algae also are very sensitive to O_2 (Fay and Cox, 1967). The photosynthetic generation of O_2 in blue-green algae subjects the vegetative cells in the light to internally generated O_2. This has led to the suggestion that N_2 fixation occurs in the heterocysts (Stewart, Haystead and Pearson, 1969), cells which are devoid to pigments of the O_2-generating photosystem II (Thomas, 1970). Fay (1970) also pointed out that the action spectrum for acetylene reduction by filaments of *Anabaena cylindrica* closely resembles the absorption spectrum of their isolated heterocysts (photosystem I). These matters are discussed in Chapter 5; since that Chapter was prepared, the observation that the non-heterocystous blue-green alga, *Plectonema boryanum,* can reduce N_2 and acetylene if placed under as anaerobic conditions as can be obtained with organisms which generate O_2 (Stewart and Lex, 1970), reopens the whole question of what algae are capable of N_2 fixation and what relation, if any, heterocyst formation has to this capability. The effect of oxygen on aerobic N_2-fixing bacteria, particularly the azotobacter, and the mechanisms whereby such organisms conduct the essentially reductive process of N_2 fixation in an aerobic environment, are discussed in Chapter 5.

4.6   KINETICS OF N_2 FIXATION

4.6.1 N_2 fixation by intact organisms

Allison and Burris (1957) followed the time course of N_2 fixation by intact *A. vinelandii*. Measurements of uptake of $^{15}N_2$ over short periods showed an initial lag of about a minute in the uptake of $^{15}N_2$ into most cellular products. The incorporation of ^{15}N exhibited a minimal lag into ammonia and products quickly formed from ammonia. Results were compatible with other data which suggested that ammonia was the first demonstrable product of N_2 fixation. Burma and Burris (1957) followed the kinetics of ammonia utilization by *A. vinelandii;* the lag in uptake of $^{15}NH_4^+$ into a variety of cellular products was brief, but there was an increased rate of uptake with time during five minutes. Bulen and coworkers (1963) also utilized ^{15}N as a tracer to follow the time course of N_2 fixation.

4.6.2 N_2 fixation in cell-free systems

Detailed studies of the kinetics of N_2 fixation have been handicapped by lack of methods for measuring the reaction continuously. It has been necessary to sample at intervals and to analyze each sample for the products of nitrogenase activity. A method for continuous ammonia analysis would be most helpful as a direct measure of N_2 fixation in contrast to the measurement of H_2 evolution as an index of nitrogenase activity (Silverstein and Bulen, 1970).

Some early studies on cell-free preparations showed an initial lag in N_2 fixation. Improved preparations have exhibited linear fixation of N_2 for periods of 30 to 60 minutes without an initial lag. This observation also it true for the reduction of acetylene, azide, cyanide, methylisocyanide and other alternative substrates for nitrogenase. When non-linear reduction has been reported, some uncontrolled factor has been limiting.

Kelly (1969a) measured rates of substrate reduction by largely purified Mo-Fe protein and Fe protein from *A. chroococcum* and found that the stoichiometric ratios of these proteins for maximum activity differed according to substrate. More Mo-Fe protein/unit of Fe protein was needed for cyanide or isocyanide reduction than for reduction of acetylene or N_2. Biggins and Kelly (1970) mentioned that acetylene could, in special circumstances, accelerate cyanide reduction by preparations of *K. pneumoniae* nitrogenase. These findings have led to speculation on the number and relative location of active sites in the nitrogenase; more detailed kinetic studies of nitrogenase action with pure proteins and with various substrates ought to clarify the position considerably.

Silverstein and Bulen (1970) examined the kinetics of H_2 evolution by preparations of the N_2-fixing complex from *A. vinelandii*. A plot of ATP concentration *vs* initial velocities established a sigmoidal curve, whereas the curve for nitrogenase concentration *vs* activity was concave upward. Computer simulation of the reactions led the authors to conclude that the Fe protein and Mo-Fe protein associated and dissociated in dynamic equilibrium, that each enzyme complex bound two ATP molecules, and that breakdown of the intermediate enzyme-ATP complex followed two pathways to yield ADP and P_i in both pathways, and H_2 plus oxidized enzyme as the products of one pathway and protons plus enzyme as the products of the other pathway. At higher concentrations of ATP the ratio of NH_3/H_2 formation increased.

4.7 METABOLIC SCHEMES FOR N_2 FIXATION

There have been many attempts to devise logical and all-embracing schemes for biological N_2 fixation. These are potentially useful, because there is a great unity among N_2-fixing agents. All have absolute requirements for ATP and a strong reducing agent, and all studied to date operate through a complex of an Fe protein and an Mo-Fe protein. Because the specific functions of these two proteins still are not clear, the schemes lack specificity.

Hoch, Schneider and Burris (1960) proposed a scheme for N_2 fixation which involved the binding of N_2 followed by its step-wise reduction by two electrons at a time; no intermediate release of any compounds was proposed before a terminal release of ammonia. The scheme in part was designed to explain the exchange reaction of H_2 in N_2-fixing agents, and it did not embrace the information derived later concerning the role of ATP and reducing agents such as ferredoxin.

Hardy and Knight (1968), Hardy and Burns (1968), and Mortenson (1968) have reviewed evidence and formulated schemes for N_2 fixation. Subsequently, Fuchsman and Hardy (1970) have studied the products of acrylonitrile reduction and have suggested that there is a cooperative binding by multiple metal sites in a step-wise reduction of N_2 and other substrates. Burris (1969) presented a scheme essentially as indicated in Fig. 1; the legend explains the proposed mechanism for N_2 fixation.

There has been much interest in deducing N_2-fixing mechanisms from inorganic and organic models; the subject has been developed in Chapters 2 and 3. For reviews of the biological process, including

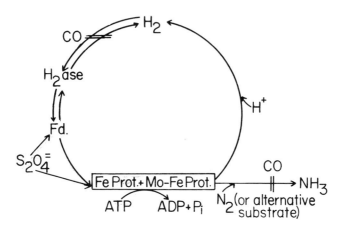

Fig. 4.1. Reaction scheme for biological N_2 fixation. The components of nitrogenase (Fe Prot. + Mo-Fe Prot.) act in conjunction to reduce N_2 to NH_3 (or to reduce alternate substrates such as C_2H_2, N_3^-, HCN, CH_3NC and N_2O). The reactions require ATP (which is converted to ADP + orthophosphate) and a reducing agent such as ferredoxin (Fd.) or dithionite (hydrosulfite, $S_2O_4^{2-}$). H_2 may be released by hydrogenase (H_2ase) through a CO-sensitive ATP-independent pathway or by nitrogenase through a CO-insensitive ATP-dependent pathway. Reduction of N_2 and other substrates is inhibited by CO.

discussion of the enzymology of N_2 fixation, the reader is referred to an outline by Postgate (1970) and more extensive ones by Burris (1966), Hardy and Burns (1968), Mortenson (1968), Stewart (1966, 1967), Yakovlev (1967), Lyubimov (1967), Nowotny-Mieczynska (1968) and Vol'pin and Shur (1969).

REFERENCES

B. A. C. ACKRELL, R. N. ASATO and H. F. MOWER, *J. Bacteriol.*, **92**, 828 (1966).

B. A. C. ACKRELL and C. W. JONES, *Biochem. J.*, **116**, 21P (1970).

R. M. ALLISON and R. H. BURRIS, *J. Biol. Chem.*, **224**, 351 (1957).

D. I. ARNON, M. LOSADA, M. NOZAKI and K. TAGAWA, *Nature*, **190**, 601 (1961).

A. N. BACH, Z. V. JERMOLIEVA and M. P. STEPANIAN, *Compt. rend. acad. sci (USSR)*, **1**, 22 (1934).

R. BACHOFEN and D. I. ARNON, *Biochim Biophys Acta*, **120** 259 (1966).

H. BEINERT and G. PALMER, *Adv. Enzymol.* **27**, 105 (1965).

J. R. BENEMANN, D. C. YOCH, R. C. VALENTINE and D. I. ARNON, *Proc. Natl. Acad. Sci. U.S.*, **64**, 1079 (1969).

A. M. BENSON, H. F. MOWER and K. T. YASUNOBU, *Arch. Biochem. Biophys.*, **121**, 563 (1967).

F. J. BERGERSEN, *J. Gen. Microbiol.*, **22**, 671 (1960).

F. J. BERGERSEN, *Nature*, **194**, 1059 (1962).

F. J. BERGERSEN, *Australian J. Biol. Sci.*, **16**, 669 (1963).

F. J. BERGERSEN, *Australian J. Biol. Sci.*, **18**, 1 (1965).

F. J. BERGERSEN, *Biochim. Biophys. Acta*, **115**, 247 (1966a).

F. J. BERGERSEN, *Biochim. Biophys. Acta*, **130**, 304 (1966b).

F. J. BERGERSEN, *Proc. Roy. Soc. B*, **172**, 401 (1969).

F. J. BERGERSEN and G. L. TURNER, *Biochim. Biophys. Acta*, **141**, 507 (1967).

F. J. BERGERSEN and G. L. TURNER, *J. Gen. Microbiol.*, **53**, 205 (1968).

F. J. BERGERSEN and G. L. TURNER, *Biochim. Biophys. Acta*, **214**, 28 (1970).

D. R. BIGGINS and M. J. DILWORTH, *Biochim. Biophys. Acta*, **156**, 285 (1968).

D. R. BIGGINS and M. KELLY, *Biochim. Biophys. Acta*, **205**, 288 (1970).

D. R. BIGGINS and J. R. POSTGATE, *J. Gen. Microbiol.*, **56**, 181 (1969).

D. C. BLOMSTROM, E. KNIGHT, Jr., W. D. PHILLIPS and J. F. WEIHER, *Proc. Natl. Acad. Sci. U.S.*, **51**, 1085 (1964).

H. L. BOHN, D. B. FENN and W. J. MOORE, *Soil Sci.*, **108**, 95 (1969).

G. BOND, *J. Exptl. Botany*, **11**, 91 (1960).

G. BOND, *Z. Allgem. Mikrobiol.*, **1**, 93 (1961).

G. BOND, *Nature*, **204**, 600 (1964).

J. H. BRUEMMER, *Biochem. Biophys. Res. Commun.*, **7**, 53 (1962).

B. B. BUCHANAN and D. I. ARNON, *Adv. Enzymol. Relat. Areas Mol. Biol.*, **33**, 119 (1970).

B. B. BUCHANAN and R. BACHOFEN, *Biochim. Biophys. Acta*, **162**, 607 (1968).

P. T. BUI and L. E. MORTENSON, *Proc. Natl. Acad. Sci. U.S.*, **61**, 1021 (1968).

P. T. BUI and L. E. MORTENSON, *Biochemistry*, **8**, 2462 (1969).

W. A. BULEN, R. C. BURNS and J. R. LeComte, *Biochem. Biophys. Res. Commun.*, **17**, 265 (1964).

W. A. BULEN, R. C. BURNS and J. R. LeComte, *Proc. Natl. Acad. Sci. U.S.*, **53**, 532 (1965).

W. A. BULEN and J. R. LeComte, *Proc. Natl. Acad. Sci. U.S.*, **56**, 979 (1966).

W. A. BULEN, J. R. LeComte and H. E. BALES, *J. Bacteriol.*, **85**, 666 (1963).

A. A. BUNDEL, V. L. KRETOVICH and V. V. BOROVIKOVA, *Mikrobiologiya*, **35**, 573 (1966).

D. BURK and H. LINEWEAVER, *J. Bacteriol.*, **19**, 389 (1930).

D. P. BURMA and R. H. BURRIS, *J. Biol. Chem.*, **225**, 287 (1957).

R. C. BURNS, Non-heme iron proteins, role in energy conversion, p. 289, Symp., Yellow Springs, Ohio (1965).

R. C. BURNS, *Biochim. Biophys Acta*, **171**, 253 (1969).

R. C. BURNS and W. A. BULEN, *Biochim. Biophys. Acta*, **105**, 437 (1965).

R. C. BURNS and W. A. BULEN, *Arch. Biochem. Biophys.*, **113**, 461 (1966).

R. C. BURNS, R. D. HOLSTEN and R. W. F. HARDY, *Biochem. Biophys. Res. Commun.*, **39**, 90 (1970).

R. H. BURRIS, *Ann. Rev. Plant Physiol.*, **17**, 155 (1966).

R. H. BURRIS, *Proc. Roy. Soc. B*, **172**, 339 (1969).

R. H. BURRIS, F. J. EPPLING, H. B. WAHLIN and P. W. WILSON, *J. Biol. Chem.*, **148**, 349 (1943).

R. H. BURRIS and C. E. MILLER, *Science*, **93**, 114 (1941).

R. H. BURRIS and P. W. WILSON, *Botan. Gaz.*, **108**, 254 (1946).

R. H. BURRIS, H. C. WINTER, T. O. MUNSON and J. GARCIA-RIVERA, Non-heme iron proteins, role in energy conversion, p. 315. Symp., Yellow Springs, Ohio (1965).

J. E. CARNAHAN, L. E. MORTENSON and J. E. CASTLE, *J. Bacteriol.*, **80**, 311 (1960).

J. E. CARNAHAN, L. E. MORTENSON, H. F. MOWER and J. E. CASTLE, *Biochim. Biophys. Acta*, **38**, 188 (1960a).

J. E. CARNAHAN, L. E. MORTENSON, H. F. MOWER and J. E. CASTLE, *Biochim. Biophys. Acta*, **44**, 520 (1960b).

A. E. CHUNG, *J. Bacteriol.*, **102**, 438 (1970).

R. M. COX and P. FAY, *Arch. Mikrobiol.*, **58**, 357 (1967).

G. DAESCH and L. E. MORTENSON, *Bact. Proc.*, 132 (1968).

J. V. DAHLEN, R. A. PAREJKO and P. W. WILSON, *J. Bacteriol.*, **98**, 325 (1969).

H. DALTON and L. E. MORTENSON, *Bact. Proc.*, 148 (1970).

H. E. DAVENPORT and R. HILL, *Biochem. J.*, **74**, 493 (1960).

H. E. DAVENPORT, R. HILL and F. R. WHATLEY, *Proc. Roy. Soc. B*, **139**, 346 (1952).

R. W. DETROY and P. W. WILSON, *Bact. Proc.*, 113 (1967).

R. W. DETROY, D. F. WITZ, R. A. PAREJKO and P. W. WILSON, *Proc. Natl. Acad. Sci. U.S.*, **61**, 537 (1968).

A. J. D'EUSTACHIO and R. W. F. HARDY, *Biochem. Biophys. Res. Commun.*, **15**, 319 (1964).

M. J. DILWORTH, *Biochim. Biophys. Acta*, **127**, 285 (1966).

M. J. DILWORTH, D. SUBRAMANIAN, T. O. MUNSON and R. H. BURRIS, *Biochim. Biophys. Acta*, **99**, 486 (1965).

R. O. D. DIXON, *Arch. Mikrobiol.*, **62**, 272 (1968).

W. DRUSKEIT, K. GERSONDE and H. NETTER, *Eur. J. Biochem.*, **2**, 176 (1967).

R. D. DUA and R. H. BURRIS, *Proc. Natl. Acad. Sci. U.S.*, **50**, 169 (1963).

R. D. DUA and R. H. BURRIS, *Biochim. Biophys. Acta*, **99**, 504 (1965).

M. C. W. EVANS and B. B. BUCHANAN, *Proc. Natl. Acad. Sci. U.S.*, **53**, 1420 (1965).

P. FAY, *Biochim. Biophys. Acta*, **216**, 353 (1970).

P. FAY and R. M. COX, *Biochim. Biophys Acta*, **143**, 562 (1967).

R. J. FISHER and P. W. WILSON, *Biochem. J.*, **117**, 1023 (1970).

R. J. FISHER and W. J. BRILL, *Bact. Proc.*, 148 (1969).

K. T. FRY and A. SAN PIETRO, *Biochem. Biophys. Res. Commun.*, **9**, 218 (1962).

W. H. FUCHSMAN and R. W. F. HARDY, *Bact. Proc.*, 148 (1970).

Y. FUJITA and J. Myers, *Arch. Biochem. Biophys.*, **111**, 619 (1965).

J. GARCIA-RIVERA and R. H. BURRIS, *Arch. Biochem. Biophys.*, **119**, 167 (1967).

N. S. GEIKO, V. I. LYUBIMOV and V. L. KRETOVICH, *Dokl. Akad. Nauk. SSSR*, **160**, 944 (1965).

K. GERSONDE and W. DRUSKEIT, *Eur. J. Biochem.*, **4**, 391 (1968).

T. V. GOGOLEVA and I. D. IVANOV, *Biol. Nauki*, 92 (1969).

F. H. GRAU and P. W. WILSON, *J. Bacteriol.*, **85**, 446 (1963).

C. T. GRAY and H. GEST, *Science*, **148**, 186 (1965).

R. I. GVOZDEV, A. P. SADKOV, V. A. YAKOVLEV and E. Ya. ALFIMOVA, *Izv. Akad. Nauk SSSR, Ser. Biol.*, 838 (1968).

K. L. HADFIELD and W. A. BULEN, *Biochemistry*, **8**, 5103 (1969).

D. O. HALL and M. C. W. EVANS, *Nature*, **223**, 1342 (1969).

W. G. HANSTEIN, J. B. LETT, C. E. McKENNA and T. G. TRAYLOR, *Proc. Natl. Acad. Sci. U.S.*, **58**, 1314 (1967).

R. W. F. HARDY and R. C. BURNS, *Ann. Rev. Biochem.*, **37**, 331 (1968).

R. W. F. HARDY and A. J. D'EUSTACHIO, *Biochem. Biophys. Res. Commun.*, **15**, 314 (1964).

R. W. F. HARDY, R. D. HOLSTEN, E. K. JACKSON and R. C. BURNS, *Plant Physiol.*, **43**, 1185 (1968).

R. W. F. HARDY and E. K. JACKSON, *Fed. Proc.*, **26**, 725 (1967).

R. W. F HARDY and E. KNIGHT, Jr., *Biochem. Biophys. Res. Commun.*, **23**, 409 (1966a).

R. W. F. HARDY and E. KNIGHT, Jr., *Biochim. Biophys. Acta*, **122**, 520 (1966b).

R. W. F. HARDY and E. KNIGHT, Jr., *Biochim. Biophys. Acta*, **139**, 69 (1967).

R. W. F. HARDY and E. KNIGHT, Jr., *Progr. Phytochem.*, **1**, 407 (1968).

R. W. F. HARDY, E. KNIGHT, Jr. and A. J. D'EUSTACHIO, *Biochem. Biophys. Res. Commun.*, **20**, 539 (1965).

P. L. HARRIS, A. A. DIAMANTIS and E. R. ROBERTS, *Biochim. Biophys. Acta*, **111**, 11 (1965).

P. L. HARRIS and E. R. ROBERTS, *Biochim. Biophys. Acta*, **111**, 15 (1965).

E. HEMMINGSEN, *Science*, **135**, 733 (1962).

E. HEMMINGSEN and P. F. SCHOLANDER, *Science*, **132**, 1379 (1960).

S. HIAI, T. MORI, S. HINO and T. MORI, *J. Biochem.*, **44**, 839 (1957).

R. HILL and F. BENDALL, *Nature*, **187**, 417 (1960).

J. W. HINKSON, *Biochemistry*, **7**, 2666 (1968).

J. W. HINKSON and W. A. BULEN, *J. Biol. Chem.*, **242**, 3345 (1967).

S. HINO and P. W. WILSON, *J. Bacteriol.*, **75**, 403 (1958).

G. E. HOCH, K. C. SCHNEIDER and R. H. BURRIS, *Biochim. Biophys. Acta*, **37**, 273 (1960).

G. E. HOCH and D. W. S. WESTLAKE, *Fed. Proc.*, **17**, 243 (1958).

T. C. HOLLOCHER and J. K. LUECHAUER, *Biochem. Biophys. Res. Commun.*, **31**, 417 (1968).

J-C. HWANG, *Diss. Abstr. B.*, **29**, 1927 (1968).

I. D. IVANOV, 'Biol.' Azot Ego Rol Zemled., Akad. Nauk SSSR, Inst. Mikrobiol., 37 (1967).

I. D. IVANOV and T. V. GOGOLEVA, *Biol. Nauki*, 107 (1968).

I. N. IVLEVA, G. I. LIKHTENSHTEIN and A. P. SADKOV, *Biofizika*, **14**, 779 (1969).

I. N. IVLEVA, A. D. SADKOV and V. A. YAKOVLEV, *Izv. Akad. Nauk SSSR, Ser. Biol.*, 688 (1969).

E. K. JACKSON and R. W. F. HARDY, *Plant Physiol.*, **42**, S-38 (1967).

D. Y. JENG, T. DEVANATHAN and L. E. MORTENSON, *Biochem. Biophys. Res. Commun.*, **35**, 625 (1969).

D. Y. JENG, T. DEVANATHAN, E. MOUSTAFA and L. E. MORTENSON, *Bact. Proc.*, 119 (1969).

D. Y. JENG, J. A. MORRIS and L. E. MORTENSON, *J. Biol. Chem.*, **245**, 2809 (1970).

D. Y. JENG and L. E. MORTENSON, *Biochem. Biophys. Res. Commun.*, **32**, 984 (1968).

C. W. JONES and E. R. REDFEARN, *Biochim. Biophys. Acta*, **143**, 340 (1967a).

C. W. JONES and E. R. REDFEARN, *Biochim. Biophys. Acta*, **143**, 354 (1967b).

K. JUNGERMANN, R. K. THAUER, E. RUPPRECHT, C. OHRLOFF and K. DECKER, *FEBS Lett.*, **3**, 144 (1969).

P. JURTSHUK, A. K. MAY, L. M. POPE and P. R. ASTON, *Can. J. Microbiol.*, **15**, 797 (1969).

S. KAJIYAMA, T. MATSUKI and Y. NOSOH, *Biochem. Biophys. Res. Commun.*, **37**, 711 (1969).

M. KELLY, *Biochem. J.*, **109**, 322 (1968).

M. KELLY, *Biochim. Biophys. Acta*, **171**, 9 (1969a).

M. KELLY, *Biochim. Biophys. Acta*, **191**, 527 (1969b).

M. KELLY, R. V. KLUCAS and R. H. BURRIS, *Biochem. J.*, **105**, 3C (1967).

M. KELLY and G. LANG, *Biochim. Biophys. Acta*, **223**, 86 (1970).

M. KELLY, J. R. POSTGATE and R. L. RICHARDS, *Biochem. J.*, **102**, 1C (1967).

I. R. KENNEDY, *Biochim. Biophys. Acta*, **130**, 285 (1966a).

I. R. KENNEDY, *Biochim. Biophys. Acta*, **130**, 295 (1966b).

I. R. KENNEDY, *Biochim. Biophys. Acta*, **222**, 135 (1970).

I. R. KENNEDY, J. A. MORRIS and L. E. MORTENSON, *Biochim. Biophys. Acta*, **153**, 777 (1968).

P. A. KETCHUM, H. Y. CAMBIER, W. A. FRAZIER III, C. H. MADANSKY and A. NASON, *Proc. Natl. Acad. Sci. U.S.*, **66**, 1016 (1970).

A. D. KIDMAN, B. A. C. ACKRELL and R. N. ASATO, *Biochim. Biophys. Acta*, **159**, 185 (1968).

B. KIRSTEINE, N. P. L'VOV, V. I. LYUBIMOV and V. L. KRETOVICH, *Dokl. Akad. Nauk SSSR*, **181**, 741 (1968).

D. KLEINER and R. H. BURRIS, *Biochim. Biophys. Acta*, **212**, 417 (1970).

R. V. KLUCAS, I. Preparation of ferredoxin. II. Investigations of N_2 exchange, p. 66. M.S. Thesis, University of Wisconsin, Department of Biochemistry (1964).

R. V. KLUCAS, B. KOCH, S. A. RUSSELL and H. J. EVANS, *Plant Physiol.*, **43**, 1906 (1968).

E. KNIGHT, Jr. and R. W. F. HARDY, *J. Biol. Chem.*, **241**, 2752 (1966).

C. J. KNOWLES and E. R. REDFEARN, *J. Bacteriol.*, **97**, 756 (1969).

C. J. KNOWLES and L. SMITH, *Biochim. Biophys. Acta*, **197**, 152 (1970).

B. KOCH and H. J. EVANS, *Plant Physiol.*, **41**, 1748 (1966).

B. KOCH, H. J. EVANS an d S. RUSSELL, *Proc. Natl. Acad. Sci. U.S.*, **58**, 1343 (1967).

B. KOCH, P. WONG, S. A. RUSSELL, R. HOWARD and H. J. EVANS, *Biochem. J.*, **118**, 773 (1970).

H. J. KOEPSELL and M. J. JOHNSON, *J. Biol. Chem.*, **145**, 379 (1942).

A. I. KRASNA, E. RIKLIS and D. RITTENBERG, *J. Biol. Chem.*, **235**, 2717 (1960).

A. I. KRASNA and D. RITTENBERG, *Proc. Natl. Acad. Sci. U.S.*, **40**, 225 (1954).

V. L. KRETOVICH, A. A. BUNDELL and N. V. BOROVIKOVA, *Mekh. Dykhaniya, Fotosin. Fiksatsii Azota*, 359 (1967).

R. L. LESTER and F. L. CRANE, *Biochim. Biophys. Acta*, **32**, 492 (1959).

C. J. LIND and P. W. WILSON, *J. Am. Chem. Soc.*, **63**, 3511 (1941).

H. L. LINDSAY, *Diss. Abstr.*, **24**, 2222 (1963).

A. LOCKSHIN and R. H. BURRIS, *Biochim. Biophys. Acta*, **111**, 1 (1965).

W. LOVENBERG and B. E. SOBEL, *Proc. Natl. Acad. Sci.*, **54**, 193 (1965).

W. LOVENBERG, W. M. WILLIAMS and E. C. BRUCKWICK, *Biochemistry*, **8**, 141 (1969).

V. I. LYUBIMOV, *Sel'skokhoz. Biol.*, **2**, 3 (1967).

V. I. LYUBIMOV, N. P. L'VOV, B. KIRSTEINE and V. L. KRETOVICH, *Izv. Akad. Nauk SSSR, Ser. Biol.*, 505 (1969).

W. E. MAGEE and R. H. BURRIS, *J. Bacteriol.*, **71**, 635 (1956).

M. C. MAHL and P. W. WILSON, *Can. J. Microbiol.*, **14**, 33 (1968).

R. MALKIN and J. C. RABINOWITZ, *Biochemistry*, **5**, 1262 (1966).

R. MALKIN and J. C. RABINOWITZ, *Ann. Rev. Biochem.*, **36**, 113 (1967a).

R. MALKIN and J. C. RABINOWITZ, *Biochemistry*, **6**, 3880 (1967b).

G. I. MATKHANOV, I. D. IVANOV, A. F. VANIN and YU. M. BELOV, *Biofizika*, **14**, 1124 (1969).

J. E. MCNARY and R. H. BURRIS, *J. Bacteriol.*, **84**, 598 (1962).

D. M. MOLNAR, R. H. BURRIS and P. W. WILSON, *J. Am. Chem. Soc.*, **70**, 1713 (1948).

L. E. MORTENSON, *Ann. Rev. Microbiol.*, **17**, 115 (1963).

L. E. MORTENSON, *Proc. Natl. Acad. Sci. U.S.*, **52**, 272 (1964a).

L. E. MORTENSON, *Biochim. Biophys. Acta*, **81**, 473 (1964b).

L. E. MORTENSON, *Biochim. Biophys. Acta*, **127**, 18 (1966).

L. E. MORTENSON, Survey of Progress in Chemistry. (A. F. Scott, ed.), p. 127. Acad. Press, New York (1968).

L. E. MORTENSON, J. A. MORRIS and D. Y. JENG, *Biochim. Biophys. Acta.* **141**, 516 (1967).

L. E. MORTENSON, J. A. MORRIS and I. R. KENNEDY, *Bact. Proc.*, 133 (1968).

L. E. MORTENSON, R. C. VALENTINE and J. E. CARNAHAN, *Biochem. Biophys. Res. Commun.*, **7**, 448 (1962).

L. E. MORTENSON, R. C. VALENTINE and J. E. CARNAHAN, *J. Biol. Chem.*, **238**, 794 (1963).

E. MOUSTAFA, *Phytochemistry*, **8**, 993 (1969).

E. MOUSTAFA and L. E. MORTENSON, *Nature*, **216**, 1241 (1967).

E. MOUSTAFA and L. E. MORTENSON, *Biochim. Biophys. Acta*, **172**, 106 (1969).

E. MOUSTAFA, L. E. MORTENSON and P. T. BUI, *Bact. Proc.*, 132 (1968).

M. M. MOZEN, The role of nitramide and nitrogen oxides in biological nitrogen fixation, Ph.D. Thesis, University of Wisconsin (1955).

M. M. MOZEN and R. H. BURRIS, *Biochim. Biophys. Acta*, **14**, 577 (1954).

M. M. MOZEN, R. H. BURRIS, S. LUNDBOM and A. I. VIRTANEN, *Acta Chem. Scand.*, **9**, 1232 (1955).

T. O. MUNSON and R. H. BURRIS, *J. Bacteriol.*, **97**, 1093 (1969).

T. O. MUNSON, M. J. DILWORTH and R. H. BURRIS, *Biochim. Biophys. Acta*, **104**, 278 (1965).

M. S. NAIK and D. J. D. NICHOLAS, *Indian J. Biochem.*, **6**, 111 (1969).

C. NEPOKROEFF and A. I. ARONSON, *Biochemistry*, **9**, 2074 (1970).

N. P. NEUMANN and R. H. BURRIS, *J. Biol. Chem.*, **234**, 3286 (1959).

D. J. NEWMAN and J. R. POSTGATE, *Eur. J. Biochem.*, **7**, 45 (1968).

D. J. D. NICHOLAS and D. J. FISHER, *Nature*, **186**, 735 (1960).

D. J. D. NICHOLAS, D. J. SILVESTER and J. FOWLER, *Nature*, **189**, 634 (1961).

A. NOWOTNY-MIECZYNSKA, *Postepy Mikrobiol.*, **7**, 289 (1968).

J. OPPENHEIM, R. J. FISHER, P. W. WILSON and L. MARCUS, *J. Bacteriol.*, **101**, 292 (1970).

J. OPPENHEIM and L. MARCUS, *Bact. Proc.*, 149 (1970).

J. G. ORMEROD, K. S. ORMEROD and H. GEST, *Arch. Biochem. Biophys.*, **94**, 449 (1961).

G. PALMER, R. H. SANDS and L. E. MORTENSON, *Biochem. Biophys. Res. Commun.*, **23**, 357 (1966).

R. A. PAREJKO and P. W. WILSON, *Can. J. Microbiol.*, **16**, 681 (1970).

C. A. PARKER and M. J. DILWORTH, *Biochim. Biophys. Acta*, **69**, 152 (1963).

R. M. PENGRA and P. W. WILSON, *J. Bacteriol.*, **75**, 21 (1958).

M. POE, W. D. WILLIAMS, C. C. McDONALD and W. LOVENBERG, *Proc. Natl. Acad. Sci. U.S.*, **65**, 797 (1970).

J. R. POSTGATE, *Nature*, **226**, 25 (1970).

J. A. RAKESTRAW and E. R. ROBERTS, *Biochim. Biophys. Acta*, **24**, 388 (1957a).

J. A. RAKESTRAW and E. R. ROBERTS, *Biochim. Biophys. Acta*, **24**, 555 (1957b).

R. REPASKE and P. W. WILSON, *J. Am. Chem. Soc.*, **74**, 3101 (1952).

M. A. RIEDERER-HENDERSON and P. W. WILSON, *J. Gen. Microbiol.*, **61**, 27 (1970).

M. ROBERG, *Jahr. Wiss. Bot.*, **83**, 567 (1936).

S. A. ROBRISH and A. G. MARR, *J. Bacteriol.*, **83**, 158 (1962).

E. D. ROSENBLUM and P. W. WILSON, *J. Bacteriol.*, **59**, 83 1950).

A. SAN PIETRO and H. M. LANG, *J. Biol. Chem.*, **231**, 211 (1958).

N-E. SARIS and A. I. VIRTANEN, *Acta Chemica Scand.*, **11**, 1438 (1957).

K. C. SCHNEIDER, C. BRADBEER, R. N. SINGH, L. C. WANG, P. W. WILSON and R. H. BURRIS, *Proc. Natl. Acad. Sci. U.S.*, **46**, 726 (1960).

R. SCHOLLHORN and R. H. BURRIS, *Fed. Proc.*, **25**, 710 (1966).

R. SCHOLLHORN and R. H. BURRIS, *Proc. Natl. Acad. Sci. U.S.*, **57**, 1317 (1967a).

R. SCHOLLHORN and R. H. BURRIS, *Proc. Natl. Acad. Sci. U.S.*, **58**, 213 (1967b).

Y. I. SHETHNA, *Biochim. Biophys. Acta*, **205**, 58 (1970a).

Y. I. SHETHNA, *Bact. Proc.*, 148 (1970b).

Y. I. SHETHNA, D. V. DER VARTANIAN and H. BEINERT, *Biochem. Biophys. Res. Commun.*, **31**, 862 (1968).

Y. I. SHETHNA, N. A. STOMBAUGH and R. H. BURRIS, *Biochem. Biophys. Res. Commun.*, **42**, 1108 (1971).

Y. I. SHETHNA, P. W. WILSON and H. BEINERT, *Biochim. Biophys. Acta*, **113**, 225 (1966).

Y. I. SHETHNA, P. W. WILSON, R. E. HANSEN and H. BEINERT, *Proc. Natl. Acad. Sci. U.S.*, **52**, 1263 (1964).

A. L. SHUG, P. B. HAMILTON and P. W. WILSON, Inorganic nitrogen metabolism. (W. D. McElroy and B. Glass, eds.) p. 344. Johns Hopkins Press, Baltimore (1956).

L. C. SIEKER and L. H. JENSEN, *Biochem. Biophys. Res. Commun.*, **20**, 33 (1965).

R. SILVERSTEIN and W. A. BULEN, *Biochemistry*, **9**, 3809 (1970).

C. SLOGER and W. S. SILVER, *Bact. Proc.*, 112 (1967).

R. V. SMITH and M. C. W. EVANS, *Nature*, **225**, 1253 (1970).

G. J. SORGER and D. TROFIMENKOFF, *Proc. Natl. Acad. Sci. U.S.*, **65**, 74 (1970).

W. D. P. STEWART, Nitrogen fixation in plants, p. 168. Athlone Press of the University of London (1966).

W. D. P. STEWART, *Science*, **158**, 1426 (1967).

W. D. P. STEWART, G. P. FITZGERALD and R. H. BURRIS, *Proc. Natl. Acad. Sci. U.S.*, **58**, 2071 (1967).

W. D. P. STEWART, G. P. FITZGERALD and R. H. BURRIS, *Proc. Natl. Acad. Sci. U.S.*, **66**, 1104 (1970).

W. D. P. STEWART, A. HAYSTEAD and H. W. PEARSON, *Nature*, **224**, 226 (1969).

W. D. P. STEWART and M. LEX, *Arch. Mikrobiol.*, **73**, 250 (1970).

G. W. STRANDBERG and P. W. WILSON, *Proc. Natl. Acad. Sci. U.S.*, **58**, 1404 (1967).

G. W. STRANDBERG and P. W. WILSON, *Can. J. Microbiol.*, **14**, 25 (1968).

N. SUZUKI and S. SUZUKI, *Sci. Reports Tohoku Univ., 4th Ser. (Biology)*, **20**, 195 (1954).

R. T. SWANK and R. H. BURRIS, *J. Bacteriol.*, **98**, 311 (1969a).

R. T. SWANK and R. H. BURRIS, *Biochim. Biophys. Acta*, **180**, 473 (1969b).

K. TAGAWA and D. I. ARNON, *Nature*, **195**, 537 (1962).

K. TAGAWA and D. I. ARNON, *Biochim. Biophys. Acta*, **153**, 602 (1968).

M. TANAKA, T. NAKASHIMA, A. BENSON, H. MOWER and K. T. YASUNOBU, *Biochemistry*, **5**, 1666 (1966).

K. B. TAYLOR, *J. Biol. Chem.*, **244**, 171 (1969).

A. TEMPERLI and P. W. WILSON, *Nature*, **193**, 171 (1962).

J. THOMAS, *Nature*, **228**, 181 (1970).

A. TISSIERES, *Biochem. J.*, **64**, 582 (1956).

A. C. TRAKATELLIS and G. SCHWARTZ, *Proc. Natl. Acad. Sci. U.S.*, **63**, 436 (1969).

G. L. TURNER and F. J. BERGERSEN, *Biochem. J.*, **115**, 529 (1969).

R. C. VALENTINE, *Bacteriol. Rev.*, **28**, 497 (1964).

J. P. VANDECASTEELE, *Diss. Abstr. B*, **29**, 4522 (1968).

J. P. VANDECASTEELE and R. H. BURRIS, *J. Bacteriol.*, **101**, 794 (1970).

A. I. VIRTANEN and S. LUNDBOM, *Acta Chem. Scand.*, **7**, 1223 (1953).

M. E. VOL'PIN and V. B. SHUR, *Priroda (Moscow)*, 29 (1969).

D. W. S. WESTLAKE, A. L. SHUG and P. W. WILSON, *Can. J. Microbiol.*, **7**, 515 (1961).

D. W. S. WESTLAKE and P. W. WILSON, *Can. J. Microbiol.*, **5**, 617 (1959).

P. W. WILSON and R. H. BURRIS, *Ann. Rev. Microbiol.*, **7**, 415 (1953).

P. W. WILSON and E. B. FRED, *Proc. Natl. Acad. Sci. U.S.*, **23**, 503 (1937).

P. W. WILSON and W. W. UMBREIT, *Arch. Mikrobiol.*, **8**, 440 (1937).

H. C. WINTER and D. I. ARNON, *Biochim. Biophys. Acta*, **197**, 170 (1970).

H. C. WINTER and R. H. BURRIS, *J. Biol. Chem.*, **243**, 940 (1968).

J. B. WITTENBERG, *Physiol. Rev.*, **50**, 559 (1970).

D. F. WITZ, R. W. DETROY and P. W. WILSON, *Arch. Mikrobiol.*, **55**, 369 (1967).

D. F. WITZ and P. W. WILSON, *Bact. Proc.*, 112 (1967).

T. YAGI, M. HONYA and N. TAMIYA, *Biochim. Biophys. Acta*, **153**, 699 (1968).

V. A. YAKOVLEV, Mekh. Dykhaniya, Fotosin. Fiksatsii Azota. (V. A. Yakovlev, ed.), p. 309. Izd. 'Nauka': Moscow, USSR (1967).

V. A. YAKOVLEV and R. I. GVOZDEV, *Zh. Evol. Biokhim. Fiziol.*, **3**, 185 (1967).

M. G. YATES, *J. Gen. Microbiol.*, **60**, 393 (1970a).

M. G. YATES, *FEBS Lett.*, **8**, 281 (1970b).

M. G. YATES and R. M. DANIEL, *Biochim. Biophys. Acta*, **197**, 161 (1970).

D. C. YOCH and D. I. ARNON, *Biochim. Biophys. Acta*, **197**, 180 (1970).

D. C. YOCH, J. R. BENEMANN, R. C. VALENTINE and D. I. ARNON, *Proc. Natl. Acad. Sci. U.S.*, **64**, 1404 (1969).

D. C. YOCH, J. R. BENEMANN, D. I. ARNON, R. C. VALENTINE and S. A. RUSSELL, *Biochem. Biophys. Res. Commun.*, **38**, 838 (1970).

D. C. YOCH, R. F. MORTENSON and E. S. LINDSTROM, *Bact. Proc.*, 133 (1968).

D. C. YOCH and R. M. PENGRA, *J. Bacteriol.*, **92**, 618 (1966).

C. S. YOCUM, *Science*, **146**, 432 (1964).

CHAPTER 5

Fixation by Free-Living Microbes: Physiology

JOHN POSTGATE

University of Sussex,
Falmer,
Sussex, BN1 9QJ, England

5.1 INTRODUCTION

5.1.1 *Aims*

The aim of this chapter is not to review exhaustively the physiology of the free-living N_2-fixing microbes. Such a task would require a book to itself and the information is already covered more than adequately in Wilson's (1958) review, Rubenchik's (1960) monograph on azotobacters and the encyclopaedic work of Mishustin and Shil'nikova (1966). Jensen (1965) has touched on many features of their physiology in an exhaustive account of their ecology. My present aim is to relate the advances over the last decade in our knowledge of the chemistry and enzymology of the N_2 fixation process to the general physiology of free-living N_2-fixing microbes; in general I shall take the information in Wilson's (1958) review for granted.

The preceding chapters have described progress made in the last decade at the purely chemical and at the enzymological levels. A most striking feature is the similarity of nitrogenases, no matter what organisms they are obtained from; a similarity emphasized by the cross-reactions which may be shown between fractions from different bacteria. The biochemical uniformity of nitrogenase means that one can make general statements about the system; three major principles emerge which are reflected in the physiology of the free-living organisms. They are:

(a) The pathway of N_2 fixation is essentially reductive. This is true even of aerobes such as *Azotobacter*, for which oxidative pathways had earlier been proposed. Not only is the pathway reductive, but at least one component of the nitrogenase system is destroyed by oxygen in all organisms so far examined.

(b) The process consumes biological energy. Despite the fact that reduction of N_2 to NH_3, coupled to oxidation of a substrate such as glucose, is exergonic at ordinary temperatures and pH values in air (Bayliss, 1956), enzyme preparations consume ATP in amounts ranging from 6 to 15 moles/mole of substrate reduced. The actual value obtained depends to some extent on the substrate (N_2, C_2H_2, CN^-, etc.) and on the ratio of the two main enzymic components of nitrogenase.

(c) Both protein components of nitrogenase contain transition metals: 'fraction 1' possesses both Fe and Mo; 'fraction 2' possesses Fe alone.

Some physiological characteristics of free-living N_2-fixing bacteria which can be related to these three main principles will be discussed separately.

5.2   THE REDUCTIVE CHARACTER OF N_2 FIXATION

5.2.1   Distribution of N_2 fixation among anaerobes compared with aerobes

The variety of microbes believed capable of fixing N_2 has fluctuated considerably during the past two decades. Nicholas (1963) and L'vov (1963) listed numerous aerobic species of microbe, including the genus *Azotomonas*, representatives of the genus *Pseudomonas* (e.g. *P. azotogensis* and *P. azotocolligans*), species of *Nocardia*, *Mycobacterium*, yeasts and blue-green algae. The advent of the sensitive acetylene test for nitrogen fixation (see page 311) has led to considerable revision of this list. Parejko and Wilson (1968) could find no evidence for fixation by available strains of *Azotomonas* species; Hill and Postgate (1969) agreed and also eliminated the two putative N_2-fixing *Nocardia* species. *P. azotocolligans* and one strain of *P. azotogensis* proved also to be non-fixers; all these organisms having in common the property of being very efficient scavengers of traces of fixed nitrogen in air or laboratory reagents. The strain of *P. azotogensis* which did fix N_2 proved (as discussed below) to be misclassified and not to fix aerobically. Campbell *et al.* (1967) isolated yeasts and *Pullularia* which apparently fixed N_2 by $^{13}N_2$ tests but one yeast strain which was closely examined by cultural and acetylene reduction tests did not do so (Dr. N. E. R. Campbell, personal communication); Millbank (1969) was unable to obtain fixation by aerobic cultures of yeasts and *Pullularia* strains which had earlier seemed to fix according to the $^{15}N_2$ test as well as to analyses for N. The status of yeasts as N_2-fixing organisms is discussed further in section 5.5.3.

At present it seems that genuine aerobic N_2-fixing microbes are rare. Among bacteria, aerobic N_2-fixers are to be found mainly in the family Azotobacteriaceae which, in the classification of De Ley (De Ley, 1968; De Ley and Park, 1966) comprises the genera *Azotobacter*, *Azomonas*, *Azotococcus*, *Beijerinckia* and *Derxia*. They are all taxonomically very similar. *Mycobacterium flavum* strain 301 and possibly one or two related species (Fedorov and Kalininskaya, 1961) are genuine fixers; Coty's (1967) *Pseudomonas methanitrificans*, a methane-oxidizing organism also fixes aerobically according to analyses, $^{15}N_2$ tests and the acetylene test but, since it forms exospores and multiplies by budding (R. Whittenbury, personal communication), it is probably misclassified in the genus *Pseudomonas*. The remainder of the aerobic bacteria which can fix are facultative anaerobes which only fix N_2 when growing anaerobically: *Bacillus macerens*, *Bacillus polymyxa* (see Grau and Wilson,

1963), *Klebsiella* species (see Hamilton, Burris and Wilson, 1964, Mahl *et al.*, 1965) and representatives of the photosynthetic genera *Rhodopseudomonas, Rhodomicrobium* and *Rhodospirillum* (Pratt and Frenkel, 1959, Schneider *et al.*, 1960) are examples. The strain called *Pseudomonas azotogensis* mentioned earlier as a genuine N_2-fixing organism was discovered by Voets and Debacher (1956); De Ley (1968) has pointed out that it is probably misclassified among the pseudomonads and Hill and Postgate (1969) showed that, contrary to the claim of its discoverers, it resembled the facultatively anaerobic group in only fixing N_2 when growing anaerobically.

Thus the ability to fix N_2 aerobically is confined to relatively few groups of bacteria among which the family Azotobacteriaciae seems predominant. Anaerobic fixation is much more widely distributed, occurring among the facultative aerobes already mentioned as well as among types of obligate anaerobe: *Clostridium pasteurianum, C. butyricum, Methanobacterium omelianskii, Desulfovibrio, Chlorobium limicola, Chromatium* (see Nicholas, 1963), *Desulfotomaculum* species (Postgate, 1970).

Among the blue-green algae aerobic fixation is common (Stewart, 1969, listed 14 N_2-fixing genera) but they appear to be more highly evolved creatures than the bacteria (van Niel, 1955); their special features are discussed in section 5.2.11.

5.2.2   The generation of reducing power
A detailed discussion of the catabolic pathways of N_2-fixing bacteria and Cyanophyceae is beyond the scope of this chapter but some aspects of the immediate metabolic links between catabolism and N_2 fixation will be commented upon. Work with cell-free preparations has shown remarkable substrate specificity in N_2 fixation. In *C. pasteurianum,* pyruvate or α-ketobutyrate are the only effective 'natural' substrates; they may be replaced by H_2 if ATP is provided, which implies that N_2 fixation in the living bacteria is intimately involved with ferredoxin and the normal pyruvic phosphoroclastic system. Pyruvate is both the source of reducing power and ATP in *Bacillus polymyxa, Klebsiella;* in Cyanophyceae it provides reducing power and light provides at least some ATP (see Chapter 4). In azotobacter preparations pyruvate is inactive and the manner in which reducing power is generated is still uncertain. Klucas and Evans (1968) obtained a soluble preparation of *A. vinelandii* which linked oxidation of β-hydroxybutyrate to acetylene reduction if benzyl viologen was added as an electron carrier; Yates and Daniel (1970) obtained partially damaged *A. chroococcum* organisms which

reduced acetylene on a small scale with 'natural' substrates such as glucose-6-phosphate, succinate or NADH, and also obtained membranous preparations which reduced acetylene on a small scale without added ATP when provided with 'natural' substrates including NADH, glucose 6-phosphate and sodium isocitrate. They also confirmed the presence of Klucas and Evans's benzyl-viologen-linked system in *A. chroococcum* and showed that it responded to substrates quoted for the membranous preparation.

Pyruvate, as mentioned earlier, functions as reductant for N_2 fixation by the blue-green algae *Anabena cylindrica* and Cox and Fay (1969) have summarized evidence that, though closely linked with photosynthesis, N_2 fixation is coupled with 'dark reactions' involving light-generated ATP but reducing power from fixed CO_2.

5.2.3   Protection of nitrogenase among aerobes

Since purified nitrogenase is, in part, inactivated by oxygen, the aerobic N_2-fixing bacteria obviously require some physiological mechanism to protect the oxygen-sensitive component of nitrogenase from damage by oxygen. The particulate nature of the enzyme as extracted in crude form from azotobacters (see p. 113), which contrasts with the soluble character of preparations obtained from *C. pasteurianum,* may be relevant to this question. The particle, as extracted from *A. vinelandii* or *A. chroococcum* by sonic disruption or decompression, is relatively insensitive to damage by oxygen, whereas the soluble extracts of *C. pasteurianum* are extremely O_2-sensitive. As soon, however, as the particulate preparations are split into their component fractions by chromatography, the oxygen-sensitivity of 'fraction 2' becomes manifest and anaerobic handling of the preparations is essential. The crude nitrogenase of *Mycobacterium flavum* 301 is also particulate but is less resistent to oxygen than comparable preparations from *Azotobacter* (Biggins and Postgate, 1969).

These findings suggest that nitrogenase, in the aerobic bacteria so far examined, can exist in a conformation in which 'fraction 2', the most oxygen-sensitive component, is protected from being damaged by oxygen. Dalton and Postgate (1969a) referred to this condition as 'conformational protection'; if this view is accepted, *M. flavum* 301 represents an organism in which 'conformational protection' is less highly developed than it is in azotobacter. I shall argue in the next section that 'conformational protection' is regulated in the living organism and that the 'conformationally protected' enzyme is also inactive towards N_2.

5.2.4 *Inhibition of aerobic N_2 fixation by oxygen: the 'phyo-enzyme'*

An empirical consequence of the sensitivity of nitrogenase to oxygen is that most aerobic N_2-fixing bacteria and those blue-green algae which have been examined, do not fix optimally at atmospheric pO_2 values. They tend to be micro-aerophilic in the sense that, though they are obligate aerobes, their fixation is most effective at sub-atmospheric pO_2 values. Fig. 5.1 illustrates typical 'bell-shaped' curves which are obtained with aerobic N_2-fixing microbes when the rate of fixation by cultures grown in air is assessed against pO_2 by the acetylene test. Drozd and Postgate (1970b) showed that, in *A. chroococcum,* the shape of the curve could be influenced by 'adapting' chemostat cultures to high pO_2 values. An example is included in Fig. 5.1; this matter is discussed more fully under 'respiratory protection' (5.2.6) below.

The state of the nitrogenase in some Azotobacteriaceae is probably under a kind of physiological control. Drozd and Postgate (1970a) showed that normal cultures of *A. chroococcum, A. vinelandii* and *Azomonas macrocytogenes* would partly or completely 'switch off' nitrogenase activity (as indicated by acetylene reduction) if cultures were vigorously shaken in air. The process was reversible: gentle shaking in air restored full activity, sometimes instantly, sometimes gradually, unless the shaking had been prolonged or the population was hypersensitive. Yates (1970a) described similar reversible inhibition by air and observed that it was accompanied by an increase in respiratory activity. Strains of *A. chroococcum* 'adapted' to atmospheres of hyperbaric pO_2 became less sensitive to the switching off process; as discussed later, they showed an elevated respiration rate and an increased content of cytochrome a_2, a terminal oxidase in this group. Among the most sensitive populations to this 'switching off' process were carbon-limited populations, which are also hypersensitive to inhibition of growth by oxygen (below).

The 'switching off' of nitrogenase activity takes less than two minutes with normal populations of *Azotobacter*. The 'switching on' process may take up to ten minutes, particularly if the population has been washed by centrifuging. By judicious choice of populations adapted to various pO_2 values, and by choosing appropriate extents of aeration, Drozd and Postgate (1970b) showed that 'switch off' and 'switch on' could be almost instantaneous. The process then recalled the reversible effects of oxygen on nitrate reduction or tetrathionate reduction in coliform bacteria (Pichinoty, 1963;

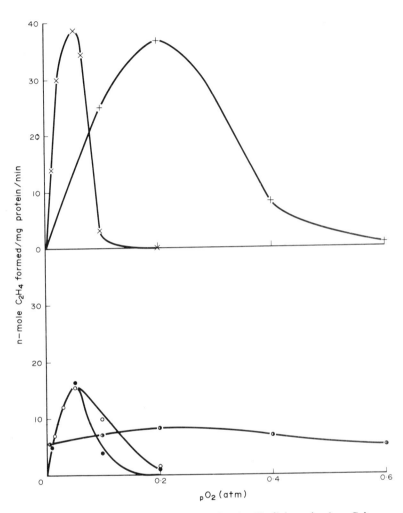

Fig. 5.1. Effect of pO_2 on acetylene reduction by N_2-fixing microbes. Cultures were exposed to atmosphere of various pO_2 values containing acetylene and rates of ethylene production were measured. Apart from the two curves with *A. chroococcum*, the actual rates have no comparative significance because the populations and test conditions differed; the general patterns of the curves are comparable. Adapted from Biggins and Postgate (1969), Stewart (1969), Drozd and Postgate (1970a).

x: *Azotobacter chroococcum* acclimatized to a pO_2 of 0.05 atm.; +: *A. chroococcum* acclimatized to pO_2 of 0.55 atm.; ○: *A. vinelandii* grown in air; ●: *Mycobacterium flavum* 301 grown in air; ◐: *Anabena flos-aquae.*

Pichinoty and Bigliardi-Rouvier, 1963), phenomena which are accounted for in terms of competition for electrons between autoxidizeable electron carriers and anaerobic respiratory processes based on tetrathionate or nitrate reduction. These processes are insensitive to inhibitors such as chloramphenicol, as is the switch-on, switch-off process in *Azotobacter*. Figure 5.2 illustrates the effect of pO_2 in causing 'switch-off' and 'switch-on' of acetylene reduction in *Azotobacter chroococcum*.

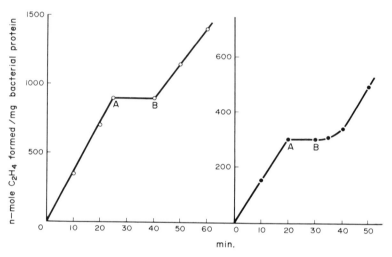

Fig. 5.2. 'Switch on' and 'Switch off' of acetylene-reducing activity induced by aeration of cultures of *Azotobacter chroococcum*. Samples of growing culture were shaken gently while reducing acetylene. Vigorous shaking at A immediately arrested acetylene reduction; return to the original shaking rate at B restored activity, either immediately or after a delay. ○: sample from a continuous culture growing at a pO_2 of 0.09 atm. and tested in air (from Drozd and Postgate, 1970b); ●: sample from a batch culture grown in air shaken at 75 strokes/min in 0.1 atm. O_2, amplitude 1.5 cm, and altered at A to 250 strokes/min at 1 cm amplitude. (Unpublished data of Mr. J. Drozd.)

An attractive anatomical mechanism for conformational protection is provided by the observation that nitrogenase may be associated with membranes in *Azotobacter vinelandii* and that the oxygen sensitivity, as well as the extent of association with membrane material, depends on the method of rupture of the bacteria (Oppenheim and Marcus, 1970; Oppenheim *et al.*, 1970). Evidence that non-specific proteins can promote association of fractions 1 and 2, and decrease oxygen sensitivity, was presented by Yates (1970b).

Prolonged aeration of sensitive strains, or exposure of insensitive strains to pure O_2, caused irreversible inhibition of ability to reduce acetylene, a result of actual damage to 'fraction 2' of the nitrogenase complex. But the existence of a reversible 'switching off' of nitrogenase in conditions of excessive aeration is probably the origin of the early belief that nitrogenase represents a class of enzymes which are only active during growth: a class designated by Burk (1934) as 'phyo-enzymes'. It is one possible reason why significant N_2 uptake by washed suspensions of azotobacters in Warburg manometers is so rarely observable even when one would expect it theoretically: the washing procedure has presumably induced the 'switching off' process. It seems reasonable to regard the 'switched off' state of the enzyme as the equivalent *in vivo* of the 'conformationally protected' form normally obtained *in vitro*.

5.2.5   Inhibition of nitrogenase synthesis by oxygen

Facultative anaerobes such as *Rhodospirillum, Klebsiella* and *Bacillus polymyxa* do not appear to synthesize nitrogenase in the presence of dissolved oxygen, though whether this is a direct effect of O_2 on synthesis or the result of oxygen inactivation of enzyme synthesized is not clear. Pengra and Wilson (1958) showed that oxygen even at a pO_2 of 0.05 atm. inhibited growth (and hence presumably nitrogenase synthesis) of a N_2-fixing *Klebsiella*; Goerz and Pengra (1961) made a similar observation with a N_2-fixing *Achromobacter**. *Rhodospirillum rubrum* can grow aerobically in the dark but then shows no nitrogenase activity. Pratt and Frenkel (1959) regarded nitrogen fixation as 'light dependent' and showed that it was inhibited by both hydrogen and oxygen. If the pO_2 at which oxygen inhibition was caused was low, the inhibition was reversed on removal of oxygen.

Strandberg and Wilson (1968), in a study of the control of nitrogenase synthesis in *Azotobacter vinelandii*, observed an adaptive lag after the culture had become depleted in repressor (ammonium or nitrate ions). Hyperbaric oxygen prolonged this lag, suggesting either that oxygen interfered directly with nitrogenase synthesis or that nitrogenase was synthesized and immediately inactivated.

5.2.6   Inhibition of aerobic growth by oxygen

Oxygen sensitivity in the growth of N_2-fixing azotobacters has been reported by many workers and was known as early as 1927. Probably

---

* Later identified as an anaerogenic *Klebsiella*.

the first explicit hypothesis was that of Parker (1954) who, with Scutt, developed an analogy between N_2-fixation and respiration (Parker and Scutt, 1960). N_2 reduction might be analogous to the oxygen reduction step of respiration and, consistent with the view, oxygen acted competitively with respect to N_2 in Parker and Scutt's growth experiments. Dilworth and Kennedy (1963) attempted to relate oxygen sensitivity to special enzymes (phosphotransacetylase and malate synthetase) which seemed particularly oxygen sensitive in *A. vinelandii*. Ivanov (1966) has taken Parker's respiration analogy sufficiently literally to refer to N_2 fixation in *Azotobacter* as 'nitrogen-respiration', a process distinct from normal aerobic respiration yet involving similar electron transport factors. Claims that cytochromes are directly involved in both processes (Ivanov, 1966; Lysenkova and Khmel, 1967; Ivanov, Denina and Gogoleva, 1967) and that N_2 fixation can be obtained anaerobically with horse-heart succinoxidase plus azotobacter cytochromes (Ivanov *et al.*, 1967) require confirmation in other laboratories before they can be accepted without reserve; Dr. M. G. Yates was not able to substantiate the latter in the writer's laboratory using beef-heart succinoxidase and cytochromes c_4 and c_5 from *A. chroococcum*. Ivanov *et al.*'s claim is particularly surprising in view of the known requirement of cell-free nitrogenase preparations for Mo and ATP; these matters are discussed further by Dr. Yates in the appendix to this volume.

There is no doubt, however, that excessive oxygen tends to inhibit growth of N_2-fixing azotobacters despite their aerobic habit. Dalton and Postgate (1969a) studied the phenomenon with both batch and continuous cultures of *A. chroococcum* and observed that sensitivity was specific to organisms fixing N_2: when provided with fixed nitrogen as the ammonium ion they showed no undue oxygen sensitivity. A particularly significant point was that nutritional status of the population exerted an important influence. Carbon- or phosphate-limited populations of *A. chroococcum* were much more sensitive than the 'normal' type of population (which, for reasons given later, may be regarded as limited by its rate of N_2 assimilation). Unpublished work in the writer's laboratory by Mr. J. Drozd shows that sulphate-limited N_2-fixing populations do not show this extra O_2 sensitivity.

5.2.7   Respiration as a protective process
Why should N_2-fixing populations of azotobacters be specially sensitive to oxygen if particulate enzyme preparations extracted

from them are not particularly sensitive? As a working hypothesis, Dalton and Postgate (1969a, b) argued that conformational protection was regulated: in the growing organism conformational protection would be controlled and, in optimal growth conditions, replaced by respiration. The high QO_2 values, for which the Azotobacteriaceae are famous (see Wilson, 1958) were regarded as expressions of an active mechanism for protection of functioning nitrogenase by respiration: access of O_2 to nitrogenase would be prevented by augmented respiration. They called this process 'respiratory protection'.

It would follow from this view that the functional enzyme in the growing organism is not the same as in the resting bacteria or in the extracted particle. Normally, when extracted in the laboratory, the organisms have 'switched off' their nitrogenase by transition to the conformationally protected state (the evidence that reversible 'switch off' can occur in living azotobacters was mentioned in section 5.2.3.). It seems likely that, in the growing, N_2-fixing organism, the nitrogenase assumes a conformation which would be susceptible to damage by oxygen, were not oxygen excluded by respiration. A consequence of this view is that cultures of N_2-fixing azotobacters show very high 'maintenance' factors: considerable proportions of the carbon-energy source are diverted from biosynthesis to 'respiratory protection'. Such high maintenance coefficients have been reported by several other groups (Phillips and Johnson, 1961; Aiba et al. 1967a, b). Where the carbon-energy source is the limiting substrate, as mentioned in the previous section, the organisms become very sensitive to oxygen. Presumably respiratory protection may break down because of a shortage of material to respire. Phosphate limitation can also lead to apparent breakdown of respiratory protection. Yates (1970) presented evidence for respiratory control through the ATP:ADP ratio in A. chroococcum; the particular oxygen sensitivity of phosphate-limited populations is readily understood if a control process of this form exists because excess O_2 would deplete the already limited adenine nucleotide pool of ADP and would thus tend to oppose respiratory protection. A physiological conflict between the two processes would occur which, in a proportion of the population, could lead to growth inhibition and even, perhaps, death.

The argument that respiration, beside its normal physiological functions, is an active protective process in nitrogen-fixing azotobacters has, like the concept of conformational protection, the status of a working hypothesis; neither can be taken as proved. Nevertheless, it will be useful to pause here to summarize the data on which the two hypotheses are based. These are:

(a) There is clear evidence for a rapid apparent 'switch off' of acetylene reduction in response to a high concentration of dissolved oxygen, which can be rapidly reversed at low dissolved oxygen concentrations. These findings could simply reflect a competition between acetylene and oxygen for electrons were it not for the fact that there is a 'memory' regarding 'switch off': it is rapid but not necessarily instantaneous (see Fig. 5.2). The rapidity of the responses and their insensitivity to chloramphenicol implies a form of control and not a sequence of damage to nitrogenase followed by resynthesis or repair of the damaged enzyme.

(b) 'Switch on' *completely* restores activity, unless the population is hypersensitive or has suffered prolonged oxygenation. In such circumstances damage to fraction 2 can be demonstrated, implying that a protected condition or protective process has been overcome.

(c) A form of nitrogenase may be extracted from azotobacters which is sedimentable and oxygen-insensitive; it resolves to given on oxygen-sensitive component on further purification. The method of extraction influences the proportion of sedimentable enzyme extracted; conditions such as glycerol lysis give less sedimentable and more oxygen-sensitive material. These facts provide evidence that the conformation of the two nitrogenase proteins, with respect to each other, to other proteins or to membrane material, influences the oxygen sensitivity of fraction 2.

(d) As described in the next section, continuous cultures adjust their rates of N_2 fixation to the dilution rate in circumstances in which all chemical components of the environment are available in excess. This observation also implies an active control process operating on nitrogenase activity.

5.2.8   *Nutritional status of 'ordinary' cultures of Azotobacters*

The view that respiration is a protective process on *Azotobacter* also accounts for what is otherwise a baffling characteristic of *A. chroococcum* or *A. vinelandii* in chemostat culture: in conventional media containing excess of all soluble nutrients, and containing free dissolved oxygen, the bacterial density is inversely related to the dilution rate. Figure 5.3. illustrates such a curve, with a 'theoretical' curve for a carbon-limited population dotted in. The deviation from theoretical at high growth rates is not relevant to the present discussion; it is attributable to selection of fast-growing variants (Dalton and Postgate, 1969b). The deviation at low growth rates is

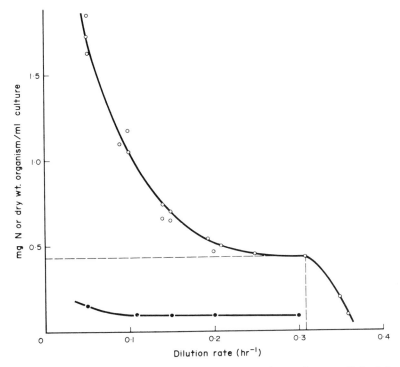

Fig. 5.3. Population densities, contents of fixed N and growth rates of *Azotobacter chroococcum* in continuous culture. ○: mg dry wt. organisms/ml culture; ● mg N/ml culture. A 'theoretical' curve for an ideal, 'maintenance'-free, carbon-limited chemostat culture of μ_{max} = 0.31 h^{-1} is dotted in (- - -). From Dalton and Postgate (1969b) including unpublished data of Dr. H. Dalton.

largely accounted for by increased contents of polyhydroxybutyrate and polysaccharide in the individual bacteria, which implies that intrinsic factors such as the rate of permeation of the carbon substrate into the cells were not limiting growth. The populations behaved exactly as if the yield were limited by a gaseous component of the environment. Since the high QO_2 values of such populations argue against oxygen as the limiting factor, one is forced to the conclusion that N_2 is somehow limiting. An elementary calculation, comparing the relative requirements of the organisms for N_2 and O_2

with their relative solubilities, shows that the physical abundance of N_2 in the solution must be adequate by nearly 100-fold. One is forced to the further conclusion that an intrinsic limitation on the rate of N_2 assimilation exists in the growing microbes, which is adjusted to the supply of other nutrients and of air. Figure 5.3 includes a plot of the steady state N/ml culture which, consistent with this view, was constant over a 3-fold range of dilution rate. Such a limitation by N_2 could arise either by decreased synthesis of nitrogenase (which seems teleologically unreasonable since all necessary materials are present in excess) or by limiting its activity. Given the evidence for control presented earlier, one might say that the organisms 'expose' as much of their nitrogenase as they can protect by respiration, and no more. Intrinsic limitations have been postulated before in continuous culture (see Elsden, 1967) and presumably operate in turbidostat cultures of any microbe. The unusual feature of the system proposed by Dalton and Postgate (1969b) is that the intrinsic limitation is under the organism's control. They termed this condition 'N_2-limited' and pointed out the corollary that any batch culture which was not O_2-limited was N_2-limited: a normal 'shake-flask' culture is N_2-limited during most of its growth and only becomes O_2-limited at high bacterial densities.

5.2.9   The role of slime

Whether the detailed view of the oxygen relations to *Azotobacter* put forward here is wholly correct requires further research; at present it has the status of a reasonable working hypothesis. Nevertheless, there is no doubt of the empirical fact that N_2-fixing azotobacters, though obligate aerobes, are unusually sensitive to oxygen and sometimes extremely so, compared with NH_4-assimilating azotobacters. Yet Azotobacteriaceae are the group of N_2-fixing bacteria which seem best adapted to aerobic existence and many workers have considered whether the slime they characteristically form is physiologically important to them. Polysaccharide formation is typical of most types of N-limited bacteria (Herbert, 1961a) and the slime formation may be nothing more than an expression of the usual nutritional status of the bacteria discussed in section 5.2.8; some evidence exists that the slime of Azotobacteriaceae is a true 'reserve' material, because it is consumed by non-growing populations (Lopez and Becking, 1968). But a slime covering would impede access to oxygen to sensitive sites and might thus have survival value; Barooh and Sen (1964) showed a correlation between slime formation and N_2 fixation in ten strains of

Beijerinckia. Anyone who has worked with *Derxia gummosa* has experienced the characteristic delay in N_2 fixation on Petri dishes which Jensen *et al.* (1960) originally described and which Tchan and Jensen (1963) attributed to a need for traces of fixed N to initiate N_2 fixation; it is as if a certain amount of slime has to be formed before N_2 fixation can begin. Once enough is formed, at random points on the dish, the characteristic yellowish growth indicative of N_2 fixation takes place. Hill and Postgate (1969) observed that *D. gummosa* fixed N_2 much more readily if cultures were initiated at low pO_2 values and took the views that polysaccharide formation provides regions of low pO_2 necessary for initiation of N_2 fixation. In these cases there is a good correlation between slime and ability to fix N_2, but evidence to the contrary is provided by the 'OP' strain of *A. vinelandii* (Bush and Wilson, 1959), a non-gummy variant much used in research, which seems to be at no disadvantage compared with the wild type in laboratory conditions. However, we have seen that ordinary laboratory cultures are 'N_2-limited'; in field conditions these organisms are likely to be limited by supplies of carbon, in which case they would be hypersensitive to oxygen inhibition and a gummy capsule which slowed diffusion of oxygen to these organisms might have selective advantages.

5.2.10   *Efficiency of aerobic nitrogen fixation*

Anabena species, Azotobacteriaceae and *Mycobacterium flavum* 301, the aerobes which have been studied from this point of view, initiate and conduct N_2 fixation most rapidly at hypobaric pO_2 values. This phenomenon has been known in azotobacters for at least 40 years; it is reflected in the 'efficiency' of fixation expressed as mg N fixed/g carbon source consumed. In most studies the nutritional status of the population was not stated explicitly, though Overbeck and Malke (1967) showed that phosphate-limited azotobacters were relatively more efficient at low pO_2 values. Dalton and Postgate (1969b) reported dramatic increase in this quotient with 'N_2-limited' *A. chroococcum* growing at low pO_2 values and a corresponding inefficiency at hyperbaric pO_2 values. Figure 5.4 illustrates this finding; these were chemostat experiments and the efficiencies obtained, even at normal pO_2 values, were well above those commonly recorded for batch cultures of Azotobacteriaceae at atmospheric pO_2 (10 to 20 mg N fixed /g; see Jensen, 1953; Stewart, 1969). They were impressively high (up to 40 mg N/g) at low pO_2 values though the population densities were low, presumably because O_2 had become the limiting nutrient. This increased efficiency may be of ecological importance: the pO_2 in the micro-environments of soils are often hypobaric; azotobacters may therefore fix more efficiently in nature, and thus contribute more to the soil nitrogen economy, than is generally realized.

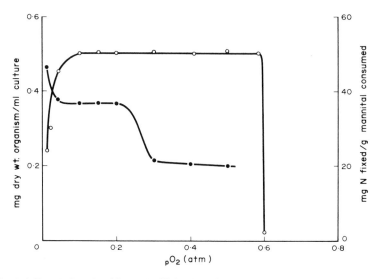

Fig. 5.4. Population densities and efficiencies of nitrogen fixation in chemostat cultures of *Azotobacter chroococcum* growing at various pO_2 values. Populations grew at D = 0.2 h^{-1} in a mannitol medium at various pO_2 values. ○: mg dry wt. organism/ml culture; ●: mg N fixed/g mannitol consumed. (After Dalton and Postgate, 1969a.)

5.2.11   The special case of the blue-green algae

Cyanophyceae capable of fixing N_2 have a special problem because their photosynthetic metabolism leads to oxygen evolution. From what has gone before, it seems highly probable that some efficient form of physiological compartmentation must exist in these organisms. The problem has occupied Fogg, Stewart, Fay and their colleagues (see Fay *et al.*, 1968) and has recently been resolved in an interesting manner. A characteristic of almost all N_2-fixing Cyanophyceae, such as members of the genus *Anabena*, is the presence of 'heterocysts': large, apparently empty cells spaced out at intervals along the algal filament which seem not to conduct normal photosynthesis. These bodies had long been suspected of involvement in nitrogen fixation and Stewart, Haystead and Pearson (1969) described the discovery, involving a pleasing element of serendipity, of their intense reducing ability: autoradiograms of *Anabena cylindrica* were a failure because the heterocysts themselves reduced the photographic emulsion. Neotetrazolium was also preferentially reduced by heterocysts and acetylene reduction was inhibited in proportion to the extent of deposition of formazan in the heterocysts. This evidence, though circumstantial, points to heterocysts as the loci of actual fixation and Stewart *et al.* (1969) successfully obtained direct evidence of fixation by selective disruption in the French press: preparations consisting mainly of heterocysts showed ATP-activated acetylene reduction with sodium dithionite; other fractions did not.

Heterocyst formation represents an extreme form of physiological separation or compartmentation which organisms require if they are to fix N_2 and to metabolize O_2 simultaneously. Stewart *et al.* (1969) ventured the prediction that only blue-green algae possessing heterocysts could fix N_2 and that reports of aerobic fixation by heterocyst-free species would prove false. In view of the phylogenetic overlap (see van Niel, 1953) between Cyanophyceae and both Athiorhodaceae (e.g. *Rhodospirillum*) and Thiorhodaceae (coloured sulphur bacteria), many of which are obligate anaerobes and some of which fix N_2, one might reasonably expect facultatively anaerobic N_2-fixing Cyanophyceae to be discovered which perform the reaction without benefit of heterocysts, as well as evolutionary intermediate forms.* In this context the report of Wyatt and Silvey (1969) of N_2 fixation by *Gleocapsa*, a coccoid blue-green alga which does not form heterocysts, is notable. A study of its oxygen sensitivity might be rewarding.

5.3 THE NEED FOR ATP IN N_2 FIXATION

5.3.1 Substrate yields of anaerobic N_2-fixing bacteria

From an energetic viewpoint, the feature which distinguishes anaerobes from aerobes is their dependence on substrate-level ATP formation: as far as we know, the fermentative anaerobes (which class excludes oxidative anaerobes such as *Desulfovibrio*) have no mechanism of ATP formation equivalent to respiratory chain phosphorylation. If the pathway of substrate fermentation is known, the moles of ATP available per mole of substrate fermented can be predicted exactly and, if the yield of organism in a substrate-limited condition is known, a figure for the yield of organism/mole ATP (Y_{ATP}) may be obtained. This figure usually lies in the range of 10.9 g dry bacterial matter/mole ATP (Stouthamer, 1969) though marked

* See also p. 149.

divergencies have been reported (see Forrest, 1969). If, in a given organism, a relatively large amount of is consumed in fixing N_2, then the yield of organism/mole substrate consumed (Y_{sub}) will be much smaller if the organism is fixing N_2 than if it is assimilating NH_3. This statement is only true (a) if the carbon-energy source is the growth-limiting substrate in both test conditions and (b) if the fermentation pathway is not altered significantly by the transition from N_2-fixing to NH_3-assimilating conditions. Senez (1962) discussed experiments with batch cultures of N_2-fixing *Desulfovibrio desulfuricans* (Le Gall and Senez, 1960) which suggested that catabolism and nitrogen fixation were uncoupled: the growth substrate (lactate) was consumed with the same rapidity whether these anaerobes were fixing N_2 or assimilating NH_3. Yet in the former case the Y_{sub} was very much lower. He proposed the elegantly simple view that N_2 fixation was limiting growth and that surplus lactate metabolism led to energy being dissipated as heat. This work was performed before the direct involvement of ATP in N_2 fixation was realized. By studying carbon-limited chemostat cultures one can, in principle, exclude such 'wastage' of energy; Daesch and Mortenson (1967) described continuous cultures of *Clostridium pasteurianum* in the two conditions and observed Y_{sub} differences equivalent, on a calculated Y_{ATP} of 10.5 g/mole, to some 20 moles ATP consumed/mole N_2 fixed. Their description left an element of doubt whether the carbon-energy source (sucrose) was truly the limiting substrate (their reference to cultures being 'limited in sucrose but to varying degrees' indicates a confusion in terminology; in normal parlance such a chemostat is either sucrose-limited or limited by something else). Professor L. Mortenson has assured the writer that the cultures were sucrose-limited in the conventional sense at some dilution rates. Crude cell-free preprations consume some 13.5 to 15 moles ATP/N_2; allowing for an error introduced by 'uncoupling' and other aspects of the 'maintenance coefficient', Daesch and Mortenson's value is of the order of magnitude to be expected from cell-free enzyme experiments. Comparable experiments by Miss S. Hill in the writer's laboratory, using anaerobically-grown *Klebseilla pneumoniae* grown in glucose-limited conditions in chemostats, have shown differences in Y_{sub} of a similar order of magnitude: if the difference in Y_{sub} with N_2 and the Y_{sub} with NH_3 is attributed entirely to ATP consumed in N_2 fixation, a value approaching 30 moles ATP consumed/N_2 fixed is obtained. The N-content of the organisms and the fermentation products did not differ greatly in the two cultural conditions compared. These findings, though they make assumptions about relations of Y_{sub} to Y_{ATP} which are not rigidly justified, indicate that physiologically intact N_2-fixing anaerobes consumed ATP in amounts similar in order to magnitude to those required by extracts. In other words, the mysterious inefficiency of cell-free N_2 fixation appears to extend to intact anaerobes.

5.3.2   Substrate yields of aerobic N_2-fixing bacteria

Y_{sub} values with aerobes such as *Azotobacter chroococcum* are, as discussed in section 5.2.7, complicated by the diversion of substrate into 'respiratory protection' of the nitrogenase. At low growth rates, the maintenace coefficients are very large. Dalton and Postgate (1969b) studied chemostat cultures of N_2-fixing and NH_3-assimilating mannitol-limited cultures of *A. chroococcum* and observed that the differences between Y_{sub} in the two conditions became progressively smaller at faster growth rates (i.e. as the maintenance coefficient declined). By extrapolating to a hypothetical zero maintenance, they obtained a difference in Y_{sub} corresponding to about 5 moles of ATP consumed/mole N_2 fixed, if the 'average' value of 10.5 for Y_{ATP} was assumed to apply to *A. chroococcum*. This assumption may not be valid; nevertheless the empirical fact is that the differences in Y_{sub} between N_2-fixing and NH_3-assimilating populations of *A. chroococcum* were very much smaller than in the two cases of anaerobes studies so far, and incompatible with so great a wastage of ATP as enzyme preparations show. There is no need to regard enzymic extracts from *A. chroococcum* as more efficient in terms of ATP consumption than preparations from anaerobes. It therefore seems likely either that Senez was right, and that Y_{sub} studies are rendered totally misleading by uncoupling, or that *A. chroococcum* possesses a physiological mechanism for recovering some of the ATP actually expended in N_2 fixation. It is conceivable that, in these bacteria, the 'N_2 pathway' includes some ATP regeneration analogous to respiratory chain phosphorylation despite the fact that the overall process leads to a net loss of ATP.

5.3.3   Phosphate and survival in Azotobacter

The intimate involvement of ATP in N_2 fixation is probably related to the lethality of oxygen to P-limited, N_2-fixing populations observed by Dalton and Postgate (1969a). Their interpretation in terms of a conflict between respiratory control and respiratory protection was mentioned in section 5.2.7; no further evidence has yet been published either to confirm or refute that proposal. But Smith and Wyss (1969) reported that *A. vinelandii* died extremely rapidly in water and that phosphate was protective; this fragility may interfere with dilution counts on Azotobacteriaceae and is, in part at least, associated with their oxygen sensitivity (Billson, Williams and Postgate, 1970). Ordinary cultures of most azotobacters are also sensitive to high phosphate concentrations: conventional media such as Burk's (Newton, Wilson and Burris, 1953) contain relatively little phosphate by bacteriological standards and azotobacters do not readily adapt to increased phosphate (unpublished experiments by the author and Miss Carole Taylor). At this stage one can only point to a correlation between the intimate involvement of ATP in N_2-fixation and the peculiar sensitivity of live azotobacters to both excess and deficiency of phosphate; more data are needed before one can propose a rationalization of N_2-fixation in these organisms taking these points into account.

5.3.4   Control of nitrogenase synthesis

N_2 fixation is a physiologically extravagant process. Though nitrogenase is sometimes formed in the presence of fixed nitrogen (by NH_3-limited *Azotobacter* or by *Klebsiella* growing with amino acids, for example) the possession of nitrogenase is presumably uneconomic when fixed nitrogen is available. The ability to synthesize nitrogenase confers selective advantage on N_2fixing microbes only in specialized environments: those from which fixed nitrogen is largely or completely absent. Nitrogenase synthesis is therefore regulated in those organisms which have been studied. NH_3 has long been known to prevent nitrogenase synthesis by most nitrogen-fixing systems, though only weakly at low concentrations, and nitrate also apparently prevents nitrogenase synthesis to some extent in azotobacters and in blue-green algae (see Wilson, 1958). Both Daesch and Mortenson (1967), using *C. pasteurianum,* and Dalton and Postgate (1969b), using *A. chroococcum,* observed that NH_3-limited continuous cultures of the organisms synthesized nitrogenase under argon atmospheres: an atmosphere of N_2 did not appear to be necessary to induce nitrogenase synthesis. Both groups of workers reported that the specific activity of such 'de-repressed' organisms was greater than that of normal N_2-grown bacteria, and so did Munson and Burris (1969), who studied N_2-limited and NH_3-limited continuous cultures of *Rhodospirillum rubrum.* The ability of NH_3-limited population to synthesize nitrogenase is of practical importance for the detection of mutants deficient in components of nitrogenase such as those of *A. vinelandii* reported by Fisher and Brill (1968).

Sorger (1968, 1969) deduced that methylamine or 2-methylalanine acted as co-repressors of nitrogenase in *A. vinelandii* because they inhibited growth of cultures fixing N_2 but not of cultures assimilating NH_3. He sought constitutive nitrogenase-positive mutants by adapting strains to 2-methylalanine; nitrate did not act as a repressor but interfered with nitrogenase action. 'Adaptive' lags in formation of nitrogenase by *Klebsiella* species after exhaustion of the NH_3 in the culture were recorded by Pengra and Wilson (1958); Yoch and Pengra (1968) observed that amino acids shortened the adaptive lag. Strandberg and Wilson (1968) found that both KNO_3 and NH_4 acetate repressed nitrogenase synthesis and found an apparent adaptive lag when cultures became NH_4-limited. Experiments on the repressive or inductive nature of the process were inconclusive. Mahl and Wilson (1968) showed that the adaptive lag of *Klebsiella* was prolonged indefinitely in a vacuum but that nitrogenase was formed on addition of N_2; this raised the question whether nitrogenase is induced by N_2 or derepressed by removal of repressor. The amount of N_2 required to induce synthesis might well be very small, and sufficient might be present in the argon or helium used by the workers who used continuous culture, whose finding of nitrogenase synthesis by ammonia-limited populations under argon would otherwise suggest a derepression mechanism. The point is very difficult to attack experimentally; at the time of writing the question is still open.

The observation of oxygen repression of nitrogenase synthesis in *Azotobacter vinelandii* (Strandberg and Wilson, 1968), a form of repression which may also be of importance in the physiology of facultative anaerobes which fix N_2 anaerobically, was dealt with in section 5.2.4.

5.4   THE NEED FOR IRON AND MOLYBDENUM

5.4.1   *Transition metals as functional components of nitrogenase*
Direct involvement of molybdenum in N_2 fixation had been suspected for many years, both in free-living and symbiotic systems (see Wilson, 1958), before Keeler, Bulen and Varner (1956) showed, with ^{99}Mo, that molybdenum tended to concentrate in a particulate fraction of *A. vinelandii*. This fraction was very similar to he particulate fraction which, in 1964, Bulen, Burns and Lecomte (1964) showed to have nitrogenase activity. To-day we realize that

Mo is a component of 'fraction 1' of the nitrogenase of all microbes so far examined. Reports that vanadium sometimes substitutes for molybdenum (Becking, 1962) must be viewed with reserve unless confirmed by isolation of an active V-containing 'fraction 1'; it is known that the V has a sparing action on Mo.* Organisms which were once thought not to require Mo, such as *Beijerinckia* or *Mycobacterium flavum* 301, have been shown by ^{99}Mo tests actually to take up Mo, though they require about $\frac{1}{100}$th the amount needed by azotobacters (Mishustin and Krylova, 1965). One can cite the analogy of *Azotobacter* species, which have an unusually high requirement for calcium which is not shown by *Beijerinckia* (Becking, 1961); *Azotobacter* species seem inefficient at assimilating trace elements, or else they are peculiarly susceptible to mineral imbalance. The position is not clear.

Fe is also an integral part of nitrogenase, being present in both fractions and associated with labile sulphide in both. Mere presence of metals in enzymes does not necessarily imply prosthetic function; however, the characteristics of the interaction of nitrogenase with analogues of N_2 add weight to the view that the metals participate directly. In Chapters 2 and 3 the authors described the discovery, in the latter part of the 1960's, of a variety of complexes formed by transition metals which retain the triply bonded N_2 group. These may be mononuclear, of the form $L_nM \leftarrow N{\equiv}N$, or binuclear, of the forms $L_nM \leftarrow N{\equiv}N \rightarrow ML_n$ (L represents a variety of ligands ranging from NH_3 to triphenyl-phosphine; M represents transition metals which may or may not be the same). The question whether complexes of this kind are models for the initial biological attack of N_2 is discussed adequately elsewhere in the book. From a physiological point of view, evidence for the direct participation of metals in nitrogenase action is rather more oblique. The reduction of analogues of N_2 such as CN^-, N_3^-, N_2O and C_2H_2, as well as the specific inhibition of such reactions (but not of hydrogen evolution) by CO, all suggest reaction at a metal site which can be blocked by carbonyl formation. The biological reduction of methyl isocyanide yields methane, methylamine and small proportions of ethane and ethylene (Kelly, Postgate and Richards, 1967; Kelly, 1968); the

---

* Personal communications from Dr. C. E. McKenna and Dr. R. W. F. Hardy, late in 1970, indicate that evidence for an active V-containing analogue of 'fraction 1' has been obtained independently in two laboratories. It is said to be present in crude cell-free extracts of *A. vinelandii* which had been grown with V in place of Mo.

methane comes from the terminal C of CH_3NC. This reaction is of special interest because pathway of reduction is characteristic of metal-bound isocyanides; the free molecule tends to become reduced to dimethylamine. Other substrate analogues such as acrylonitrile and allene give reduction products which suggest that they have become bound to a metal (Hardy and Burns, 1968). The nature of the metal is still a matter of speculation; studies using Mössbauer (γ-resonance) spectra with ^{57}Fe-labelled nitrogenase from *Klebsiella pneumoniae* (Kelly and Lang, 1971) have provided direct evidence that Fe is involved in the activation of the nitrogenase complex but are inconclusive on the question whether an Fe-substrate bond is formed.

Substrates such as acetylene and methyl isocyanide are reduced by intact, whole organisms; indeed, the value of the now widespread acetylene test depends on the fact that, in all systems tested so far, the substrate reaches the appropriate, presumably metal, site in the living organism. Substrates which attack other enzymes preferentially would not be expected to be reduced and this is probably the reason why, according to unpublished data obtained in the writer's laboratory by Dr. M. G. Yates, intact *Azotobacter chroococcum* does not reduce cyanide though extracts do. Other differential effects may be expected when nitrogen analogues, either inhibitors or alternative substrates, are tested with intact organisms and compared with enzyme preparations; an obvious instance arises from the CO-insensitivity of the ATP-activated hydrogen evolution reaction from nitrogenase (see Chapter 4). One would expect hydrogen evolution in intact organisms to be inhibited by CO if the reaction involves conventional hydrogenase but not if it involves nitrogenase. Relatively few physiological studies along such lines have been reported.

5.4.2   Cytochromes and non-haem iron enzymes

Though historically the work of Carnahan *et al.* (1960) on cell-free N_2 fixation by *C. pasteurianum* was a seminal influence in the discovery and characterization of non-haem iron proteins such as ferredoxin, rubredoxin and flavodoxin in microbes, whether N_2-fixing or not, these can not at present be said to have a *specific* role in nitrogen fixation by *C. pasteurianum* because they are present and, as far as one knows, functional in non-nitrogen-fixing populations. They appear also to be present in species of anaerobe which do not fix N_2.

Claims that the cytochrome-based electron transport systems of aerobic N_2-fixing microbes are directly involved in fixation should be regarded with reserve; such factors as oxygen transfer rate in the culture medium or QO_2 of the organisms may well account for differences reported in the cytochrome-contents of N_2-fixing *versus* NH_3-grown organisms or the degrees of reduction of these factors (Lysenkova and Khmel, 1967; Knowles and Redfearn, 1968); for the special case of *A. vinelandii,* Daniel and Erikson (1969) have observed that the pattern of the respiratory chain is influenced by the carbon-energy source: organisms grown in a glucose medium contained much less cytochrome a_2 than those grown with mannitol. Dalton and Postgate (1969b) found that nutritional status regarding carbon or phosphate had effects as great as changes of N-source in chemostat cultures of *A. chroococcum.* However, the work of Jones, Redfearn and their colleagues indicates that *A. vinelandii* has an unusual type of respiratory chain in that it may well be doubly branched (Jones and Redfearn, 1967a). It includes normal transport factors such as cytochromes, quinones, non-haem iron components, etc., but can be separated into two classes of sub-cellular fraction (red or green) in which different branches of the chain appear to predominate (Jones and Redfearn, 1967b). Essentially the view of this school is that succinate or NADH pass electrons, by independent routes, into a common part of the respiratory chain and this chain later branches again. One branch leads *via* cytochrome c_4 and c_5 to a terminal oxidase involving cytochromes o and a; the other passes *via* cytochrome b to cytochrome a_2. There is some suggestion that the system shows differences in P/O ratios according to which pathway is functioning (Ackrell and Jones, 1970) though some ATP is always generated. If respiratory protection is really a part of the physiology of azotobacters as suggested in section 5.2.7, then the facility to control within limits the amounts of respiratory chain phosphorylation would enable the organism to operate respiratory protection without drastically altering its ATP-ADP economy. Since, however, respiration seems never to be wholly uncoupled in the living bacteria, conflict between respiratory control and respiratory protection would still be expected in phosphate- or carbon-energy-limited conditions and could still account for hypersensitivity to oxygen among such populations.

Obviously, as in the rest of the organisms' physiology, the respiratory chain of *Azotobacter* must be indirectly involved in N_2 fixation, but at the time of writing there is no convincing evidence that any electron transport factor of the conventional respiratory chain more positive than NAD is directly involved in the fixation process.

Benemann *et al.* (1969) and Yoch *et al.* (1969) have, in fact, succeeded in linking the particulate nitrogenase of *Azotobacter vinelandii* to 'photosystem 1' of spinach chloroplasts by way of electron carriers derived from the azotobacter itself. One effective carrier resembles a ferredoxin and the other (azotoflavin) is a flavoprotein (see pp. 126-7); if these are the 'natural' electron carriers to nitrogenase, the electron transport pathway is far from conventional.

5.5 CONCLUSION: SOME UNRESOLVED PROBLEMS

5.5.1 *The hydrogenase-nitrogenase correlation*

There is an element of irony in the fact that one of the earliest striking features of N_2 fixation is still unresolved. The correlation of hydrogenase with nitrogenase (Wilson, 1958) may be summarized in two statements: (i) all N_2-fixing systems, including symbiotic ones, possess hydrogenase but not all microbes which possess hydrogenase fix N_2; (ii) hydrogen generally, but not always, inhibits N_2 fixation by growing organisms. The second of these propositions has been somewhat undermined by Parker and Dilworth's (1963) argument that, in *A. vinelandii*, H_2-N_2 antagonism was an experimental artefact: to obtain a high pH_2 in the atmosphere it is necessary to have a low pN_2 which, in itself, could limit fixation rates. The first proposition, however, is undoubtedly true, and the early demonstration of ATP-activated hydrogen evolution in nitrogenase preparations, described in Chapter 4, provides at least a partial explanation for the universality of hydrogenase in these organisms. But *Clostridium pasteurianum*, *Desulfovibrio* and all other N_2-fixing bacteria so far examined also contain conventional (non-ATP-activated) hydrogenases. *C. pasteurianum*, indeed contains so much that the ATP-activated hydrogen evolution reaction could only be demonstrated under an atmosphere of CO, which inhibits conventional hydrogenase but not the ATP-activated hydrogen evolution reaction of nitrogenase (Mortenson, 1966). Hydrogenase is also present in the aerobic *Azotobacter* species, whether the organisms have fixed nitrogen or grown with urea (Bulen *et al.* 1967), as well as in organisms such as *Escherichia coli* which do not fix nitrogen. In *Azotobacter* the question at once arises that, if its nitrogenase evolves hydrogen when extracted from the organism, why do not the intact bacteria evolve hydrogen? An attractive suggestion made to the writer by Dr. Bulen, that the conventional hydrogenase of *Azotobacter* serves to re-oxidize any hydrogen evolved by the nitrogenase, is not substantiated by experiment: CO, which inhibits the conventional hydrogenase (Hyndman, Burris and Wilson, 1953), does not cause *Azotobacter* to evolve H_2. To the writer, the fact that live, N_2-fixing *Azotobacter* does not evolve H_2 implies that, in the living organism the site of fixation is effectively anhydrous: the hydrogen ion may be unable to reach it.

The question of the physiological role of conventional hydrogenase and its curious correlation with nitrogenase is still open.

5.5.2 Thermophily among N_2-fixing bacteria

Nitrogenase preparations from *Azotobacter* withstand heating for 10 min at 60° (Bulen *et al.*, 1965; Hardy and Knight, 1966; Kelly, 1966); some members of the Azotobateriaceae resist heating to 100° for up to 7 min, though whether cyst formation is involved is doubtful (see Garbosky and Giambi, 1963). Yet the literature contains no report of a truly thermophilic N_2-fixing organism (if one accepts as criterion of thermophily the ability to grow at 45°C and above). Galovacheva and Kalininskaya (1968) isolated a N_2-fixing bacterium, *Pseudomonas ambigua*, from a hot spring, but it proved to be mesophilic. Perhaps thermophilic organisms have not been sought systematically; but one would expect examples to have turned up at least occasionally during the decades over which thermophily has been studied, if they actually exist. Their biochemistry would have features of obvious interest.

5.5.3 Restriction of N_2 fixation in eukaryotes

In section 5.2.1 the point was made that N_2 fixation is commonest among anaerobes such as butyric clostridia, *Desulfovibrio*, Thiorhodaceae, methane bacteria: organisms which seem to have relatively primitive metabolisms (Klein and Cronquist, 1967). The process becomes increasingly rare among facultative aerobes and obligate aerobes and, though Millbank's publication (1969) did not exclude the possibility of fixation by yeasts in anaerobic or microaerophilic conditions, he has excluded the possibility in later work (Dr. J. Millbank, personal communication). It now seems likely that reports of N_2 fixation by yeasts and their eukaryotic relatives were mistaken. N_2 fixation, therefore, seems restricted to prokaryotes and is, moreover, most widespread among the most primitive prokaryotes. If these organisms truly represent the primitive denizens of this planet, and if one accepts conventional views of the absence of N_2 and O_2, but presence of CH_4 and NH_3, in the primitive atmosphere, one is forced to the paradoxical view that N_2 fixation is characteristic of to-day's representatives of creatures which had no need to fix N_2. One is tempted to the view that the earliest evolutionary function of what is now the nitrogenase was something quite different. Was nitrogenase developed from a system evolved for detoxification of HCN or removal of C_2H_2, both of which are known components of the primitive atmosphere? Or do the relationships between bacterial ferredoxins and photo-pyridine nucleotide reductases suggest that nitrogenase bears an evolutionary relationship to the photosynthetic apparatus? Perhaps further speculations on these lines would be fruitless at present, but the questions are cogent to understanding the evolution and distribution of ability to fix dinitrogen.

REFERENCES

B. A. C. ACKRELL and C. . JONES, *Biochem. J.*, **116**, 21P, (1970).

S., AIBA, S., NAGAI, Y. NISHIZAWA and M. ONODERA, *J. gen. appl. Microbiol. Tokyo*, **13**, 73, (1967a).

S. AIBA, S. NAGAI, Y. NISHIZAWA and M. ONODERA, *J. gen. appl. Microbiol. Tokyo*, **13**, 85 (1967b).

P. P. BAROOH and A. SEN, *Arch. Mikrobiol.*, **48**, 311. (1964).

N. S. BAYLISS, *Aust. J. biol. Sci.*, **9**, 369 (1956).

J. H. BECKING, *Plant and Soil*, **14**, 287 (1961).

J. H. BECKING, *Plant and Soil*, **15**, 171. (1962).

J. R. BENEMANN, D. C. YOCH, R. C. VALENTINE and D. I. ARNON, *Proc. Natl. Acad. Sci., U.S.A.*, **64**, 1079 (1969).

D. R. BIGGINS and J. R. POSTGATE, *J. gen. Microbiol.*, **56**, 181 (1969).

S. BILLSON, K. WILLIAMS and J. R. POSTGATE, *J. appl. Bact.*, **33**, 270 (1970).

W. A. BULEN, R. C. BURNS and J. R. LECOMTE, *Proc. Natl. Acad. Sci. U.S.A.*, **53**, 532 (1965).

W. A. BULEN, J. R. LECOMTE, R. C. BURNS and J. HINKSON, Non-heme iron proteins. (A. San Pietro, ed.). Antioch Press: U.S.A. (1965).

D. BURK, *Ergebn. Enzymforsch.*, **3**, 23 (1934).

J. A. BUSH and P. W. WILSON, *Nature (Lond.)*, **184**, 381 (1959).

N. E. R. CAMPBELL, R. DULAR, H. LEES and K. G. STANDING, *Can. J. Microbiol.*, **13**, 587 (1967).

J. E. CARNAHAN, L. E. MORTENSON, H. F. MOWER and J. E. CASTLE, *Biochim. biophys. Acta*, **44**, 520 (1960).

V. F. COTY, *Biotechn. Bioengng.*, **9**, 25 (1967).

R. M. COX and P. FAY, *Proc. Roy. Soc. B*, **172**, 357 (1969).

G. DAESCH and L. E. MORTENSON, *J. Bact.*, **96**, 346 (1967).

H. DALTON and J. R. POSTGATE, *J. gen. Microbiol.*, **54**, 463 (1967a).

H. DALTON and J. R. POSTGATE, *J. gen. Microbiol.*, **56**, 307 (1969b).

J. De LEY, *Antonie van Leeuwenhoek*, **34**, 66. (1968).

J De LEY and I. W. PARK, *Antonie van Leeuwenhoek*, **32**, 6 (1966).

R. M. DANIEL and S. K. ERICKSON, *Biochim. biophys. Acta*, **180**, 63 (1969).

M. J. DILWORTH and I. R. KENNEDY, *Biochim. biophys. Acta.* **67**, 240 (1963).

J. DROZD and J. R. POSTGATE, *J. gen. Microbiol.* **60**, 427 (1970a).

J. DROZD and J. R. POSTGATE, *J. gen. Microbiol.* **63**, 63 (1970b).

S. R. ELSDEN, Microbial physiology and continuous culture. (E. O. Powell, C. G. T. Evans, R. E. Strange and D. W. Tempest, eds.), p. 255. H.M.S.O. London (1967).

P. FAY, W. D. P. STEWART, A. E. WALSBY and G. E. FOGG, *Nature (Lond.)*, **220**, 810 (1968).

M. V. FEDOROV and T. A. KALININSKAYA, *Mikrobiologiya*, **30**, 9 (1961).

R. J. FISHER and W. J. BRILL, *Biochim. biophys. Acta*, **184**, 99 (1968).

W. W. FORREST, Microbial Growth. (P. Meadow and S. J. Pirt, eds.), p. 65, *19th Symp. Soc. gen. Microbiol.* C.U.P. London (1969).

A. J. GARBOSKY and N. GIAMBI, *Ann. Inst. Pasteur*, **105**, 203 (1963).

R. S. GOLOVACHEVA and T. A. KALININSKAYA, *Mikrobiologiya*, **37**, 941 (1968).

R. D. GOERZ and R. M. PENGRA, *J. Bact.*, **81**, 568 (1961).

F. H. GRAU and P. W. WILSON, *J. Bact.*, **85**, 2 (1963).

I. R., HAMILSON, R. H. BURRIS and P. W. WILSON, *Proc. Natl. Acad. Sci., U.S.A.*, **52**, 637 (1964).

R. W. F. HARDY and E. KNIGHT, *Biochim. biophys. Acta*, **122**, 520 (1966).

R. W. F. HARDY and R. C. BURNS, *Ann. Rev. Biochem.* **37**, 331 (1968).

D. HERBERT, Microbial Reaction to environment. (G. G. Meynell and H. Gooder eds.), p. 391, *11th Symp. Soc. gen. Microbiol.* C.U.P. London (1961a).

D. HERBERT, *Soc. chem. Ind. Monograph*, (12) 21 (1961b).

S. HILL and J. R. POSTGATE, *J. gen. Microbiol.*, **58**, 277 (1969).

L. A. HYNDMAN, R. H. BURRIS and P. W. WILSON, *J. Bact.*, **65**, 522 (1953).

I. D. IVANOV, *Abs. 9th. Intern. Congr. Microbiol.*, *Moscow*, p. 276. C2/19, (1966).

I. D. IVANOV, N. S. DEMINA and T. V. GOGOLEVA, *Mikrobiologiya*, **36**, 8 (1967).

I. D. IVANOV, G. I. MATHANKOV, Y. M. BELOV and T. V. GOGLEVA, *Mikrobiologiya*, **36**, 205 (1967).

H. L. JENSEN, *3rd Intern. Congr. Microbiol.*, *Rome*, **6** (sect. 18) 245 (1953).

H. L. JENSEN, Soil Nitrogen. (W. V. Bartholemew and F. E. Clark, eds.), p. 440, *Amer. Soc. agron.*, Monograph No. 10 (1965).

H. L. JENSEN, E. J. PETERSEN, P. K. DE and R. BHATTACHARYA, *Arch. Mikrobiol.*, **36**, 182 (1960).

C. W. JONES and E. R. REDFEARN, *Biochim. biophys. Acta*, **143**, 340 (1967a).

C. W. JONES and E. R. REDFEARN, *Biochim. biophys. Acta*, **143**, 354 (1967b).

R. F. KEELER, W. A. BULEN and J. E. VARNER, *J. Bact.*, **72**, 394 (1956).

M. KELLY, *9th Intern. Congr. Microbiol.*, *Moscow*, p. 277, C2/20 (1966).

M. KELLY, *Biochem. J.*, **107**, 1 (1968).

M. KELLY and G. LANG, *Biochim. biophys. Acta*, **223**, 86 (1971).

M. KELLY, J. R. POSTGATE and R. L. RICHARDS, *Biochem. J.*, **102**, 1c (1967).

R. M. KLEIN and A. CRONQUIST, *Q. Rev. Biol.*, **42**, 105 (1967).

R. V. KLUCAS and H. J. EVANS, *Plant Physiol.*, **43**, 1458 (1968).

C. J. KNOWLES and E. R. REDFEARN, *Biochim. biophys. Acta*, **162**, 348 (1968).

J. LeGALL and J. O. SENEZ, *C. R. Acad. Sci., Paris*, **250**, 404 (1960).

R. LOPEZ and J. H. BECKING, *Microbiol. Espanol*, **21**, 53 (1968).

N. P. L'VOV *Izvestia. Acad. Naak. U.S.S.R. (Ser. biol.), (2), 270 (1963).*

L. L. LYSENKOVA and I. A. KHMEL, *Mikrobiologiya*, **36**, 905 (1967).

M. C. MAHL and P. W. WILSON, *Can. J. Microbiol.*, **14**, 33 (1968).

M. C. MAHL, P. W. WILSON, M. A. FIFE and W. H. EWING, *J. Bact.*, **89**, 1482 (1965).

J. W. MILLBANK, *Arch. Mikrobiol.*, **68**, 32 (1969).

E. N. MISHUSTIN and N. B. KRYLOVA, *Mikrobiologiya*, **34**, 683 (1965).

E. N. MISHUSTIN and V. K. SHIL'NIKOVA, Biological fixation of atmospheric nitrogen. Moscow: U.S.S.R. Acad. Sci. (1966). (English translation 1970, Macmillan, London).

L. E. MORTENSON, *Biochim. biophys. Acta*, **127**, 18 (1966).

T. O. MUNSON and R. H. BURRIS, *J. Bact.*, **97**, 1092 (1969).

J. W. NEWTON, P. W. WILSON and R. H. BURRIS, *J. biol. Chem.*, **204**, 445 (1953).

D. J. NICHOLAS, Symbiotic associations. (P. S. Nutman and B. Mosse, eds.), p. 92, *13th Symp. Soc. gen. Microbiol.* C.U.P. London (1963).

C. B. van NIEL, A century of progress in the natural sciences, p. 89. Calif. Acad. Sci., San Francisco (1955).

J. OVERBECK and H. MELKE, *Z. allg. Mikrobiol.*, **73**, 197 (1967).

J. OPPENHEIM and L. MARCUS, *J. Bact.*, **101**, 286 (1970).

J. OPPENHEIM, R. J. FISHER, L. MARCUS and P. W. WILSON, *J. Bact.*, **101**, 292 (1970).

R. A. PAREJKO and P. W. WILSON, *J. Bact.*, **95**, 143 (1968).

C. A. PARKER, *Nature (Lond.)*, **173**, 780 (1954).

C. A. PARKER and M. J. DILWORTH, *Biochim. biophys. Acta*, **69**, 152 (1963).

C. A. PARKER and P. B. SCUTT, *Biochim. biophys. Acta*, **38**, 230 (1960).

R. M. PENGRA and P. W. WILSON, *J. Bact.*, **75**, 21 (1958).

D. H. PHILLIPS and M. J. JOHNSON, *J. Biochem. microbiol., Technol. Engng.*, **3**, 277 (1961).

F. PICHINOTY, *Ann. Inst. Pasteur*, **104**, 394 (1963).

F. PICHINOTY and J. BIGLIARDI-ROUVIER, *Biochim. biophys. Acta*, **67**, 366 (1963).

J. R. POSTGATE, *J. gen. Microbiol.*, **63**, 137 (1970).

D. C. PRATT and R. W. FRENKEL, *Plant Physiol.* **34**, 333 (1959).

L. I. RUBENCHIK, Azotobacter and its use in agriculture. *Translated version.* Office of Technical Services, U.S. Dept. of Commerce, Washington D.C., U.S.A. (1963).

K. C. SCHNEIDER, C. BRADBEER, R. N. SINGH, L. C. WANG, P. W. WILSON and R. H. BURRIS, *Proc. Natl. Acad. Sci., U.S.A.*, **46**, 726 (1963).

J. C. SENEZ, *Bact. Rev.*, **26**, 95 (1962).

D. D. SMITH and D. WYSS, *Antonie van Leeuwenhoek*, **35**, 84 (1969).

G. J. SORGER, *J. Bact.*, **95**, 1721 (1968).

G. J. SORGER, *J. Bact.*, **98**, 56 (1969).

W. D. P. STEWART, *Proc. Roy. Soc. B.*, **172**, 367 (1969).

W. D. P. STEWART, A. HAYSTEAD and H. W. PEARSON, *Nature (Lond.)*, **224**, 226 (1969).

A. H. STOUTHAMER, Methods in Microbiology. (J. R. Norris and D. W. Ribbons, eds.), p. 629. Acad. Press, London (1969).

G. W. STRANDBERG and P. W. WILSON, *Can. J. Microbiol.*, **14**, 25 (1968).

Y. T. TCHAN and H. L. JENSEN, *Proc. Linnean Soc. New South Wales*, **88**, 329 (1963).

J. P. VOETS and J. DABACHER, *Naturwiss*, **43**, 40 (1956).

P. W. WILSON, Encyclopedia of Plant Physiology. (W. Ruhland, ed.), p. 9. Springer-Verlag, Berlin (1958).

P. W. WILSON, *Proc. Roy. Soc. B.*, **172**, 319 (1969).

J. T. WYATT and J. K. G. SILVEY, *Science*, **115**, 908 (1969).

M. G. YATES, *J. gen. Microbiol.*, **60**, 393 (1970).

M. G. YATES, *FEBS Letters*, **8**, 281 (1970).

M. G. YATES and R. M. DANIEL, *Biochim. biophys. Acta*, **197**, 161 (1970).

D. C. YOCH, and R. M. PENGRA, *J. Bact.*, **92**, 618 (1966).

D. C. YOCH, J. R. BENEMANN, R. C. VALENTINE and D. I. ARNON, *Proc. Natl. Acad. Sci., U.S.A.*, **64**, 1404 (1969).

CHAPTER 6

Physiological Chemistry of Symbiotic Nitrogen Fixation by Legumes

HAROLD J. EVANS and STERLING A. RUSSELL

Department of Botany and Plant Pathology,
Oregon State University,
Corvallis, Oregon

6.1   INTRODUCTION

The beneficial effects of including legumes in crop rotations were realized by early Greek, Roman and Chinese agriculturists centuries before the elementary principles of biological N_2 fixation were established (Fred, Baldwin and McCoy, 1932; Stewart, 1966).

Although the occurrence of nodules on roots of legumes had been described (Malpighi, 1675), and evidence for N_2 fixation by leguminous plants in field plots had been reported (Boussingault, 1838), the basic biological aspects of symbiotic N_2 fixation by legumes were not established until the results of the classical experiments of Hellriegel and Wilfarth (1888) were reported. They assumed that nodules on the roots of *Pisum sativum* were caused by bacteria and that nodulation enabled the plants to fix N_2. Furthermore, both non-legumes and leguminous plants without nodules were considered to be dependent upon fixed nitrogen compounds in the soil. Experiments in which legumes and non-legumes were cultured in sterile soil with and without non-sterile extracts of garden soil provided conclusive evidence that the non-sterile extract was needed for nodulation and that nodulation was essential for normal growth without adequate combined nitrogen. These experiments were quickly confirmed by other workers and were better understood when Beijerinck (1888) isolated pure cultures of nodule-forming bacteria. He named his first isolate *Bacillus radicola,* which is now known as *Rhizobium leguminosarum.*

An understanding of a few of the basic aspects of symbiotic N_2 fixation, and the realization of the economic importance of symbiotic N_2 fixation to the agricultural industry, has stimulated an enormous amount of research activity in this field. Progress has been discussed in the monographs by Fred, Baldwin and McCoy (1932), Wilson (1940), Stewart (1966), and in reviews by Carnahan and Castle (1963), Virtanen and Meittinen (1963), Burris (1966) and Dixon (1969). Burris (1965) stated that about 13 000 species of leguminous plants had been described and that most of them were believed to possess N_2-fixing capabilities. According to Allen and Allen (1949), however, a relatively small proportion of known leguminous species have been examined for nodulation, and of those that have, 89% were nodulated. The importance of legumes as contributors of fixed nitrogen to the soils of the world has been discussed by Donald (1960); he suggested that they contribute the major portion of the estimated total biological fixation of 10^8 tons of N_2 per year. As a result of improvements in methods for detecting biological N_2 fixation, several additional N_2-fixing organisms have been identified during the past decade (Burris, 1966) and others have been excluded (see Chapter 5). The economic importance of N_2 fixation by different kinds of N_2-fixing organisms, therefore, needs to be re-evaluated.

The importance of symbiotic N_2 fixation by legumes has demanded the development of methods for culturing *Rhizobium* species and determining their effectiveness as symbionts for the different groups of legumes. From these investigations, leguminous species have been classified into cross-inoculation groups, each of which is relatively compatible with a particular *Rhizobium* species. Some of the major *Rhizobium* species and their corresponding cross-inoculation groups of legume hosts have been listed by Erdman (1948). As pointed out by Nutman (1956, 1963), the relationship of *Rhizobium* species and strains to leguminous hosts is relatively complex. Within each *Rhizobium* species numerous strains have been identified that differ in effectiveness not only with the various leguminous species within a cross-inoculation group, but also with different legume varieties within a particular species. Further complications are introduced by the fact that the N_2-fixing effectiveness of *Rhizobium*-legume associations are influenced by environmental factors. An effective symbiotic N_2-fixing relationship between a particular strain of *Rhizobium* and a specific variety of legume represents a delicate balance between the genetic constitution of two different kinds of organisms.

In this chapter we intend to consider briefly the general aspects of the symbiotic association between *Rhizobium* and leguminous plants. This is needed in our opinion as a basis for the main theme of the chapter. This will be followed by a discussion of the requirements and, where known, the biochemical roles of the mineral elements that appear to be specifically associated with symbiotic growth and N_2 fixation by legumes. After this, the properties and physiological role of leghemoglobin will be discussed, and then a considerable portion of the chapter will be devoted to the biochemistry of the nitrogenase system. Finally, our present understanding of electron transport systems in legume nodules and possible relationships of them to the nitrogenase system will be considered.

6.2   THE SYMBIOTIC ASSOCIATION

A consideration of the physiology and biochemistry of N_2 fixation by legumes requires familiarity with recent concepts concerning the initiation and development of the symbiotic association between *Rhizobium* and leguminous species. Research in this area has been reviewed by Nutman (1956) and by Dixon (1969). The essential anatomy involved in the infection of a *Medicago sativa* root by *Rhizobium meliloti* is illustrated by Thornton and Nicol (1936).

(Fig. 6.1). In a large number of leguminous species, the root hairs are the sites of infection. Exceptions are the aquatic legumes, where rhizobia are reported to enter the root tissue through epidermal cells (Schaede, 1940), and *Arachis hypogea* where the point of lateral root

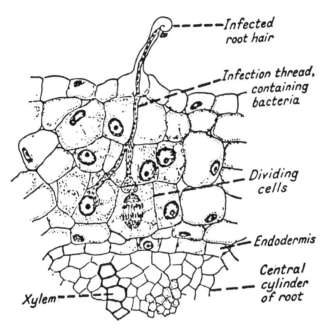

Fig. 6.1. An illustration of the infection of a root of *Medicago sativa* by *Rhizobium meliloti* (after Thornton and Nicol, 1936).

emergence serves as site for entry of the bacteria (Allen and Allen, 1940). In the initial stages of infection in most legumes, *Rhizobium* cells are concentrated around root hairs apparently in response to the excretion of substances of plant origin. The interaction of the root hairs with effective *Rhizobium* cells results in elongation, curling and branching of root-hair cells (Thornton and Nicol, 1936; Haak, 1964; Kefford, Brockwell and Zwar, 1960).

Rhizobium cells in the immediate vicinity of the legume-root hair are reported to excrete indole acetic acid (IAA), and to participate in the conversion of tryptophan from the plant into IAA. The IAA formed appears to influence the lengthening of root hairs, but the branching and curling of root hairs that is associated with the interaction are considered to be more specific reactions than can be accounted for by IAA alone. Fahraeus and Ljunggren (1968) proposed that the secretion of a polysaccharide by *Rhizobium* cells induced the legume-root cells to synthesize a pectinase which participates in cell wall degradation and entry of rhizobia into the root cells. This interpretation, however, has been questioned (Lillich and Elkan, 1968) and further research in the area is needed. Although evidence exists for a role of pectinase in the entry of rhizobia into the root hairs, the specific types of enzymes involved in the postulated enzymatic degradation of cell walls have not been identified.

In his review, Dixon (1969) suggested that the crude polysaccharide material reported to be excreted by rhizobia may contain a nucleic acid that functions in the inductive synthesis of cell wall-degrading enzymes. He speculates that a nodule-inducing principle from the bacteria may change the metabolism of root hairs so that a growth promoting substance such as IAA may function in the infection process. The information available is consistent with the idea that DNA in the polysaccharide excreted by rhizobia may be the nodule-inducing principle that causes the host tissue to synthesize IAA. The IAA, then, may influence the formation of cell wall-softening enzymes that, presumably, are necessary for initial infection. The specificity of *Rhizobium* species toward certain leguminous plants may be related to the type of DNA postulated to be present in the crude polysaccharide material that is excreted by the rhizobia, rather than to the polysaccharide *per se*.

At least two mechanisms have been suggested for the entry of rhizobia into the root hairs. Dart and Mercer (1964) proposed that small coccoid forms of rhizobia enter gaps in the cellulose microfibrils, and Nutman (1956) has suggested that the bacteria enter the root hairs as a result of an invagination that accompanies root-hair curling. The precise mechanism of entry of rhizobia however has not been established. From recent reviews it would appear that an infection thread of plant origin is formed as the rhizobia enter root hairs. The cellulose wall of the infection thread is continuous with the cell wall of the host and the initial growth of the infection thread seems to be oriented by the nucleus of the root-hair cell. The bacteria within the infection thread proliferate as the infection thread penetrates the cortical region of the root. When the infection thread reaches one of the few tetraploid cells within the cortical region of the root, bacteria are released and the tetraploid and adjacent diploid cells are induced to meristematic activity. As the plant cells divide and the nodule structure develops, the bacterial cells divide and eventually single bacterial cells or groups of two or three or more cells are enclosed in a membrane (Goodchild and Bergersen, 1966). In some nodulated legumes the bacteria that are released from the infection thread develop into pleomorphic forms. A large proportion of the bacteria within the nodule lose their capacity to divide and to grow and are referred to as bacteroids (Almon, 1933).

6.3   SPECIFIC MINERAL REQUIREMENTS

In general, the mineral requirements for non-nodulated leguminous plants are not essentially different from the requirements of non-legumes. There is considerable evidence, however, that the mineral requirements for the symbiotic association between *Rhizobium* species and leguminous plants are quantitatively and in some cases qualitatively different from the requirements of the leguminous plants *per se*. It is the requirements, and the biochemical roles of those minerals that seem to be specifically related to the symbiotic association that will be discussed in this section.

6.3.1   Molybdenum

The biological importance of molybdenum was not appreciated until Bortels (1930) observed that traces of the element markedly stimulated growth and N_2 fixation of cultures of *Azotobacter chroococcum*. Bortels (1937) reported that *Medicago sativa, Trifolium pratense,* and *Glycine max*. Merr, cultured in sand without combined nitrogen responded strikingly to small amounts of molybdenum but less to the addition of traces of vanadium. There are indications (Anderson, 1956; Stewart, 1966) of partial substitution of vanadium for molybdenum in the growth of N_2 fixing microorganisms, but the evidence for substitution in symbiotic N_2 fixation by legumes is not conclusive. Bortels's observations were followed by the important discovery of Anderson (1942) that a few ounces of molybdenum supplied to *T. subterraneum* and *M. sativa* cultured in an ironstone soil in Australia resulted in significant increases in dry weights. The initial observations of molybdenum deficiency in legumes in Australia were made on relatively acid soils. Further experimentation revealed that the addition of lime to some acid soils increased the availability of molybdenum. Anderson's early discovery of molybdenum deficiency in the field in Australia was followed by reports of molybdenum deficiency in field-grown legumes in other locations throughout the world.

In his early work, Anderson clearly indicated that responses of legumes of molybdenum additions were dependent upon the nitrogen status of soils. Symptoms of nitrogen deficiency on the foliage of molybdenum-deficient *Pisum sativum,* and scattered patterns of nodule distribution upon the roots of deficient plants were observed by Mulder (1948). In contrast, he reported that leaves of plants with adequate molybdenum were dark green in color, and nodules were concentrated in large aggregates on the major root stems. Characteristic symptoms of molybdenum-deficient *M. sativa* cultured in an acid soil from New Jersey (Fig. 6.2) were not essentially different from symptoms of nitrogen deficiency (Evans, Purvis and Bear, 1951). Although major symptoms of molybdenum deficiency in symbiotically grown legumes appear to be identical with symptoms of nitrogen deficiency, molybdenum is needed also by legumes for the utilization of nitrates (Evans, Purvis and Bear, 1950; Stout and Meagher, 1948). The magnitude of the requirement, however, appeared to be somewhat greater for N_2 fixation than for nitrate utilization. A molybdenum requirement for growth of legumes on ammonium nitrogen has been difficult to establish (Hewitt, 1958). If it is required at all, the quantity needed is definitely less than that required for either N_2 fixation or nitrate utilization.

Molybdenum is an essential constituent of the nitrate reductase that is formed in tissues of leguminous plants supplied with nitrate (Evans and Hall, 1955). In this aspect of metabolism the biochemical role of molybdenum in legumes and non-legumes is the same. Evidence to be discussed in a subsequent section of this chapter indicates that the additional molybdenum required for the N_2-fixing process *per se* can be logically explained by evidence that molybdenum is a constituent of the nitrogenase complex. This information is consistent with and provides a logical explanation for the observations and implications of the early nutritional research.

Before the properties of cell-free nitrogenase were established, Cheniae and Evans (1960) observed that the bacteroids from nodules of *G. max* and several other symbiotically grown legumes contained a very active nitrate reductase. The activities of the enzyme in a series of samples collected at intervals during the growth period were positively correlated with leghemoglobin contents of nodules and N_2 fixation by the leguminous plants. Furthermore, the effectiveness of *Rhizobium japonicum* strains in symbiotic N_2 fixation with *G. max* was positively correlated with the specific activities of nitrate reductase in preparations of nodule bacteroids (Cheniae and Evans,

1957). When these observations were made, it was known that molybdenum was an essential constituent of nitrate reductase and was involved in some unknown way in N_2 fixation. There was insufficient biochemical information, however, for an adequate interpretation of the relationship between nitrogenase and nitrate reductase. Recent information to be discussed in section 6.5.4 suggests that nitrogenase and nitrate reductase may share a common molybdo-protein subunit. This exciting new information provides a feasible explanation for some of the puzzling early observations.

6.3.2   Cobalt
6.3.2.1   Requirements
The first association of a nutritional cobalt requirement with N_2 fixation was made by Holm-Hanson *et al.* (1954), who showed that several species of blue-green algae responded to cobalt additions. Those species capable of fixing atmospheric N_2 were reported to exhibit a greater response than those lacking N_2-fixing capabilities. That blue-green algae required cobalt was confirmed by Buddhari (1960), but he observed no consistent effect of the form of nitrogen supplied on the magnitude of the cobalt response. The observation of Levin, Funk and Tendler (1954) that the vitamin B_{12} content of legume nodules was up to 34 times greater than that in adjacent root tissues, stimulated investigations of the possibility that cobalt was essential for legumes when cultured in symbiosis with *Rhizobium*.

Beginning in 1958, Evans and associates refined procedures for the removal of traces of cobalt from culture vessels, water, and reagent salts, and developed special cultural techniques for studying cobalt requirements of legumes. For use in these experiments, macronutrient salts were purified by extraction with 1-nitroso-2-naphthol (Ahmed and Evans, 1961). The purities of micronutrient salts that were utilized were established by spectrographic methods. Experiments were conducted in growth chambers where contaminating dust was removed from the air by filtration. Nodulated *G. max* plants cultured by these methods in a purified nutrient solution lacking combined nitrogen responded strikingly to cobalt additions. In comparison with control cultures without cobalt, the addition of 0.1 μg cobalt per liter of nutrient solution resulted in a 12-fold increase in the dry weight of shoots and a 22-fold increase in their nitrogen content. Plants supplied with cobalt appeared to be vigorous and were dark green in color. In contrast, the symptoms of nitrogen deficiency of plants lacking cobalt were extremely severe and at least 25 per cent

Fig. 6.2. The effect of sodium molybdate on the growth of *Medicago sativa* in Norton silt loam at pH 5.5 Culture No. 2 was not treated with $Na_2MoO_4 \cdot 2H_2O$, whereas culture No. 6 received 20 mg of $Na_2MoO_4 \cdot 2H_2O$ (after Evans, Purvis and Bear, 1951).

(*To face page 198*)

Fig. 6.3. The effect of the addition of 0.1 μg of cobalt per liter (0.1 ppb.) of nutrient solution on the growth of *Glycine max* in symbiosis with *Rhizobium japonicum*. Details of the experiments including purification of the nitrogen-free nutrient solution were described by Ahmed and Evans (1961).

of them failed to survive (Fig. 6.3). The response to cobalt was specific and was not replaced by a series of other trace elements.

During the same period that these experiments were in progress Delwiche *et al.* (1961) demonstrated that *M. sativa* cultured in symbiosis with *R. meliloti* required cobalt for growth and N_2 fixation, and Halsworth *et al.* (1960) reported that symbiotically cultured *T. subterraneum* also responded to cobalt additions. In all cases, symptoms of cobalt deficiency of symbiotically cultured legumes were indistinguishable from symptoms of nitrogen deficiency.

Using the refined techniques that reproducibly produced cobalt deficiency in nodulated legumes, Ahmed and Evans (1961) consistently failed to demonstrate a cobalt requirement for non-nodulated soybean plants supplied with nitrate nitrogen. Furthermore, Kliewer and Evans (unpublished results, 1963) observed no significant effect of cobalt on the growth of tomato plants in a purified nutrient solution containing nitrates. Delwiche *et al.* (1961) also reported that *M. sativa* supplied with nitrates did not respond to added cobalt. In contrast, Bolle-Jones and Mallikarjuneswara (1957) previously had observed a small but statistically significant increase in growth of *Hevea brasiliensis* and *Lycopersicum esculentum* from additions of traces of cobalt. More recently, Wilson and Nicholas (1967) have cultured *Triticum durum* and non-nodulated *T. subterraneum* in purified nutrient solutions containing nitrate nitrogen and have indicated that cobalt deficiency resulted in a chlorosis of young leaves and a small but statistically significant decrease in dry weight. The consistency of the response and number of replicate cultures was not reported. They could not identify cobamide compounds in normal *T. subterraneum* tissues, but evidence for the presence of other cobalt complexes was presented. In the present authors' opinion the question of the essentiality of cobalt for non-nodulated legumes and for non-legumes remains open. If cobalt is essential for higher plants *per se* the quantity required must be much less than that needed for symbiotic N_2 fixation by legumes.

On the basis of the results obtained with nodulated and non-nodulated legumes, Lowe, Evans and Ahmed (1960) and Lowe and Evans (1962a) investigated the cobalt requirements of cultured *Rhizobium* species. Several different *Rhizobium* species cultured in purified medium responded to traces of cobalt when nitrogen was supplied either as nitrate or ammonium compounds. Maximal growth was observed at a concentration of about 0.04 μg cobalt per l. The effect of cobalt on growth was specific and could not be replaced by a series of twelve trace elements that presumably were removed from the medium by the purification procedure.

6.3.2.2 Effects of deficiency In addition to symptoms of nitrogen deficiency on the foliage, the nodules from cobalt deficient legumes were soft and flaccid and contained decreased leghemoglobin contents (Ahmed and Evans, 1961). Nodules from cobalt deficient *M. sativa* fixed $^{15}N_2$ at a considerably slower rate than those from normal plants (Delwiche *et al.,* 1961). A biological assay of nodules revealed a striking effect of cobalt on the vitamin B_{12} content. Nodules from plants supplied with 0, 1.0 and $50\mu g$ of cobalt per l contained, respectively, 0.13, 0.65 and 519 ng of vitamin B_{12} per g of fresh nodules. Further research by Kleiwer and Evans (1963a) proved that the major vitamin B_{12} component of soybean nodules and cultured *R. meliloti* was 5,6-dimethylbenzimidazolylcobamide coenzyme (DBC coenzyme) (Fig. 6.4). Analyses of tissues (Kleiwer and Evans, 1963b) from a variety of normal leguminous species showed that nodules contained from 30 to 64 ηmoles DBC coenzyme per g fresh nodules, but roots, stems or leaves contained no detectable amount of this coenzyme. Furthermore, the DBC coenzyme content (Table 6.1) of *R. meliloti* cultured in purified nutrient media was markedly influenced by the cobalt supply, ranging from less than 1 $n$mole per g fresh cells where no cobalt was added up to 109 $n$moles per g cells from a medium containing 200 ng of cobalt per l nutrient medium (Kleiwer and Evans, 1962).

6.3.2.3 Role in propionate metabolism A search (DeHertogh, Mayeux and Evans, 1964a, b) for metabolic pathways in nodule bacteroids and cultured *Rhizobium* cells in which B_{12} coenzyme might function resulted in the identification of the following enzymic reactions:

Propionate $+ ATP + CoA \leftrightarrows$ Propionyl-CoA $+ AMP +$ pyrophosphate 1.
Propionyl-CoA $+ O_2 + ATP \leftrightarrows$ Methylmalonyl-CoA (a) $+ ADP +$ 2.
orthophosphate
Methylmalonyl-CoA (a) $\leftrightarrows$ Methylmalonyl-CoA (b)   3.
Methylmalonyl-CoA (b) $\leftrightarrows$ Succinyl-CoA   4.

The enzymes catalyzing these reactions are: 1. acetyl-CoA kinase, 2. propionyl-CoA carboxylase and 3. methylmalonyl-CoA racemase and 4. methylmalonyl-CoA mutase. Normal *Rhizobium* cells from pure cultures and nodule bacteroids not only contain the enzymes necessary for reactions 1, 2, 3 and 4, but they exhibit a capacity to oxidize propionate. In contrast, *R. meliloti* cultured in media lacking

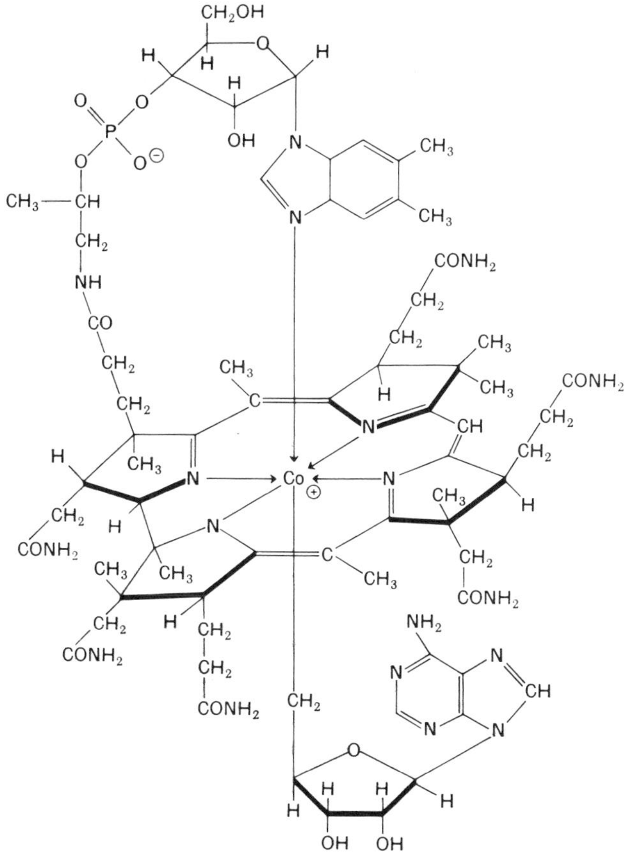

Fig. 6.4. The structure of 5,6-dimethylbenzimidazolylcobamide coenzyme (DBC), the major B_{12} coenzyme in bacteroids or cultured rhizobia (Kliewer and Evans, 1961a).

TABLE 6.1. Effect of cobalt on the growth and on the contents of nitrogen and B_{12} coenzyme in *Rhizobium meliloti* (after Kliewer and Evans, 1963a).

Co concentration in medium	Fresh weight of cells	Total N	B_{12} coenzyme content
μg/liter	mg/flask	mg/flask	nmoles/g fresh cells
0.00	547	118	0.0*
0.02	790	188	12.9
2.00	1034	272	45.0
200.00	430	87	109.2

* Less than 1 nmole.

sufficient cobalt failed to oxidize this substrate but oxidized glutamate at a normal rate (Fig. 6.5).

Exposure of extracts of normal *R. meliloti* cells to visible light, a procedure reported to destroy B_{12} coenzyme, resulted in loss of methylmalonyl mutase activity. The addition of the benzimidazolyl-cobamide coenzyme (BC) or DBC coenzyme to the light-inactivated

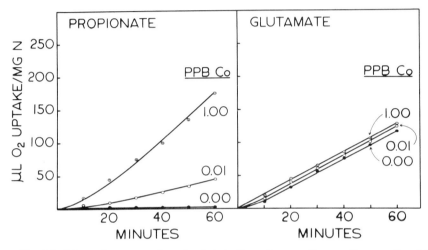

Fig. 6.5. Effect of concentration of cobalt in the culture medium on the capacity of *R. meliloti* to oxidize propionate and glutamate. Cells were cultured in media containing 0.0, 0.01 or 1.0 μg cobalt per liter (indicated as ppb). Cells were harvested and their capacities to oxidize propionate or glutamate determined in a Warburg respirometer (after DeHertogh *et al.*, 1964a).

preparation resulted in reversal of a major portion of the inactivation caused by light. As shown in Table 6.2, extracts of cobalt deficient cells showed no capacity to convert propionyl-CoA to methyl-malonyl-CoA but extracts from cells supplied with an adequate amount of cobalt converted methylmalonyl-CoA to succinyl-CoA at a rapid rate. The addition of the DBC coenzyme to inactive extracts from the deficient cells restored methylmalonyl-CoA mutase activity. From these results it was concluded that the apoenzyme for methylmalonyl-CoA mutase was present in deficient cells. The lack of enzyme activity in extracts of deficient cells was caused by cobalt deficiency, which prevented the synthesis of an adequate quantity of cobamide coenzyme.

Although there is no conclusive evidence to show that propionate oxidation is an essential pathway in the metabolism of *Rhizobium* and nodule bacteroid cells, Jackson and Evans (1966) established that extracts from soybean nodules convert lactate to propionate, and that propionate is converted to succinyl-CoA *via* a series of reactions involving methylmalonyl mutase. Furthermore, when propionate labelled with ^{14}C is supplied to nodule slices, the radio-

TABLE 6.2. Effect of different concentrations of cobalt in growth medium on activity of propionyl-CoA carboxylase and methylmalonyl-CoA mutase in cell-free extracts of *R. meliloti*.

Propionyl-CoA carboxylase and methylmalonyl-CoA mutase activities were measured as described by DeHertogh *et al.* (1964b). The quantity of DBC coenzyme added to each reaction mixture (final volume, 1.1 ml) was 1.65 × 10^{-3} μmole.

Concentration of cobalt added to culture medium	Propionyl-CoA carboxylase activity	Methylmalonyl-CoA mutase activity	
		−DBC	+DBC
(μg/liter)	(μmoles $H^{14}CO_3$-fixed/mg protein/h)	(μmoles $H^{14}CO_3$-fixed into permagnanate stable compounds/mg protein/h)	
None	1.83	0.00	1.22
0.01	2.75	0.05	1.15
1.00	2.41	1.07	1.07

activity is incorporated into the heme moiety of leghemoglobin at a rate equal to or greater than the rate of incorporation into heme of a variety of radioactive citric acid cycle intermediates. All the evidence was consistent with the conclusion that incorporation of propionate into the heme of leghemoglobin proceeded through the sequence of reactions involving methylmalonyl mutase. The propionate pathway may serve as an additional mechanism for the input into the citric-acid cycle of essential intermediates under conditions where key intermediates such as α-ketoglutarate are continuously removed for the synthesis of glutamate, a known early product of N_2 fixation (Kennedy, 1965). It is of interest that no evidence for the presence of methylmalonyl mutase could be found in tissues of legumes *per se*.

6.3.2.4 Role in nucleotide reduction Cobalt deficiency in *Rhizobium* causes the development of abnormally elongated cells, a symptom that suggests a lesion in the cell division process (Lowe and Evans, 1962; Cowles, Evans and Russell, 1969) (Fig. 6.6). Since a B_{12} coenzyme requirement had been established (Blakeley, 1965) for the nucleotide reductase system in *Lactobacillus leichmannii,* the mechanism of conversion of ribonucleotides to deoxyribonucleotides in *Rhizobium* species and nodule bacteroids was examined. This research (Cowles and Evans, 1968; Cowles, Evans and Russell, 1969) established that the activity of nucleotide reductase in extracts of nodule bacteroids and several *Rhizobium* species was dependent upon B_{12} coenzyme. In this regard, the nucleotide reductase from *Rhizobium* was similar in certain respects to the nucleotide reductase from *L. leichmannii,* but was different from the nucleotide reductases from *E. coli* and most other organisms where B_{12} coenzyme is not required for activity. It is apparent that cobalt deficient *Rhizobium* cells contain insufficient B_{12} coenzymes to allow the normal function of the nucleotide reductase system. Experiments in which the effects of cobalt deficiency on the activity of nucleotide reductase in *R. meliloti* revealed that the deficient cells responded by synthesizing a 5- to 10-fold greater quantity of the apoenzyme of nucleotide reductase than cells supplied with adequate cobalt. Cobalt deficiency apparently results in a derepression of the synthesis of the apoenzyme of nucleotide reductase. The evidence seems convincing that abnormal cell elongation associated with cobalt deficiency in *Rhizobium* is caused by a deficiency of nucleotide reductase. This deficiency results from an inadequate supply of an appropriate B_{12} coenzyme that is essential for nucleotide reductase activity and deoxynucleotide synthesis.

Although the search has not been exhaustive to date, nucleotide reductase and methylmalonyl mutase are the only enzymes identified in *Rhizobium* cells or nodule bacteroids that require B_{12} coenzyme. Preliminary investigations into the mechanism of ribonucleotide reduction in legume tissues *per se* have not identified a B_{12} coenzyme dependent reaction. At present there is no convincing evidence that higher plants synthesize B_{12} coenzymes or utilize B_{12} coenzyme requiring enzymes in their metabolic processes. It seems clear that cobalt is indispensible, for normal metabolic processes of rhizobia regardless of whether these organisms are maintained in cultures or in nodules in symbiotic association with legumes. Since normal metabolic processes of rhizobia are indispensible for the maintenance of the symbiotic association between *Rhizobium* species and legumes, cobalt is clearly essential for symbiotic N_2 fixation.

6.3.3   Calcium

According to Hallsworth (1958), vigorous growth of many nodulated legumes requires a medium with a higher calcium content and greater

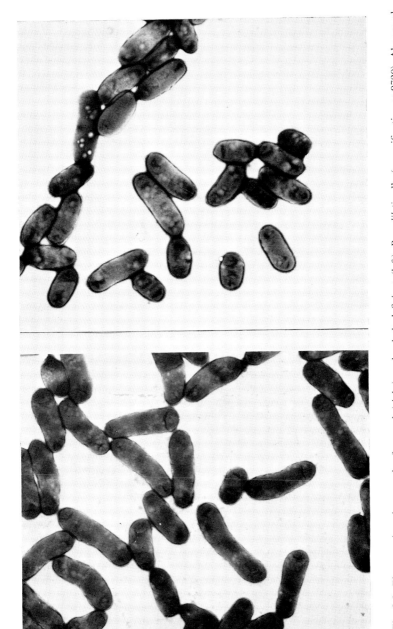

Fig. 6.6. Electronmicrophotograph of normal (right) and cobalt-deficient (left) *R. meliloti* cells (magnification × 9720). Abnormal elongation of cells in cobalt-deficient media resulted regardless of whether deoxynucleosides were or were not added (after Cowles *et al.*, 1969).

(*To face page 204*)

alkalinity than is needed for the culture of most legumes and non-legumes with combined nitrogen. In addition to supplying calcium, the common practice of adding calcium carbonate to acid soils used for the culture of inoculated legumes may produce several other beneficial effects. Increased alkalinity from calcium carbonate or other lime results in decreased solubility in the soil of elements such as manganese and aluminium that are toxic when present in excessive concentrations. The rise in soil pH that accompanies additions of lime to acid soils increases the solubility of molybdenum, an element specifically involved in the N_2-fixing process. A supply of calcium carbonate may increase the soil bicarbonate content and influence growth of legumes because bicarbonate or CO_2 is essential for growth of free-living rhizobia and is required for optimum nodulation of legumes (Mulder and van Veen, 1960; Lowe and Evans, 1962b).

A specific effect of calcium on nodulation recently has been identified by Lowther and Loneragan (1968). They found that nodulation of *T. subterraneum* was inhibited in a medium with a pH of 5.2 and a calcium concentration of 0.5 mM. At the same pH an increase in the calcium concentration to 8 mM resulted in effective nodulation. From experiments in which the calcium supply was altered at different stages of the nodulation sequence, Lowther and Loneragan concluded that the calcium requirement for nodulation of *T. subterraneum* was greater during the first day of infection of root hairs by *Rhizobium* than at other stages of the nodulation process. Munns (1968) presented evidence indicating that the same stage of the nodulation process that was sensitive to calcium deficiency also was sensitive to acidity. When the concentration of calcium was sufficiently low to inhibit nodulation, Lowther and Loneragan (1968) found that root-hair curling, which is associated with infection, was also inhibited. The high calcium requirement for initiation of nodulation by *T. subterraneum* was not related to growth or survival of *Rhizobium,* nor to growth of lateral or tap roots or to the development of root hairs. The specific calcium requirement for symbiotic N_2 fixation was consistently associated with the infection of legume roots by rhizobia.

Although several workers (Dixon, 1969; Lillich and Elkan, 1968) have studied the role of pectinases in the nodulation process, to the authors' knowledge there is no pertinent information available on the calcium requirements for specific pectin degrading enzymes associated with the infection process. It is of considerable interest that an enzyme present in culture filtrates of *Erwinia carotovora* and *Bacillus polymyxa* split pectic substances through the action of poly-galacturonic-trans-eliminase, the activity which is absolutely dependent upon calcium (Starr and Moran, 1962). No one has identified polygalacturonic-trans-eliminase in *Rhizobium* species or legume nodules, but Nagel and Anderson (1965) have concluded that it is now well established that the principal mechanism involved in bacterial degradation of pectic substances is the trans-eliminative cleavage of glycosidic linkages. A thorough search for a calcium requiring trans-eliminase in *Rhizobium* cells obviously is overdue.

6.3.4   *Iron*

Iron is an essential constituent of the various cytochromes that participate in respiration and photosynthesis in both leguminous and non-leguminous species regardless of the form of nitrogen that is utilized by the plants. Iron in a non-heme form is an essential constituent of chloroplast ferredoxin and on the basis of research with organisms other than higher plants, non-heme iron would be expected to function in the respiration of both leguminous and non-leguminous plant tissues as a component of succinic dehydrogenase and coenzyme-Q-reductase (Malkin and Rabinowitz, 1967).

Although convincing nutritional evidence is not available, there is ample circumstantial evidence of a biochemical nature to support the view that the iron requirement for legumes growing in symbiosis with *Rhizobium* is greater than that for growth of legumes on combined forms of nitrogen. Nodules of legumes contain a relatively high concentration of leghemoglobin which has an iron content of about 0.3% (Virtanen, 1947; Virtanen and Miettinen, 1963). On the basis of the positive correlation between leghemoglobin contents of legume nodules and N_2-fixing effectiveness, leghemoglobin very likely plays an important indirect role in the symbiotic N_2 fixation process. Klucas *et al.* (1968) established that both components of the bacteroid nitrogenase contain non-heme iron. Another non-heme iron protein has been identified in nodule bacteroids (Koch *et al.*, 1970) and it seems probable that it functions as a link in the transfer of electrons from substrates to nitrogenase. In addition, Bergersen (1963b) has shown that non-heme iron granules associated with plastids are present in nodules, and he believes that this ferretin-like protein serves as a storage form of iron for the nodules. Although elucidation of the precise roles of all the iron proteins present in nodules requires further study, the biochemical evidence that is available strongly suggests that the iron requirements for growth and symbiotic N_2 fixation by legumes in general must be greater than the requirements for growth of non-nodulated legumes.

6.3.5 Copper

Copper is essential for both legumes and non-legumes. Limited evidence indicates a specific requirement for symbiotic N_2 fixation. According to Hallsworth (1958) and Hallsworth, Wilson and Greenwood (1960), the copper supply has a definite effect on the manner in which nodules are distributed on the roots of *T. subterraneum*. When copper was added at 10 times the normal concentration (0.064 ppm) nodules developed in dense clusters on main roots, whereas at concentrations below 0.064 ppm a more sparse distribution was apparent. When *T. subterraneum* was grown with adequate calcium an increase in the copper content in the nutrient solution from 0.0064 to 0.064 ppm or higher greatly stimulated N_2 fixation. Hallsworth's group suggested that the response might be related to a role of copper in leghemoglobin synthesis but no experimental evidence to support this suggestion was presented.

In another study, Yates and Hallsworth (1963) observed that copper deficient non-nodulated *T. subterraneum* supplied with nitrates accumulated amino acids. In contrast, the addition of copper at several concentrations above that ordinarily supplied in normal nutrient solutions resulted in accumulation of soluble amino acids in symbiotically grown *T. subterraneum*. The increased copper supply to the nodulated clover resulted in an increase in the contents of glutamic acid, α-alanine and γ-amino-*n*-butyric acid. The rate of incorporation of ^{14}C labeled glucose into soluble amino acids and protein of isolated nodules was correlated positively with the copper supply. They suggested that copper may be directly involved in the synthesis of γ-amino-*n*-butyric acid. Although the copper requirement for symbiotically grown *T. subterraneum* appears to be somewhat higher than that for growth of clover on combined nitrogen, sites of action of copper specifically related to the N_2-fixing process have not been clearly identified.

6.4 LEGHEMOGLOBIN

6.4.1 Association with N_2 fixation

A unique characteristic of the symbiotic association between *Rhizobium* and leguminous species is the presence of a red pigment in the root nodules. Kubo (1939) identified the compound as a hemoglobin, a conclusion that has been confirmed by Virtanen (1945) and Keilin and Wang (1945). The association of leghemoglobin contents of nodules with N_2-fixing capacities was discovered by Virtanen, Erkama and Linkola (1947), and the early work on the subject has been reviewed in detail by Virtanen (1947). Leghemoglobin has been identified in the nodules of inoculated *Pisum*, *Vicia*, *Lathyrus*, *Phaseolus*, *Trifolium*, *Melilotus*, *Glycine*, *Caragena* and other species, and a positive relationship between N_2-fixing effectiveness and leghemoglobin contents was indicated consistently. In an investigation of the importance of leghemoglobin, Virtanen (1945) pointed out that vigorous N_2 fixation began in *P. sativum* at about the time that leghemoglobin synthesis was initiated, and that a series of conditions that inhibited N_2 fixation, including insufficient O_2 and inadequate light, resulted in decreased leghemoglobin in nodules and a parallel decrease in the rate of N_2 fixation. In an investigation of *Rhizobium* strains exhibiting different effectiveness in symbiosis with *P. sativum*, Virtanen *et al.* (1947) observed a rough parallel between leghemoglobin content of nodules induced by the different strains and the quantities of N_2 fixed. Similar positive correlations have been reported by several researchers (Graham and Parker, 1961; Jordan and Garrard, 1951). Convincing evidence has been presented (Virtanen and Miettinen, 1963) that the initiation of N_2 fixation and its decline when legumes become scenescent closely coincided with the appearance of leghemoglobin and its degradation to bile pigments. Bergersen (1962a) has emphasized that the correlation between N_2 fixation and leghemoglobin contents is not necessarily direct because an excellent positive correlation can be made between the vigor of N_2 fixation and the volume of active nodule tissue. Nevertheless, the consistent association of leghemoglobin with N_2 fixing effectiveness strongly suggests an important role.

6.4.2 *Properties*

The spectral properties of leghemoglobin and its derivatives are not appreciably different from those of mammalian hemoglobin. Characteristic absorption spectra of ferroleghemoglobin, oxyferroleghemoglobin and ferrileghemoglobin have been reported (Kulo, 1939; Hamilton, Shug and Wilson, 1957), and the prosthetic group of the protein has been identified as protoheme (Ellfolk and Sievers, 1965). Crude preparations of leghemoglobin contain endogenous phenoloxidase that apparently participates in the catalysis of the conversion of ferroleghemoglobin to ferrileghemoglobin (Keilin and Wang, 1945). Ferrileghemoglobin may be reduced back to ferroleghemoglobin by the addition of sodium hydrosulfite. In a study of the O_2 equilibrium of leghemoglobin, Appleby (1961) found that a crude preparation and also purified components were one-half oxygenated at an O_2 pressure of 0.05 mm of Hg. The affinity of leghemoglobin for O_2, therefore, is greater than that of other hemoglobins.

Ellfolk (1960) has successfully separated leghemoglobin from *G. max* nodules into two main components by use of DEAE-cellulose chromatography, ammonium sulfate fractionation and electrophoresis. After measurement of the sedimentation and diffusion coefficients, partial specific volume and molar frictional ratios, he calculated molecular weights of 16 800 and 15 400 for the electrophoretically fast and slow components respectively. From amino-acid analyses, mean molecular weights were calculated that were in excellent agreement with the data obtained from physical methods (Ellfolk, 1961). Amino-acid analyses revealed that the fast moving component contained an N-terminal glycine, whereas an N-terminal valine was identified in the electrophoretically slow moving component. Other differences in amino-acid content were relatively minor. On the basis of chromatographic separation of peptides from tryptic digestion of leghemoglobins, Ellfolk (1962) concluded that the synthesis of the two components very probably was directed by different genetic information.

Ehrenberg and Ellfolk (1963) have studied the paramagnetic susceptibilities of leghemoglobin derivatives. Both the acidic and alkaline forms of ferrileghemoglobin were temperature-dependent equilibrium mixtures of high- and low-spin states. The acetate complex of ferrileghemoglobin, however, was a pure high-spin form. In a recent study Applely (1969a) has found that 70% of the leghemoglobin in *G. max* nodules occurs in the high-spin oxyleghemoglobin form provided that appropriate precautions are taken to prevent autoxidation during preparation.

6.4.3 Biosynthesis

Neither *Rhizobium* species in pure culture nor non-nodulated leguminous plants synthesize leghemoglobin. A hemoglobin component was recently identified (Appleby, 1969b) in pure cultures of *Rhizobium japonicum* but it is unrelated to leghemoglobin. In the symbiotic association in legume nodules, leghemoglobin is reported (Goodchild and Bergersen, 1966) to be located inside membrane envelopes that surround individual bacteroids or groups of bacteroids. On the other hand Dart (1968) has presented evidence, that the pigment is located outside the membrane envelopes. The role of the plant and the bacteroids in the synthesis of this pigment is of considerable fundamental importance.

In regard to the mechanism of heme biosynthesis, Richmond and Solomon (1955) used radioactive glycine and acetate and concluded that the pathway of heme biosynthesis in *G. max* nodules was similar to the established pathway for heme synthesis in mammalian tissues (Shemin and Russell, 1963). Jackson and Evans (1966) demonstrated that· radioactive propionate was incorporated into the heme moiety of leghemoglobin of *G. max* nodules, and presented evidence that the enzymes necessary for heme synthesis from this precursor were located in the bacteroids. Cutting and Schulman (1969) have carefully separated *G. max* nodules into soluble and bacteroid components (free of plant mitochondria) and obtained convincing evidence that bacteroids and not the soluble protein fraction were the major sites of conversion of radioactive δ-amino-levulinic acid into the heme moiety of leghemoglobin.

In an effort to identify the site of biosynthesis of the globin moiety of leghemoglobin, Dilworth (1969) cultured *Ornithopus sativus* and *Lupinus luteus* in symbiotic association with *Rhizobium lupini,* an organism which was known to nodulate both of these species effectively. Isolation of the leghemoglobins from the two species by chromatographic procedures revealed two distinctly different elution patterns. In contrast, nodules from *L. luteus* inoculated with two different strains of *R. lupini* produced only one type of leghemoglobin. Dilworth concluded, therefore, that the genetic specification for leghemoglobin composition is determined by the plant.

Cutting and Schulman (1968) have investigated the site of biosynthesis of leghemoglobin by using antibodies to highly purified leghemoglobin components. *G. max* infected with *R. japonicum* showed no evidence of cross-reactive material in bacteroids, leaves, stems or roots but contained cross-reactive material in the host cytoplasm of nodules. A cell-free protein-synthesizing system prepared from nodules induced by either effective or ineffective *R. japonicum* strains and established to be free of bacteroids, produced cross-reactive material when supplied with radioactive amino acids. They concluded that the host plant cells were the major sites of synthesis of the protein moiety of leghemoglobin. From this research they state that leghemoglobin biosynthesis in nodulated legumes is partitioned, and concluded that the heme moiety was synthesized by the bacteroids, whereas the synthesis of the globin moiety was directed by cells of the host.

6.4.4   *Physiological role*
Early researchers (Kubo, 1939; Keilin and Wang, 1945) believed that

leghemoglobin functioned in the transport of O_2 to the bacteroid tissue of legume nodules. This conclusion was based to some extent on the similarities of the gross chemical properties of hemoglobins from legume nodules and mammalian sources. On the other hand, Smith's (1949a, b) experiments failed to show a positive relationship between leghemoglobin content and O_2 uptake by detached nodules. He observed that the addition of sufficient CO to block O_2 transport by leghemoglobin failed to influence the respiratory rate of nodules. Also, Burris and Wilson (1952) found no positive correlation between the leghemoglobin content of nodules and rates of O_2 uptake, but they did observe that the addition of swine hemoglobin to a cell suspension of *R. leguminosarum* stimulated the respiratory rate and the CO inhibited the increase. Little and Burris (1947) also showed that the addition of leghemoglobin increased the respiratory rate of soybean nodule bacteroids under conditions of low O_2 tension. The failure to establish a definite role of leghemoglobin in the respiration process of nodules has resulted in a search for other explanations.

On the basis of a series of investigations Bauer (1960a, b) and associates (Bauer and Mortimer, 1960; Abel, Bauer and Spence, 1963), Ehrenberg and Ellfolk (1963) and Abel (1963) postulated that leghemoglobin was the actual site of N_2 complexation and reduction. Abel and Bauer (1962) observed striking differences in the properties of leghemoglobin that were dependent upon the methods of isolation and purification. Leghemoglobin extracted by 3 M ammonium sulfate at pH 9.0 under either H_2 or argon retained its 'native form' and could be stored under H_2 at $-10°C$ for long periods. The heme moiety of leghemoglobin prepared under these conditions was believed to be buried in the molecule and protected from oxidation. In contrast, preparations made without appropriate precautions were reported to be subject to oxidation and partial denaturation. Leghemoglobin isolated by the procedure that prevented denaturation was reported (Abel, Bauer and Spence, 1963) to preferentially complex N_2 rather than O_2. The complex failed to dissociate under low partial pressure of N_2 and was relatively stable.

On the basis of these investigations Abel (1963) postulated that the reduced forms of the two leghemoglobin components were linked together in a unique way that allowed N_2 to bind between Fe atoms of the heme moieties of the two oriented components. In this environment two electrons from iron atoms of the leghemoglobin molecules and two protons from the medium were postulated to be transferred to bound N_2 to form a bound diimide. Electrons from metabolic processes in the bacteroids and protons from the medium theoretically provided the energy for continued reduction of N_2, bound as N_2^{2-}, by way of a diimide stage and a postulated bound hydrazine intermediate to ammonia. As the sequence of reductions proceeded, Abel postulated that the conformation of the leghemoglobin was altered to accomodate the expected increase in N—N bond lengths that would accompany the reduction steps.

The existence of a leghemoglobin N_2 complex has been challenged by Appleby (1969c). A mass spectrometric analysis of gases released from a preparation presumed to be a leghemoglobin-N_2-complex (Abel and Bauer, 1962) identified O_2 rather than N_2 as the major product. Furthermore, chromatography of the 'leghemoglobin-N_2-complex' revealed a mixture of leghemoglobin and cytochrome c. The spectrum of the mixture according to Appleby would account for the spectrum that Abel and Bauer associated with a leghemoglobin-N_2-complex. From Appleby's results it would appear that Abel's hypothetical scheme for a role of leghemoglobin in N_2 fixation is based upon an artifact.

On the basis of recent investigations to be discussed subsequently, the nitrogenase system from nodule bacteroids contains non-heme iron and molybdenum but no leghemoglobin. Leghemoglobin apparently plays no direct role in N_2 reduction, but the possibility that it is involved in the transport of O_2 or of N_2 has not been ruled out.

Bergersen (1960a) proposed a hypothetical sequence of reactions to explain the mechanism of N_2 fixation by legume nodules including the role of leghemoglobin. The hypothesis was based on several observations including the report by Bergersen and Briggs (1958) that groups of bacteroids within nodules were surrounded by a membrane, and that leghemoglobin was located in solution within the membrane envelops. This interpretation, however, has been questioned by Dart (1968). On the basis of kinetic experiments in which G. max nodules were exposed to $^{15}N_2$ then fractionated into bacteroids, soluble proteins, and membranes, Bergersen (1960b) concluded that membranes contained the greatest concentration of ^{15}N and that membranes were the sites of N_2 reduction. The hypothesis also was based on observations by Hamilton et al. (1957), Shug, Hamilton and Wilson (1956) and Bergersen and Wilson (1959) from which they claimed that nodule bacteroid preparations reduced ferrileghemoglobin to ferroleghemoglobin and that ferroleghemoglobin was oxidized to ferrileghemoglobin in presence of N_2. Bergersen (1960a) proposed, therefore, that cytochromes in the plasma membranes of the bacteroids were reduced by bacteroid metabolic processes, and that electrons from the reduced cyto-chromes were transferred *via* an unknown intermediate, to ferrileg-hemoglobin to form ferroleghemoglobin. With the aid of another unknown component the ferroleghemoglobin was postulated to be oxidized by N_2, which resulted in the production of NH_3. In this scheme photosynthetic products from the legume provided the reducing power for bacteroids and also the intermediate carbon skeletons necessary for the synthesis of amino acids.

As pointed out in section 6.5.1, washed bacteroids apparently free of leghemoglobin retained their N_2-fixing capacity, and the activity of purified nitrogenase from nodule bacteroids was not stimulated by addition of leghemoglobin to the nitrogenase reactions. A role of leghemoglobin in N_2 fixation as indicated in Bergersen's scheme therefore, is no longer tenable and has been discarded by him (Bergersen and Turner, 1967).

As a result of these developments the old idea of a unique role of leghemoglobin as an O_2 carrier in nodule metabolism has gained support. Legume nodules were reported to contain a low concentra-tion of oxyleghemoglobin (Bergersen, 1962a; Appleby, 1961). More recently however, up to 20% of the total leghemoglobin in young nodules has been identified as oxyleghemoglobin (Appleby, 1969d). A marked enhancement in O_2 transport through solutions by mammalian hemoglobin has been demonstrated (Hemmingsen and Scholander, 1960; Hemmingsen, 1962), and on the basis of similar-ities in properties the same type of facilitated O_2 transport by leghemoglobin would be expected. From this information, and the known sensitivity of bacteroid nitrogenase to O_2 damage (Koch, Evans and Russell, 1967a), it has been argued (Bergersen and Turner, 1967; Yocum, 1964) that leghemoglobin tenaciously binds O_2 and transports it through the dense nodular tissue to respiratory sites in a unique way that does not damage the O_2 sensitive nitrogenase in the bacteroids. The facilitated diffusion could provide the necessary O_2 for oxidative phosphorylation (see section 6.7) by the dense mass of bacteroids. This hypothesis may be considered attractive, but it suffers from a lack of supporting experimental evidence and, there-fore, requires further investigation.

6.5   THE NITROGENASE SYSTEM

6.5.1   *Investigations with intact nodules*
Much information was obtained on the properties of nitrogenase before researchers were able to fractionate nodules and investigate

their components. Wilson, Umbreit and Lee (1937, 1938) established that H_2 was a specific inhibitor of N_2 fixation by nodulated red clover plants, and suggested a relationship between N_2 fixation and hydrogen metabolism. Initial efforts to identify a hydrogenase in nodules were negative, but Hoch, Little and Burris (1957) and Hoch, Schneider and Burris (1960) using the mass spectrometric technique clearly demonstrated H_2 evolution and also N_2-dependent HD formation from D_2. Bergersen (1963a) reported that both HD formation from D_2, and N_2 fixation by soybean nodules responded to O_2 and CO in similar ways. Bergersen also observed that N_2 was a competitive inhibitor of H_2 evolution and that H_2 competitively inhibited N_2 fixation by nodules. On the basis of evidence to be discussed subsequently (6.5.3.5), the inhibition of H_2 evolution by N_2 may be interpreted on the basis of a competition between N_2 and H^+ for the ATP and nitrogenase-dependent transfer of electrons from endogenous donors. Inhibition of N_2 fixation by H_2 apparently results from a competition between N_2 and H_2 for a binding site on the nitrogenase complex.

Burris (1956) has reviewed the early work in which the optimum conditions for nitrogen fixation by excised nodules were established and effects of metabolic inhibitors on fixation were determined. From investigations with inoculated red clover plants a Michaelis constant for N_2 of 0.05 atm. was estimated (Bergersen, 1963a). The comparable value for excised soybean nodules was 0.025 atm. (Burris, Magee and Bach, 1955). Excised nodules required a pO_2 of about 0.5 atm. for optimum fixation and exhibited an extreme sensitivity to CO (Burris, 1956). At supra-optimal O_2 partial pressures (0.5 to 0.8 atm.) Bergersen (1962b) observed that O_2 was a competitive inhibitor of N_2 fixation. Intact nodules incubated under optimal conditions continued to incorporate $^{15}N_2$ for about 2 h (Aprison, Magee and Burris, 1954). Slicing nodules reduced the rate of fixation to about 25% of the normal and crushing nodules resulted in complete loss of activity (Burns, 1956).

After the development of the acetylene assay, the conditions required for optimum acetylene reduction by *G. max* nodules were investigated by Koch and Evans (1966). More detailed investigations, including a study of the procedures for handling nodules, the most favorable environmental conditions for optimum acetylene reduction, and correlations between rates of acetylene reduction and N_2 fixation by nodulated legumes have been conducted by Schwinghammer *et al.* (1970), Sprent (1969) and Hardy *et al.* (1968). The evidence seems convincing that the acetylene reduction assay, when utilized with appropriate precautions, provides a convenient, inexpensive and reliable method for the assessment of nitrogenase activity in biological materials.

Aprison, McGee and Burris (1954) determined the distribution of ^{15}N in nodule components after exposure of intact nodules to $^{15}N_2$ and observed the greatest concentration in glutamic acid. Using an exposure to $^{15}N_2$ of 1 min, Bergersen (1965) concluded that 94% of the fixed nitrogen was recovered as ammonia in the soluble fraction of soybean nodules. These results were consistent with the early general conclusions of the Wisconsin group (Burris, 1956) that ammonia was the first identifiable intermediate in the N_2 fixing process.

Recently Pate et al. (1969) have conducted an investigation of the vascular transport system of legume root nodules. Amino-acid analyses of tissue fluids or sap from bleeding nodules, roots, nodule bacteroid tissue, and nodule cortex of Vicia faba have revealed that the nodule bleeding sap from severed vascular tissue, in comparison with the other fluids, was unusually rich in amides, particularly asparagine. Pate et al. suggested that specialized transport cells were involved in an active uptake and secretion of nitrogenous compounds from the bacteroid tissue where N_2 fixation takes place, and that amides are major nitrogenous compounds that are transported from the nodules. Mr. Peter Wong, from our laboratory (unpublished experiments, 1969) has conducted amino-acid analyses of the bleeding sap from G. max root nodules and obtained results in good agreement with those of Pate et al.

Although much insight into the mechanism of nitrogen fixation has been obtained from investigations (Burris, 1956; Bergersen, 1969) with intact nodules, the early difficulties in demonstrating N_2 fixation by breis or extracts of nodules delayed the elucidation of a considerable portion of the detailed biochemistry of the N_2-fixing process.

6.5.2 Bacteroids as the site of activity

Bergersen's (1960b) proposal that the membrane envelopes which surround bacteroids were the sites of N_2 fixation was based on limited data and therefore, was subjected to re-examination. Kennedy (1965) exposed O. sativus nodules to $^{15}N_2$ then fractionated them in an isotonic medium. Analyses revealed that the greatest concentration of the isotope was retained in bacteroids and not in the membrane envelopes. Similar experiments by Kennedy, Parker and Kidby (1966), in which nodules were crushed at $-20°C$

then fractionated in undiluted nodule sap, also indicated that bacteroids rather than membranes were the sites of fixation. When insufficient precautions were taken to prevent amino acids from leaching out of bacteroids, the greatest ^{15}N concentration was found in the soluble fractions of nodules. Kinetic experiments (Klucas and Burris, 1967) similar to those conducted by Bergersen, but without appropriate steps to prevent loss of amino acids from bacteroids, showed that the soluble fraction of nodules contained the greatest enrichment of ^{15}N.

Bergersen (1967) developed a special press which enabled him to crush soybean nodules under anaerobic conditions and to prepare a brei that retained the capacity to fix N_2. From the active breis, maintained under argon, Bergersen and Turner (1967) collected the bacteroids by centrifugation, washed them and, in agreement with the conclusions of Kennedy et al. (1966), found that the bacteroids were the sites of N_2 fixation. In these experiments, it was necessary to avoid exposure of macerated nodules to O_2 during preparation, but bacteroids, like intact nodules, required O_2 for N_2 reduction. Excessive exposure of isolated bacteroids to O_2 (i.e. 15 minutes in air), resulted in a striking loss in activity (Bergersen and Turner, 1968). Both N,2 fixation and H_2 evolution by washed bacteroids was stimulated by succinate, fumarate and pyruvate and, like nodules, they exhibited a great sensitivity to CO (i.e. a K_i for CO of 0.00057 atm.). Neither N_2 fixation nor H_2 evolution by bacteroid preparations was dependent upon an ATP-generating system. More recently Dixon (1968) reported an uptake of H_2 accompanied by O_2 uptake by washed bacteroids from the nodules of Pisum sativum. This observation is of considerable interest because catalysis of H_2 uptake is a characteristic reaction of hydrogenase and not a reaction directly involving nitrogenase (see 6.5.3.5).

As a result of some major developments in methodology, our laboratory in 1966 embarked upon a program to investigate the nitrogenase system in legume nodules. The primary objective was to isolate active bacteroids and to prepare and characterize cell-free nitrogenase from them. The acetylene test proved to be of great value in the assay of nitrogenase activity in nodules and bacteroids (Koch and Evans, 1966; Koch, Evans and Russell, 1967b). Another method that proved to be of major importance in the successful isolation of active bacteroids and cell-free extracts involved the use of insoluble polyvinylpolypyrrolidone (PVP) for the removal of phenolic compounds from plant tissues (Loomis and Battaile, 1966). Phenolic compounds that are released after maceration of plant tissues react with peptide bonds to form phenolpeptide bond complexes. This type of reaction is reported to cause the inactivation of many plant enzyme systems. If phenolic compounds in plant tissues are maintained in a reduced state by the addition of ascorbate or other approriate reductants they may react with polyvinylpoly-pyrrolidone to form a relatively insoluble complex that may be removed from the macerated plant material by filtration. The reactions reported to be involved are outlined in Fig. 6.7. A

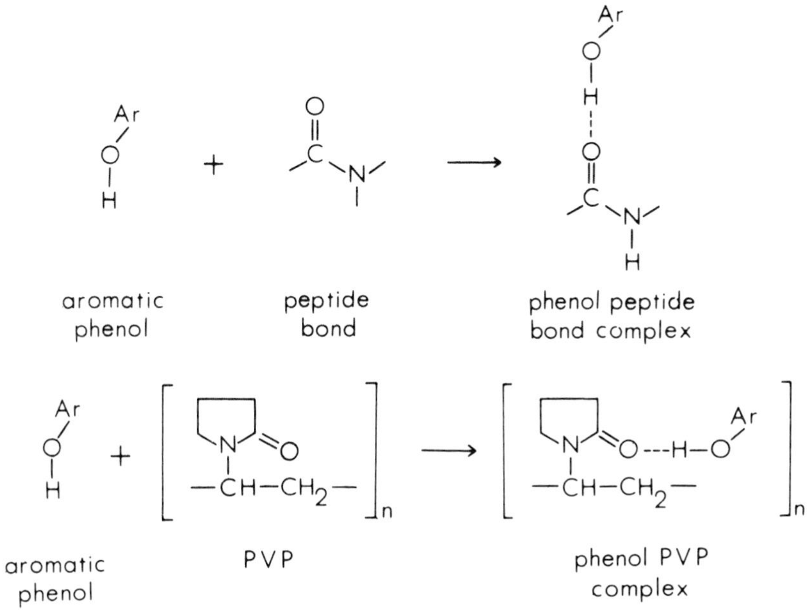

aromatic phenol   peptide bond   phenol peptide bond complex

aromatic phenol   PVP   phenol PVP complex

Fig. 6.7. Proposed reactions of an aromatic phenol with a peptide bond and with polyvinylpolypyrrolidone (based upon evidence presented by Loomis and Battaile, 1966).

procedure was developed in which soybean nodules were placed in a blender vessel and mixed with buffered polyvinylpolypyrrolidone and ascorbate while gassing under a stream of N_2 or A. The brei from the macerated nodules was filtered through bolting cloth and the bacteroids were collected and washed by centrifugation. It was necessary to conduct all of these steps under anaerobic conditions. Both the brei and the washed bacteroids prepared under these conditions retained N_2-fixing capacities (Table 6.3). The bacteroids required O_2 for N_2 fixation, but did not respond to leghemoglobin additions. The nitrogenase activity in washed bacteroids, however, was stimulated by the addition of substrates, including succinate and β-hydroxybutyrate. These results independently confirmed those of Kennedy *et al.* (1966) and of Bergersen and Turner (1967). The evidence that bacteroids are the site of nitrogen fixation is considered conclusive.

TABLE 6.3.   The effect of polyvinylpolypyrrolidone (PVP) and ascorbate in the preparation medium on the acetylene reducing capacity of a *G. max* nodule brei.

Each reaction mixture contained 10 ml of nodule brei prepared from 2.5 g of *G. max* nodules. The gas volume (22 ml) above the mixture was composed of 0.25 atm. of O_2, 0.65 atm. of A and 0.1 atm. of acetylene (after Koch, Evans and Russell, 1967a, b).

Preparation of brei	Ethylene produced (μmoles/h)
With PVP	7.50
Without PVP	0.08
With PVP and 0.3 M ascorbate*	14.11
Without PVP, with 0.3 M ascorbate	0.48

* Further investigation revealed that a concentration of 0.2 M sodium ascorbate was optimum for the preparation.

6.5.3   Cell-free extracts

6.5.3.1   Preparation   The availability of a reproducible method for the preparation of active bacteroids, and of a rapid and sensitive assay technique for nitrogenase, provided the essential conditions necessary for demonstrating N_2 fixation by cell-free extracts of legume nodules (Koch, Evans and Russell, 1967a; Evans, Koch and Klucas, 1970). A large quantity of biological material was essential and the choice was nodulated *G. max* cultured in a nitrogen-free nutrient medium under greenhouse conditions. Sufficient cultures were maintained to supply 300 to 400 g nodules per week from plants that were four to five weeks-old. In the early experiments all bacteroid preparations were made with fresh nodules, but it was discovered that nitrogenase remained active in nodules that were frozen and stored at −70°C. Bacteroids prepared by the procedure referred to in section 6.5.2 were used for the preparation of cell-free extracts.

All steps in the preparation of cell-free extracts were carried out under strictly anaerobic conditions. To accomplish this, all buffers or other solutions utilized were made anaerobic by sparging with purified N_2 or A. Extracts or other materials were transferred by use of hypodermic syringes and certain operations were conducted in a glove box filled with O_2-free N_2. In each operation about 25 g of bacteroids collected from 150 g-batches of nodules were suspended

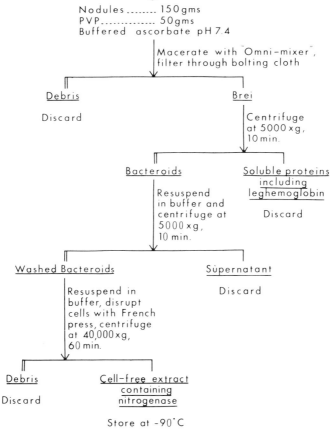

Nodules .... 150 gms
PVP .... 50 gms
Buffered ascorbate pH 7.4

Macerate with Omni-mixer,
filter through bolting cloth

Debris — Discard

Brei

Centrifuge at 5000 xg, 10 min.

Bacteroids

Resuspend in buffer and centrifuge at 5000 xg, 10 min.

Soluble proteins including leghemoglobin — Discard

Washed Bacteroids

Resuspend in buffer, disrupt cells with French press, centrifuge at 40,000 xg, 60 min.

Supernatant — Discard

Debris — Discard

Cell-free extract containing nitrogenase

Store at -90°C

Fig. 6.8. A diagram illustrating the procedure for the preparation of a cell-free extract containing nitrogenase from legume-root nodules. The entire operation must be conducted under anaerobic conditions (after Klucas *et al.*, 1968).

in 25 ml of 0.1 M tris (tris(hydroxymethyl)amino methane) buffer at pH 8.5 and placed in a pre-cooled French press. The bacteroids were ruptured by forcing them through the orifice of the press at a pressure of 8 tons per square inch. The ruptured bacteroids were collected under a stream of N_2, centrifuged, and the dark brown supernatant liquid collected and used as a source of nitrogenase. The essential features of the preparation of the extract are indicated in the flow diagram (Fig. 6.8) and have been described in detail (Shug, Hamilton and Wilson, 1956; Evans, Koch and Klucas, 1970).

Bergersen and Turner (1968) have used a procedure for preparation of extracts that is similar to the one initially developed in our laboratory, with the exception that a soluble rather than insoluble polyvinylpolypyrrolidone was employed. This procedure yielded a crude nitrogenase preparation with a maximum specific activity (nmoles N_2 reduced per min per mg of protein) of 0.067. With this low activity it was necessary to continue to use the $^{15}N_2$ assay. More recently Bergersen and Turner (1970) have utilized the method recommended by Koch *et al.* (1970) and have obtained crude extracts with higher specific activities.

6.5.3.2 Requirements for activity   In the initial experiments, crude extracts were assayed for both N_2 fixation and acetylene reduction. More recently, Koch, Klucas and Evans (unpublished results, 1970) have observed that the non-specificity for substrate can be extended to cyanide and azide, but these substrates are not particularly suitable for a convenient assay. Reaction mixtures utilized for routine assays of N_2 or acetylene reduction were similar in most respects to those described by Bulen's group (See chapter 4) in which $Na_2S_2O_4$ was employed as the electron donor and creatine phosphate and creatine phosphokinase utilized to maintain an ATP supply. The components of the reaction mixture for both assays are shown in Table 6.4. After the reactions were terminated, the synthesis of ammonia was measured by Nesslerization as described by Dilworth *et al.* (1965). Crude extracts ordinarily contain endogenous NH_4^+ and also may produce NH_4^+ from hydrolytic reactions that take place while reaction mixtures are incubated. To avoid this problem, crude extracts to be assayed for ammonia synthesis were passed through an anaerobic Bio-Gel column to remove compounds of low molecular weight. Crude extracts also were assayed by the acetylene reduction method, which in addition to having extreme sensitivity has the advantage that endogenous NH_4^+ does not interfere with measurement of the product of the reaction (Table 6.4). In early experiments, cell-free N_2 fixation was confirmed by use of $^{15}N_2$ and mass spectrometric analyses of products. This procedure also is not complicated by endogenous ammonia in reactions. It has been utilized extensively by Bergersen (1965, 1969) and by Burris and co-workers (Burris, 1956, 1966).

In the early experiments with cell-free extracts, we reported that the specific activity of a crude soybean bacteroid extract in a complete reaction was 1.8 nmoles N_2 reduced per minute per mg of protein. After gaining experience in preparing bacteroid nitrogenase, over 200 crude extracts with specific activities of 9 to 13 have been

TABLE 6.4. Essential reaction components for cell-free N_2 fixation and acetylene reduction.

The complete reaction mixture for N_2 fixation, in a final volume of 1.5 ml, contained 100 μmoles of TES buffer at pH 7.4, 10 μmoles of $MgCl_2$, 50 μmoles of creatine phosphate, 7.5 μmoles of ATP, and 0.5 mg of creatine phosphokinase. All reagents were adjusted to pH 7.4 with KOH. Each bottle (21 ml volume) was flushed 5 times with high-purity N_2 and 40 μmoles of $Na_2S_2O_4$ was injected into each bottle. The reaction was initiated by the addition of an extract containing bacteroid nitrogenase (1.8 mg protein) prepared by heat treatment and protamine sulfate precipitation.

The complete reaction in which acetylene reduction was determined was identical with that described for N_2-fixation, except that the gas phase in the reaction vessels was composed of 0.1 atm. of acetylene and 0.9 atm. of A. When N_2 or acetylene was omitted, A was added as the gas phase. Other methods utilized were those described by Klucas *et al.* (1968). (Unpublished data of Koch, Evans and Russell, 1970).

Reaction	N_2 fixed $n$moles/min/ mg protein)	C_2H_2 formed ($n$moles/min/ mg protein)
Complete (N_2-fixation)	12.2	—
Without N_2 gas	0.0	—
Complete (C_2H_2 reduction)	—	37.8
Without C_2H_2	—	0.0
Without ATP-generating system*	0.6	0.0
Without creatine phosphate	0.4	3.2
Without ATP	0.7	0.0
Without creatine phosphokinase	1.2	2.4
Without $MgCl_2$	3.3	4.7
Without $Na_2S_2O_4$	0.2	0.0

* The ATP-generating system consisted of creatine phosphate, creatine phosphokinase and ATP.

prepared routinely within the past two years. The production of NH_4^+ was convincingly dependent upon active nitrogenase, $Na_2S_2O_4$, an ATP-generating system, N_2 and anaerobic conditions. Acetylene reduction assays of crude extracts or partially purified extracts (Table 4) produced similar results, with the exception that the ratio, in moles of acetylene reduced to N_2 reduced, was about three to one as expected. Reactions containing relatively crude extracts (Table 6.4) were not consistently completely dependent upon creatine phosphokinase and Mg^{2+}. The low activity observed without creatine phosphate was due to the small amount of ATP that was added to the mixture. Determination of the activity of the partially purified bacteroid nitrogenase under a series of different partial pressures of N_2 indicated a maximum velocity near 0.25 atm. and one-half maximum velocity at 0.053 atm. of N_2. Saturation curves for the major reactants used in the $Na_2S_2O_4$ assay have established optimum concentrations of reactants that are similar to those reported for the nitrogenase from *Azotobacter vinelandii* (Bulen, Burns and LeComte, 1965). Under optimum conditions the reaction proceeded at a near linear rate for at least 1 h. Although the nodule bacteroid nitrogenase utilizes ATP for N_2 reduction, the reaction is strongly inhibited by ATP concentrations above 4 μmoles per ml (Bergersen, 1969) and, therefore, a continuous low concentration of ATP must be maintained in assays by an ATP-generating system (Koch *et al.*, 1967a).

Cell-free extracts of bacteroids from nodules of *G. max, O. sativus* and *Lupinus* sp. (Evans, 1970) have been prepared with activities in the $Na_2S_2O_4$ assay that approach those of crude extracts of *Azotobacter*. It seems obvious that the $Na_2S_2O_4$ assay procedure may be applied toward the study of nitrogenase in bacteroid extracts from a variety of different legumes.

6.5.3.3 Purified components Further purification of the bacteroid nitrogenase was accomplished by use of adaptations of procedures described by Bulen and LeComte (1966). Maintenance of anaerobic conditions has proven to be absolutely essential. In our procedure, nucleic acids were removed from a crude extract by protamine sulfate precipitation. The resulting supernatant liquid containing most of the nitrogenase was further purified on a DEAE-cellulose column by use of a step-wise elution with different buffered NaCl and $MgCl_2$ solutions. This fractionation resulted in the elution of a major component with 0.035 M $MgCl_2$ which contained 44 *n*moles of non-heme iron and 2.2 *n*moles of molybdenum per mg of protein (fraction 1). A second component (fraction 2) was eluted from the column at a concentration of 0.1 M $MgCl_2$. It contained 50 *n*moles of non-heme iron per mg of protein, and no greater concentration of molybdenum than was found in the crude extract. Both fractions also contained acid-labile sulfur. The fractions were not considered sufficiently pure to justify expression of the Fe, Mo and acid-labile sulfur contents on the basis of moles of protein.

In a variation of the purification procedure, the protamine sulfate precipitation step was omitted and instead the nitrogenase was precipitated with polypropylene glycol. The protein was dissolved in buffer and chromatographed on DEAE-cellulose. A summary of the specific activities for N_2-fixation and acetylene reduction at various steps in the procedure are presented in Table 6.5. These data

TABLE 6.5.   Relationship between the reduction of acetylene and N_2 by extracts and fractions of nodule bacteroids.

The extracts of bacteroids were prepared and assayed for acetylene reduction and for nitrogenase activities as described by Klucas *et al.* (1968). When fractions 1 and 2 were assayed together, 0.05 ml of each fraction (0.29 mg protein from fraction 1 and 0.20 mg from fraction 2) was utilized and the results expressed on the basis of the protein in fraction 2 (from Klucas *et al.*, 1968).

Type of extract	Rate of reduction		Ratio of rate of reduction C_2H_2 : N_2
	C_2H_2	N_2	
	*n*moles per mg protein per min		
Crude	35.2	11.8	2.98:1
25–55% PPG* ppt	76.1	27.4	2.78:1
Fraction 1	0.0	0.0	—
Fraction 2	74.4	19.3	3.85:1
Fractions 1 and 2	607.4	192.6	3.15:1

* Polypropylene glycol (P-400).

illustrate the approximate 3 to 1 ratio of the rates of acetylene reduction to N_2 fixation and provide no conclusive evidence of altered ratios associated with fractionation. The data in Table 6.5 also show no activity in fraction 1, a very low activity in fraction 2 alone, and a striking increase in activity from a combination of fractions 1 and fractions 2. In some experiments, complete resolution of fractions 1 and 2 was accomplished and each fraction from such experiments was titrated against the other to determine optimum proportions for activity. When the quantity of either one of the fractions was held constant and the other systematically increased in assays, typical enzyme saturation curves were obtained. From the shapes of these curves (Fig. 6.9) the functional unit of the combined fractions appeared to involve equivalent amounts of each unit. This point requires reinvestigation, however, when the fractions of higher purities are available.

In considering the purity of fractions of the bacteroid nitrogenase, assays of either one of the two fractions must be conducted under conditions where the other fraction is added in excess and specific activity expressed on the basis of protein content of the fraction that limits the reaction rate. Purification and determination of the extract of purity of the two individual fractions has been complicated by the fact that both are O_2 labile and easily denatured when subjected to electrophoresis or analytical ultracentrifugation. Klucas *et al.* (1968) have purified the nitrogenase on a DEAE-cellulose without dissociation of the two fractions into subunits. When this preparation was subjected to ultracentrifugation, two main components were observed with apparent S values of 8 and 4.5 respectively. From the

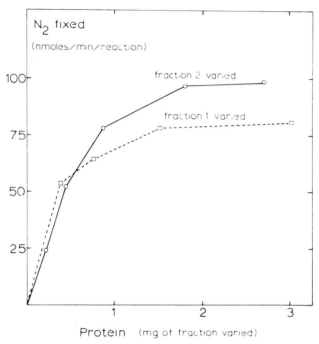

Fig. 6.9. Nitrogenase activity as influenced by different combinations of fractions 1 and 2 of soybean bacteroid nitrogenase. When fraction 2 was varied, all reactions contained a constant amount (1.5 mg of protein) of fraction 1. When fraction 1 was varied all reactions contained a constant amount (0.9 mg protein) of fraction 2. Details of the experiment were described by Klucas *et al.* (1968).

areas under the Schlieren patterns, 88% of the entire protein in the purified preparation was estimated to be accounted for by the two major components. Tentative estimates of the molecular weights of the 8S and 4.5S components were 150 000 and 55 000 respectively (unpublished results R. V. Klucas, Burton Koch and H. J. Evans, 1968). Bergersen and Turner (1970) have subjected a concentrated *G. max* nodule extract to chromatography on Sephadex G-200 and also have obtained two fractions. Component 1 was estimated to have a molecular weight of about 180 000 and to contain 9 gram-atoms of iron and 1 gram-atom of molybdenum per molecule of protein. They estimated the molecular weight of component 2 as 51 000 and reported that it contained 1 gram-atom of Fe per molecule of protein. Since no evidence regarding the homogeneity or purity of the fractions was presented, the data indicating molecular proportions of Fe, Mo and protein have little meaning. The molecular weights of fractions of bacteroid nitrogenase are within the general range of molecular weights of fractions 1 and 2 proteins respectively that have been purified from free-living nitrogen fixing microorganisms (Hardy and Burns, 1968). A more precise investigation of the physical properties of the G. max bacteroid nitrogenase components must be delayed until homogenous fractions are available.

6.5.3.4 *Inhibitors* The extreme sensitivity of G. max bacteroid nitrogenase to O_2 has been emphasized in a discussion of the preparation of cell-free extracts (6.5.3.1). In assays in which $Na_2S_2O_4$ was utilized as the electron donor, anaerobic conditions in reaction mixtures were maintained efficiently by a reaction of $Na_2S_2O_4$ with O_2. Considerable difficulty has been encountered in an adequate removal of O_2 from reaction mixtures in which electron donors such as NADH have been investigated (Klucas and Evans, 1968).

In addition to O_2 lability, Bergersen and Turner (1968) claimed that cell-free nitrogenase from G. max nodule bacteroids was cold labile. Crude preparations with low specific activities (0.076 or less) were reported to lose 50% of their activities in 30 min when stored in an ice bath. In three years' experience, cold lability of G. max bacteroid nitrogenase has not been observed in our laboratory. In a series of experiments 1-ml samples of crude bacteroid nitrogenase (specific activity of 12) stored under A at 0° for 20 h with 4 mg of $Na_2S_2O_4$ retained 96% of the original activity. Comparable samples stored at 0° for 20 h without $Na_2S_2O_4$ retained an average of 62% of their activities. One-ml samples of bacteroid nitrogenase containing 4 mg of $Na_2S_2O_4$ and stored for 20 h at 25°C under A retained 90% of their original activities whereas samples without $Na_2S_2O_4$ retained 83% of the original activities. It is the authors' opinion that the claims of cold lability are more logically explained as O_2 lability associated with an increased solubility of O_2 at lower temperatures.

Bacteroid nitrogenase is extremely sensitive to H_2 and CO. At a pN_2 of 0.25 atm., and an equal partial pressure of H_2, the rate of ammonia synthesis was inhibited by 67%. From a Lineweaver-Burk plot of data obtained from an experiment in which reactions were run at a series of increasing partial pressures of N_2, both in the presence and absence of 0.05 atm. H_2, Koch *et al.* (1967a) calculated a K_i for H_2 of 0.106 atm. This inhibition appeared to be competitive. Bergersen and Turner (1970) have studied CO inhibition of N_2 fixation by cell-free extracts of nodule bacteroids and reported a K_i for CO of 0.23 mm of Hg. The inhibition also was competitive.

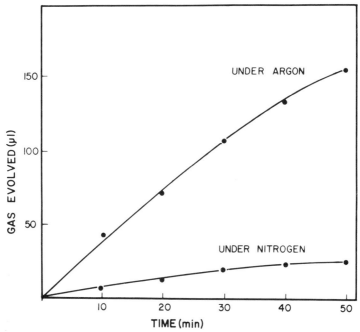

Fig. 6.10. Nitrogenase-dependent evolution of gas under either N_2 or argon. The complete reaction mixture (3 ml) in a Warburg vessel contained chromatographed bacteroid extract (12 mg of protein), sodium hydrosulfite, buffer, and the ATP-generating system. Prior to the addition of the extract, flasks were flushed thoroughly with N_2 or A as indicated. After 50 min NH_3 synthesis was measured after microdiffusion, by Nesslerization (after Koch *et al.*, 1967a).

6.5.3.5 Hydrogen evolution and HD exchange　The availability of active nitrogenase in cell-free extracts of bacteroids provided an opportunity to reinvestigate early reports of Hoch *et al.* (1960) Bergersen, (1963a) and Wilson, Umbreit and Lee (1937) that hydrogen metabolism in some way was related to N_2 fixation. Experiments were designed (Koch *et al.*, 1967a) to identify and measure quantities of gases evolved during N_2 fixation (Fig. 6.10). Complete reactions containing $Na_2S_2O_4$, an ATP-generating system, buffer, and active extract were placed in Warburg flasks and gassed with N_2. Comparable reactions were conducted with the exception that N_2 was replaced by an atmosphere of A. The complete reaction under N_2 synthesized NH_3 and evolved some H_2 both of which utilized about 11.2 μeq. of reductant. In contrast, a comparable reaction under A failed to fix N_2 but evolved a quantity of H_2 requiring about 13.6 μeq. of reductant. Both N_2 fixation and H_2 evolution in these experiments were dependent upon both reductant and the ATP-generating system. It was concluded from these results that the magnitude of ATP-dependent reductant utilization either with or without N_2 fixation was roughly the same. Since some H_2 was evolved in presence of N_2, some inefficiency was exhibited in the transfer of electrons from donors to N_2. In this respect cell-free nitrogenase behaves like intact nodules. All of the factors that control the coupling of electron donor systems to nitrogenase have not been established. This problem has been discussed by Burris (1969).

Hydrogen gas fails to function as an electron donor for the nitrogenase system in crude cell-free extracts of *G. max* nodules bacteroids (Koch *et al.*, 1967a), and all efforts to identify a cell-free hydrogenase have been unsuccessful. From the work of Dixon (1968), however, a hydrogenase is present in bacteroids from the nodules of *Pisum sativum*. Turner and Bergersen (1969) have shown that relatively crude extracts from *G. max* nodule bacteroids catalyzed the formation of HD from D_2. Some activity was observed without the addition of substrate, but a marked enhancement in the exchange resulted from the addition of $Na_2S_2O_4$, an ATP-generating system and N_2. These results are in general agreement with those reported by Jackson, Parshall and Hardy (1968), who conducted similar experiments with *Azotobacter* extracts (see p. 138). The results are consistent with the conclusion that an exchange occurs between D_2 and hydrogen of a nitrogenase-bound diimide or hydrazine intermediate of N_2 reduction. The nature of the reaction that is responsible for catalysis of HD production from D_2 in the absence of N_2, reductant, and an ATP supply remains to be clarified. It seems apparent that the properties of nitrogenase from nodules are not essentially different from those of the enzyme system from other sources.

6.5.4 *Relation to nitrate reductase*

Recent information (Nason *et al.*, 1970) regarding the subunit structure and *in vitro* complementation of *Neurospora crassa* nitrate reductase has provided considerable insight into the possible relationship of nitrate reductase to nitrogenase (see 6.3.1). Both genetic and biochemical evidence has shown that *Neurospora crassa* assimilatory nitrate reductase is composed of at least two subunits. The synthesis of one of these is induced by nitrate in a mutant (nit 1) lacking the capacity to grow on nitrate. This component is characterized biochemically by a capacity to catalyze the reduction of cytochrome *c* by NADPH. The other subunit is a constititutive component of wild-type *Neurospora crassa* and certain mutants, and is characterized biochemically by a capacity for catalysis of nitrate reduction by $FADH_2$ or reduced methylviologen when this subunit is combined with the inducible subunit. Recently Ketchum *et al.* (1970) (and private communication from Dr. A. Nason) have shown that the molybdenum-containing constitutive subunit from *Neurospora crassa* can be replaced by a molybdenum-containing subunit from the following sources: bovine milk xanthine oxidase, intestinal xanthine oxidase, rabbit liver aldehyde oxidase, chicken liver xanthine oxidase, fraction 1 protein from *A. vinelandii* nitrogenase or fraction 1 protein from *G. max* bacteroid nitrogenase. Combination of a molybdenum-containing constitutive subunit from any one of the various sources with the inducible subunit from *N. crassa* (nit 1) results in reconstitution of nitrate reductase with physical properties indistinguishable from those of wild-type *Neurospora* nitrate reductase.

From this information, it seems highly probable that the significant positive correlation between nitrate reductase activities in nodule bacteroids and nitrogen-fixing capacities of nodulated legumes (see 6.3.1) may be explained by assuming that a constitutive subunit of bacteroid nitrate reductase (containing molybdenum and perhaps non-heme iron) and fraction 1 protein of bacteroid nitrogenase are identical. A combination of fraction 1 protein from bacteroid nitrogenase with a nitrate inducible subunit in bacteroids would be expected to lead to the synthesis of nitrate reductase. On the other hand, the combination of a molybdenum-containing constitutive subunit with an N_2-inducible subunit (presumably fraction 2 of bacteroid nitrogenase) would be expected to lead to the synthesis of nitrogenase. The observed correlation between N_2 fixation and nitrate reductase activities could be explained if the quantity of the constitutive molybdenum-containing subunit limited the synthesis of both nitrate reductase and nitrogenase. Under conditions where ample nitrate is supplied one might postulate that the nitrate inducible subunit would be synthesized in sufficient quantity to combine with all the available constitutive molybdenum subunit and consequently an inhibition of nitrogenase synthesis would result. This type of control mechanism might explain the well known inhibitory effect of nitrate on N_2 fixation. Under conditions where nodulated legumes are grown in absence of added fixed nitrogen, sufficient traces of nitrate are continuously supplied through microbial oxidation of nitrogenous compounds from the plant to induce the synthesis of nitrate reductase. Under such conditions the limited nitrate supply apparently is not sufficient to curtail the synthesis of nitrogenase seriously. Nitrate reductase activity induced by traces of nitrate may reflect the capacity or in some cases the genetic capability to synthesize the constitutive molybdenum subunit that may limit the expression of both nitrogenase and nitrate reductase activities. It is now possible to test these postulations.

6.6 ELECTRON TRANSPORT

6.6.1 Oxidative pathways in bacteroids and cultured rhizobia

There are marked differences in the metabolism of cultured *Rhizobium* cells and bacteroids from nodules. One of the most significant is the capability of bacteroid cells to synthesize nitrogenase and transfer electrons to N_2, a property that has not been detected in cultured cells. In addition to differences in morphology, ribosome content and capacity to grow in ordinary culture media, (Almon, 1933; Klucas and Burris, 1967) cultured cells and bacteroids differ to some extent in their abilities to oxidize substrates (Tuzimura and Miguro, 1960). Both cultured *Rhizobium japonicum* cells (on glucose or succinate) and nodule bacteroids oxidized organic acids and fructose 1,6-diphosphate, but failed to utilize a series of hexose and triose phosphates. Tuzimura and Watanabe (1964) have reported that glucose was oxidized slowly by cells cultured on succinate, but bacteroid cells failed to oxidize glucose, fructose, sucrose, or mannitol. Kidby and Parker (personal communication), however, have reported that bacteroids from *Lupinus* sp. regained the capacity to oxidize glucose after a pre-incubation period of 7 to 10 h in which the glucose transport system apparently was restored. They believe that sucrose is the major substrate supporting N_2 fixation by bacteroids. Thorne and Burris (1940) observed relatively minor differences in the utilization of a series of carbohydrates and organic acids by cultured *R. leguminosarum* or *R. japonicum* cells as compared to bacteroids from legumes inoculated with these two organisms.

A series of investigations (Appleby and Bergersen, 1958; Falk, Appleby and Porra, 1959; Tazimura and Watanabe, 1964) have indicated that cytochrome patterns in cultured *Rhizobium* cells were strikingly different from those in nodule bacteroids. In a thorough

TABLE 6.6.   Hemoprotein concentrations in *R. japonicum* cultured cells and in nodule bacteroid cells (from Appleby 1969b, e).

Hemoproteins	Pigment contents (μmoles heme/g of protein)	
	Cultured cells	Bacteroids
Total cytochromes		
Cytochrome c	0.41	0.94
Cytochrome b	0.45	0.42
Cytochrome $a + a_3$	0.15	0.00
CO-reaction pigments		
Cytochrome a_3	0.09	0.00
Cytochrome o	+ *	0.00
Cytochrome c (552)	0.00	0.24
P-450	0.00	0.11
P-428	+ *	+ *
P-420	0.00	++ *
Rhizobium hemoglobin	++ *	0.00

* The symbols + or ++ denote low or high concentration of pigments whose absorption bands are partially masked by other pigments.

investigation Appleby (1969b, e) recently has shown that cultured *R. japonicum* cells (Table 6.6) contain cytochromes a and a_3, cytochrome o (cytochrome b, 559.5) and *Rhizobium* hemoglobin, and that bacteroids lack all three of these pigments. In contrast, cytochrome c (552), P-450, and P-420 were identified in nodule bacteroids but were not found in cultured cells. As mentioned previously, most researchers (Bergersen and Briggs, 1958) believe that nodule bacteroids are located inside membrane envelopes and are surrounded by a high concentration of leghemoglobin. A role of this pigment in the transfer of O_2 to the bacteroids which lack cytochrome oxidase has been suggested (Bergersen and Turner, 1967). *Rhizobium* hemoglobin at a relatively low concentration has been identified (Appleby, 1969b) in cultured cells but this pigment is antigenically unrelated to leghemoglobin.

Appleby has outlined an electron transport scheme for both cultured *R. japonicum* cells and nodule bacteroids. In cultured cells, a major pathway of electron transport from dehydrogenases through cytochrome b (556), cytochrome c, cytochromes a and a_3 and then to O_2 was proposed. As an alternate major pathway in cultured cells Appleby (1969b) suggested that electrons may be transferred from dehydrogenases to cytochrome o (cytochrome b, 559.5) and then to O_2. Evidence for some minor electron pathway also was presented. In another phase of the investigation Appleby (1969e) has diagrammed major routes of electron transport in bacteroids which include a sequence from particle bound dehydrogenases through cytochrome c (552) to O_2 or from particle bound dehydrogenases through cytochrome b (556) to cytochrome c and then through P-502 to O_2. Other pathways were suggested in which electrons were transferred via pigments P-420 and P-428. In considering the role of P-450, Appleby suggests reduction of the pigment by soluble dehydrogenases and direct oxidation by O_2. In several organisms P-450 plays a major role in O_2 activation prior to hydroxylation (Mason, 1965). Since P-450 was reported (Hernandez, Mazel and Gillette, 1967) to catalyze the anaerobic reduction of azo-compounds to amines, Appleby has discussed the possibility that P-450 may play a role in the reduction of intermediates of N_2 fixation. If P-450 functioned in this manner, one would expect that CO inhibition of N_2 fixation would be reversed by light, but unpublished results from our laboratory have provided no indication of light reversal of the CO inhibition of N_2 fixation.

Tuzimura and Watanabe (1964) have reported that *Rhizobium* cells cultured with insufficient O_2 exhibit cytochrome patterns that are similar to those observed in bacteroid cells from nodules. Despite these similarities no convincing evidence has been presented that cultured *Rhizobium* cells synthesize nitrogenase, and the factors involved in the control of the synthesis of the enzyme remain obscure. The convincing evidence that bacteroids require O_2 for N_2 fixation, (Bergersen and Turner, 1967) and the established (Koch *et al.*, 1967a) high magnitude of the ATP requirement for cell-free nitrogenase activity, provides strong circumstantial evidence that the necessary ATP supply for N_2 fixation *in vivo* must be derived from oxidative phosphorylation coupled to the electron transport chain.

6.6.2 Pathways to nitrogenase

  6.6.2.1 NADH as a donor In the early investigations of the properties of nitrogenase in cell-free extracts of nodule bacteroids, $Na_2S_2O_4$ was used exclusively as the electron donor. Natural electron donors such as pyruvate, α-ketobutyrate and H_2, all of which were effective for the nitrogenase from *Clostridium pasteurianum*, failed to serve as donors for the cell-free bacteroid system. Information from several sources led to the postulation that a continuous supply of NADH might serve as a source of electrons for nitrogenase dependent acetylene reduction. Hayward *et al.* (1959) had reported that both cultured *Rhizobium* cells and bacteroids accumulated poly-β-hydroxybutyrate. Wong and Evans (1969) and Evans and Russell (1965) confirmed these observations, and in addition have identified a very active NAD specific β-hydroxybutyrate dehydrogenase and also an NADH dehydrogenase in bacteroid extracts. The NADH dehydrogenase catalyzed the transfer of electrons to low potential dyes such as benzylviologen or methyl viologen. On the basis of this information a hypothetical scheme in which electrons were transferred from β-hydroxybutyrate through NAD, benzylviologen and then to nitrogenase was tested. As illustrated (Fig. 6.11), the addition to a reaction mixture of active bacteroid nitrogenase, an ATP-generating system, benzylviologen, a catalytic amount of either NADH or NAD and β-hydroxybutyrate resulted in the catalysis of nitrogenase-dependent acetylene reduction at a relatively rapid rate. When β-hydroxybutyrate was omitted a strikingly decreased rate of the reaction occurred. The omission of the ATP-generating system or benzylviologen resulted in little or no activity. From this evidence, the sequence of reactions outlined in Fig. 6.12 (with benzyl viologen replacing intermediate 'X' in the scheme) appeared to be functional. In other experiments the β-hydroxybutyrate dehydrogenase and β-hydroxybutyrate in Fig. 6.12 were effectively replaced by substrate concentrations of NADH provided that benzylviologen was included in reaction mixtures. Obviously, any effective NADH-generating system could be utilized in the complete reaction sequence. Although the crude extract contained a very powerful NADH dehydrogenase, and the activity of this dehydrogenase was correlated with the rate of acetylene reduction in a series of experiments, an absolute requirement for NADH dehydrogenase has not been established. A thorough search for a natural component to replace 'X' in the scheme has met with only limited success. Some evidence (Evans, 1970) has been obtained showing that a combination of a crude extract of bacteroids and a rather high concentration of FMN or FAD consistently substituted for 'X' in the scheme. In these experiments, however, both the extract and the bacteroid nitrogenase preparations were impure. Essential components in these extracts are labile and purification attempts conducted so far have resulted in loss of activity. Some progress has been made in the resolution of electron transport factors for bacteroid nitrogenase by

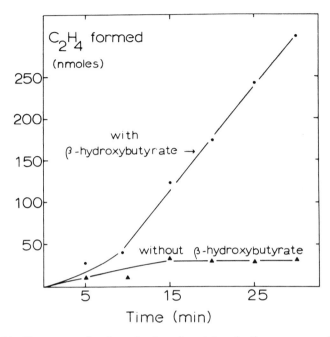

Fig. 6.11. Time course for the reduction of acetylene in the presence and absence of β-hydroxybutyrate. The complete system (in a final volume of 2.0 ml) contained: an ATP-generating system (75 μmoles of creatine phosphate, 7.5 μmoles of ATP, 10 μmoles of MgCl$_2$, 0.2 mg creatine phosphokinase) 200 μmoles of DL-β-hydroxybutyrate, 0.5 μmole of benzylviologen, 1 μmole of NADH and 120 μmoles of N-tris-(hydroxymethyl)-2-aminomethane sulfonic acid (TES) buffer at pH 7.5 and bacteroid nitrogenase (6.8 mg protein). The gas phase was composed of 0.1 atm. acetylene and 0.9 atm. A. Reactions without β-hydroxybutyrate or the ATP-generating system or without active nitrogenase produced little or no ethylene (after Klucas and Evans, 1968).

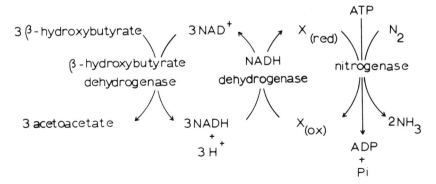

Fig. 6.12. A scheme of electron transport from β-hydroxybutyrate to nitrogenase. In most of the experiments conducted benzylviologen has been utilized in place of unknown 'X'.

use of a different type of assay procedure. This will be discussed in he following section (6.6.2.2).

   6.6.2.2   Coupling to photosystem I   Recently two factors have been shown (Benemann *et al.*, 1969; Yoch *et al.*, 1969) to function in the light-dependent transfer of electrons from photosystem I of spinach chloroplast fragments to *Azotobacter* nitrogenase. One of these apparently is a flavo-protein and has been referred to as 'azotoflavin', and the other is a non-heme iron protein (see p. 126) with properties similar to ferredoxin. In the presence of ascorbate, 2,6-dichlorophenolindophenol (DCIP), an ATP-generating system, active nitrogenase, and chloroplast fragments, either factor catalyzed the light-dependent reduction of acetylene to ethylene. As a consequence of these discoveries, Yoch *et al.* (1970) conducted experiments with an extract of *G. max* nodule bacteroids from our laboratory, and reported evidence for a factor in the extract that catalyzed the light-dependent transfer of electrons from photosystem I to *Azotobacter* nitrogenase. The addition of the bacteroid factor to a reaction containing ascorbate, DCIP, photosystem I, an ATP-generating system, and crude bacteroid nitrogenase resulted in limited stimulation of the photochemical reduction of acetylene to ethylene.

   Recently Koch *et al.* (1970) have purified and characterized a non-heme iron protein from *G. max* nodule bacteroids that functions effectively in the light-dependent transfer of electrons from ascor-

TABLE 6.7.   Essential components for the photochemical reduction of acetylene.

The complete reaction mixture in a final volume of 1.5 ml contained: a photochemical reducing system (consisting of chloroplast fragments, 2,6-dichlorophenolindophenol (DCIP) and ascorbate); bacteroid nitrogenase; an ATP-generating system, and buffer. The specific activity of the purified bacteroid non-heme iron protein (0.13 mg protein) was 130 (after Koch *et al.*, 1970).

Reaction system	Ethylene produced (*n*moles/20 min)
Complete	
Complete except chloroplasts, DCIP and ascorbate	336.0
omitted	0.4
Complete except nitrogenase omitted	0.3
Complete except ATP-generating system omitted	0.4
Complete except bacteroid non-heme iron protein	
omitted	2.9
Complete except incubated in dark	1.0

bate *via* photosystem I to a purified bacteroid nitrogenase. Both the non-heme iron protein and the nitrogenase utilized in the assay system were purified sufficiently to convincingly establish the necessary components of the system (Table 6.7). The sequence of electron transport is illustrated in Fig. 6.13. The purified non-heme

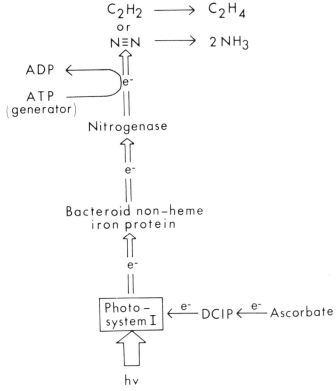

Fig. 6.13. An electron transport scheme from ascorbate, via photosystem I, to nitrogenase. The assay for non-heme iron protein is based on this series of reactions.

iron protein contained 0.31 μatoms of non-heme iron and 0.35 μatoms of acid-labile sulfur per mg of protein. From sedimentation velocity measurements an S_{20w} value of 1.3 was calculated, and by use of assumed values for the partial specific volume and diffusion coefficient of the protein, a tentative molecular weight of 9400 was estimated. Although the non-heme iron protein functioned efficiently in the transfer of electrons from photosystem I to nitrogenase, it has proven to be ineffective as an electron transport factor in the phosphoroclastic breakdown of pyruvate. The bacteroid non-heme iron protein also failed to function in the photoreduction of NADP in presence of spinach chloroplasts. On the basis of this behaviour the nodule factor is different from the ferredoxin isolated from *Azotobacter*, (pages 126-7) and has been referred to as a non-heme iron protein rather than a ferredoxin.

The non-heme iron protein from bacteroids is relatively labile and loses activity as a result of repeated freezing and thawing. During such operations the normal greenish-brown color of the protein faded, acid-labile sulfur content decreased, and the activity of the protein in the photochemical acetylene reduction assay also decreased. Some of the lost activity was restored by treating the preparation with sodium sulfide, mercaptoethanol and ferric ammonium sulfate according to the procedure described by Malkin and Rabinowitz (1966).

A large number of experiments have been conducted to determine whether or not the non-heme iron protein will function as a carrier in the transfer of electrons from NADH to nitrogenase to reduce acetylene to ethylene. In the experiments conducted to date, results have not been definitive because some of the extracts contain more than one component and purification has resulted in loss of activity. Further research is essential, therefore, before details of the sequence of electron transport of nitrogenase are established.

6.7   CONCLUSIONS

Many aspects of our present understanding of the biochemistry of symbiotic N_2 function are diagrammatically illustrated in Fig. 6.14. A similar scheme with fewer details regarding postulated and established electron transport pathways has been presented by Bergersen (1969). The usual assay for nitrogenase is outlined in which $Na_2S_2O_4$ and an artificial ATP-generating system are utilized as sources of energy for the nitrogenase reaction. Creatine phosphate and creatine phosphokinase have been utilized as an ATP-generating system in the usual assay, and also in experiments where the reductant was supplied as (a) NADH via benzylviologen, (b) NADH via flavin nucleotide and protein factors, (c) as electrons from reactions involving photosystem I, or (d) from H_2 through *C. pasteurianum,* hydrogenase and ferredoxin. Although the provision of reductant to nitrogenase through photosystem I has no *in vivo* significance *per se,* it has proven to be an exceedingly useful tool for the assay of the non-heme iron protein and flavoprotein (azotoflavin) from *Azotobacter* and for the assay of similar protein factors from nodule bacteroids. It seems probable that the non-heme iron protein and the bacteroid flavoprotein described by Koch *et al.* (1970) may participate in the transfer of electrons from NADH, to nitrogenase. All efforts to demonstrate requirements for these factors in coupled reactions so far have produced inconclusive results.

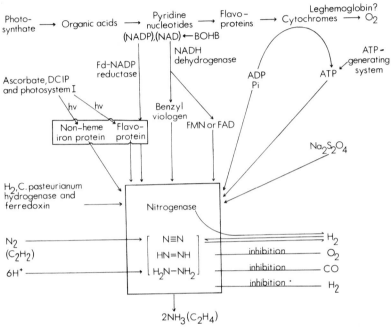

Fig. 6.14. A schematic representation of N_2 fixation and related processes in legume-root nodules. Reactions demonstrated to occur by use of cell-free systems are indicated by solid lines. An exception is oxidative phosphorylation coupled to electron transport which is based on inhibitor evidence (Bergersen and Turner, 1967). The arrow from NADP to non-heme iron protein and flavoprotein (both enclosed by a solid line) indicates that both components are essential for electron transport from NADP to nitrogenase.

It seems obvious that photosynthate from the legume component of the symbiotic association provides the substrates required by nodule bacteroids for the generation of the necessary ATP and reductant needed for N_2 fixation. The carbon skeletons that are essential for the synthesis of amides and amino acids by N_2 fixing bacteroids also must be derived from the utilization of sucrose or other photosynthate. Since bacteroids or intact nodules require O_2 for N_2 fixation, it seems highly probable that the *in vivo* supply of ATP for N_2 fixation must be derived from oxidative phosphorylation rather than glycolysis. The possibility exists however, that factors other than the supply of ATP limits N_2 fixation under anaerobic conditions. In view of the facts that the N_2 fixation reaction is anaerobic and purified nitrogenase is exceedingly O_2 labile, it has not been possible to conduct *in vitro* experiments in which the components required for both N_2 reduction and oxidative phosphorylation were included in the same reaction. Perhaps O_2 tenaciously bound to leghemoglobin is transported to the respiratory apparatus under *in vivo* conditions in a manner that does not interfere with the O_2 sensitive nitrogenase system.

In the lower part of the diagram, the scheme represents N_2 bound to nitrogenase and shows reduction occurring in two-electron steps. Bound intermediates at the oxidation-reduction stage of diimide and hydrazine are postulated. Although there is no direct evidence for these intermediates, alternate nitrogenase substrates such as acetylene and cyanide apparently are reduced in two-electron steps by the nitrogenase from free-living microorganisms (Hardy and Burns, 1968) and, therefore, an analogous stepwise reduction of N_2 by bacteroid nitrogenase seems reasonable. Also, as illustrated in the diagram, is the capacity of nitrogenase to catalyze the ATP-dependent transfer of electrons from reductant to hydrogen ions to form H_2, a reaction that proceeds at a rapid rate when nitrogenase acceptor substrates are absent. The susceptibility of nitrogenase to inhibition by O_2, CO and H_2 and the possibility of H_2 exchange with the hydrogen on postulated bound intermediates of N_2 reduction are shown.

Despite the progress that has been made many aspects of symbiotic N_2 fixation are not understood. For example, what is the role of ATP in the N_2 reduction reaction? What are the roles of iron and molybdenum in the binding and reduction of N_2? How many, and what kind of factors are involved in the flow of electron from dehydrogenases to nitrogenase, and what determines whether electrons are transferred to O_2 through the electron transport chain, or to N_2 via the electron transport sequence to nitrogenase? What controls the synthesis of bacteroid nitrogenase in the relatively anaerobic environment of the nodule, and why have all efforts to induce cultured rhizobia to synthesize the enzyme failed? What role, if any, is played by leguminous plants in provision of genetic information for the synthesis of components of the nitrogenase system? Although our understanding of symbiotic nitrogen fixation is far from complete, considerable progress has been made in the elucidation of the biochemistry of symbiotic N_2 fixation, and it seems highly probable that many of our questions will be clarified as research continues.

REFERENCES

K. ABEL, *Phytochemistry*, **2**, 429 (1963).

K. ABEL and N. BAUER, *Arch. Biochem. Biophys.*, **99**, 8 (1962).

K. ABEL, N. BAUER and J. T. SPENCE, *Arch. Biochem. Biophys.*, **100**, 339 (1963).

S. AHMED and H. J. EVANS, *Proc. Natl. Acad. Sci. U.S.A.*, **47**, 24 (1961).

O. N. ALLEN and E. K. ALLEN, *Bot. Gaz.*, **102**, 121 (1940).

E. K. ALLEN and O. N. ALLEN, *Proc. Soil Sci. Soc. Amer.*, **14**, 179 (1949).

L. ALMON, *Zentr. Bakteriol. Parasitenk. Abt. II*, **87**, 289 (1933).

A. J. ANDERSON, *J. Aust. Inst. Agr. Sci.*, **8**, 73 (1942).

A. J. ANDERSON, Inorganic Nitrogen Metabolism. (W. D. McElroy and B. Glass, eds.), p. 3. Johns Hopkins Press, Baltimore (1956).

C. A. APPLEBY, *Biochim. Biophys. Acta*, **60**, 226 (1961).

C. A. APPLEBY, *Biochim. Biophys. Acta.*, **189**, 267 (1969a).

C. A. APPLEBY, *Biochim. Biophys. Acta.*, **172**, 88 (1969b).

C. A. APPLEBY, *Biochim. Biophys. Acta.*, **180**, 202 (1969c).

C. A. APPLEBY, *Biochim. Biophys. Acta.*, **188**, 222 (1969d).

C. A. APPLEBY, *Biochim. Biophys. Acta.*, **172**, 71 (1969e).

C. A. APPLEBY and F. J. BERGERSEN, *Nature*, **182**, 1174 (1958).

M. H. APRISON, W. E. McGEE and R. H. BURRIS, *J. Biol. Chem.*, **208**, 29 (1954).

N. BAUER, *Nature*, **188**, 471 (1960a).

N. BAUER, *J. Phys. Chem.*, **7**, 833 (1960b).

N. BAUER and R. G. MORTIMER, *Biochim. Biophys. Acta.*, **40**, 170 (1960).

M. W. BEIJERINCK, *Botan. Ztg.*, **46**, 725 (1888).

J. R. BENEMANN, D. C. YOCH, R. C. VALENTINE and D. I. ARNON, *Proc. Natl. Acad. Sci. U.S.A.*, **64**, 1079 (1969).

F. J. BERGERSEN, *Bacteriol Rev.*, **24**, 246 (1960a).

F. J. BERGERSEN, *J. Gen. Microbiol.*, **22**, 671 (1960b).

F. J. BERGERSEN, *Nature*, **194**, 1059 (1962a).

F. J. BERGERSEN, *J. Gen. Microbiol.*, **29**, 113 (1962b).

F. J. BERGERSEN, *Aust. J. Biol. Sci.*, **16**, 669 (1963a).

F. J. BERGERSEN, *Aust. J. Biol. Sci.*, **16**, 916 (1963b).

F. J. BERGERSEN, *Aust. J. Biol. Sci.*, **18**, 1 (1965).

F. J. BERGERSEN, *Biochim. Biophys. Acta.*, **130**, 304 (1967).

F. J. BERGERSEN, *Proc. Roy. Soc.*, **172**, 401 (1969).

F. J. BERGERSEN and M. J. BRIGGS, *J. Gen. Microbiol.*, **19**, 482 (1958).

F. J. BERGERSEN and G. L. TURNER, *Biochim. Biophys. Acta.*, **141**, 507 (1967).

F. J. BERGERSEN and G. L. TURNER, *J. Gen. Microbiol.*, **53**, 205 (1968).

F. J. BERGERSEN and G. L. TURNER, *Biochim. Biophys. Acta.*, **214**, 28 (1970).

F. J. BERGERSEN and P. W. WILSON, *Proc. Natl. Acad. Sci. U.S.A.*, **45**, 1641 (1959).

R. L. BLAKLEY, *J. Biol. Chem.*, **240**, 2173 (1965).

E. W. BOLLE-JONES and V. R. MALLIKARJUNESWARA, *Nature*, **179**, 738 (1957).

H. BORTELS, *Arch. Mikrobiol.*, **1**, 333 (1930).

H. BORTELS, *Arch. Mikrobiol.*, **8**, 13 (1937).

J. B. BOUSSINGAULT, *Ann. Chim. et. Phys.*, *2nd Ser.*, **67**, 5 (1838).

W. BUDDHARI, Ph.D. dissertation. Cobalt as an Essential Element for Blue-Green Algae. Univ. of Calif., Berkeley. (1960).

W. A. BULEN and J. R. LeCOMTE, *Proc. Natl. Acad. Sci. U.S.A.*, **56**, 979 (1966).

W. A. BULEN, R. C. BURNS and J. R. LeCOMTE, *Proc. Natl. Acad. Sci. U.S.A.*, **53**, 532 (1965).

R. H. BURRIS, Inorganic Nitrogen Metabolism (W. D. McElroy and Bently Glass, eds.) p. 316. Johns Hopkins Press, Baltimore (1956).

R. H. BURRIS, Plant Biochemistry (J. Bonner and J. E. Varner, eds.), p. 961. Academic Press, New York (1965).

R. H. BURRIS, *Ann. Rev. Plant Physiol.*, **17**, 155 (1966).

R. H. BURRIS, *Proc. Roy. Soc.*, **172**, 339 (1969).

R. H. BURRIS and P. W. WILSON, *Biochem. J.*, **51**, 90 (1952).

R. H. BURRIS, W. E. MAGEE and M. K. BACH, *Ann. Acad. Sci. Fenn. Ser.*, AII, **60**, 190 (1955).

JAMES E. CARNAHAN and JOHN E. CASTLE, *Ann. Rev. Plant Physiol.*, **14**, 125 (1963).

G. M. CHENIAE and H. J. EVANS, *Biochim. Biophys. Acta.*, **26**, 654 (1957).

G. M. CHENIAE and H. J. EVANS, *Plant Physiol.*, **35**, 454 (1960).

J. R. COWLES and H. J. EVANS, *Arch. Biochem. Biophys.*, **127**, 770 (1968).

J. R. COWLES, H. J. EVANS and S. A. RUSSELL, *J. Bacteriol.*, **97**, 1460 (1969).

J. A. CUTTING and H. M. SCHULMAN, *Fed. Proc. Fed. Amer. Soc. Exp. Biol.*, **27**, 3100 (1968).

J. A. CUTTING and H. M. SCHULMAN, *Biochim. Biophys. Acta.*, **192**, 486 (1969).

P. J. DART, Proc. 4th European Reg. Conf. Electr. Micro., p. 69 (1968).

P. J. DART and F. V. MERCER, *Arch. Mikrobiol.*, **47**, 334 (1964).

C. C. DELWICHE, C. M. JOHNSON and H. M. REISENAUER, *Plant Physiol.*, **36**, 73 (1961).

A. A. DeHERTOGH, P. A. MAYEUX and H. J. EVANS, *J. Bacteriol.*, **87**, 746 (1964a).

A. A. DeHERTOGH, P. A. MAYEUX and H. J. EVANS, *J. Biol. Chem.*, **239**, 2446 (1964b).

M. J. DILWORTH, *Biochim. Biophys. Acta.*, **184**, 432 (1969).

M. J. DILWORTH, D. SUBRAMANIAN, T. O. MUNSON and R. H. BURRIS, *Biochim. Biophys. Acta.*, **99**, 486 (1965).

R. O. D. DIXON, *Arch. Mikrobiol.*, **62**, 272 (1968).

R. O. D. DIXON, *Ann. Rev. Microbiol.*, **23**, 137 (1969).

C. M. DONALD, *J. Aust. Inst. Agr. Sci.*, **26**, 319 (1960).

A. EHRENBERG and N. ELLFOLK, *Acta. Chem. Scand.*, **17**, 343 (1963).

N. ELLFOLK, *Acta. Chem. Scand.*, **14**, 1819 (1960).

N. ELLFOLK, *Acta. Chem. Scand.*, **15**, 545 (1961).

N. ELLFOLK, *Acta. Chem. Scand.*, **16**, 831 (1962).

N. ELLFOLK and G. SIEVERS, *Acta. Chem. Scand.*, **19**, 268 (1965).

L. W. ERDMAN, Farmers' Bull. 2003. U.S. Department of Agri., p. 1 (1948).

H. J. EVANS, How Crops Grow—A Century Later. (James G. Horsfall, ed.), p. 110. *Conn. Agr. Exp. Sta. Bull.* 708 (1970).

H. J. EVANS and N. S. HALL, *Science*, **122**, 922 (1955).

H. J. EVANS and S. RUSSELL, *Plant Physiol. Suppl.*, **40**, iii (1965).

H. J. EVANS, B. KOCH and R. KLUCAS, Methods in Enzymology (Anthony San Pietro ed.). Academic Press, New York (1970).

H. J. EVANS, E. R. PURVIS and F. E. BEAR, *Plant Physiol.*, **25**, 555 (1950).

H. J. EVANS, E. R. PURVIS and F. E. BEAR, *Soil Sci.*, **71**, 117 (1951).

G. FAHRAEUS and H. LJUNGGREN, The Ecology of Soil Bacteria. (T. R. G. Gray and D. Parkinson, ed.) Liverpool University Press (1968).

J. E. FALK, C. A. APPLEBY and R. J. PORRA, Sym. Soc. Exp. Biol. (H. K. Porter, ed.), *13*, p. 73. Academic Press, New York (1959).

E. B. FRED, I. L. BALDWIN and E. McCOY, Root Nodule Bacteria and Leguminous Plants. University of Wisconsin, Madison (1932).

D. J. GOODCHILD and F. J. BERGERSEN, *J. Bacteriol.*, **92**, 204 (1966).

P. H. GRAHAM and G. A. PARKER, *Aust. J. Sci.*, **23**, 231 (1961).

A. HAAK, *Zentr. Bakteriol. Parasitenk. Abt. II*, **117**, 343 (1964).

E. G. HALLSWORTH, Nutrition of Legumes (E. G. Hallsworth, ed.), p. 183. Butterworths Scientific Publications, London (1958).

E. G. HALLSWORTH, S. B. WILSON and E. A. N. GREENWOOD, *Nature*, **187**, 79 (1960).

P. B. HAMILTON, A. L. SHUG and P. W. WILSON, *Proc. Natl. Acad. Sci. U.S.A.*, **43**, 297 (1957).

R. W. F. HARDY and R. C. BURNS, *Ann. Rev. Biochem.*, **37**, 331 (1968).

R. W. F. HARDY, R. D. HOLSTEN, E. K. JACKSON and R. E. BURNS, *Plant Physiol.*, **43**, 1185 (1968).

A. C. HAYWARD, W. G. C. FORSYTH and J. B. ROBERTS, *J. Gen. Microbiol.*, **20**, 510 (1959).

H. HELLREIGEL and H. WILFARTH, Beilage. zu der Ztschr. Ver. Rubenzucker-Ind. dtsch. Reichs. (1888).

E. HEMMINGSEN, *Science*, **135**, 733 (1962).

E. HEMMINGSEN and P. F. SCHOLANDER, *Science*, **132**, 1379 (1960).

P. H. HERNANDEZ, P. MAZEL, J. R. GILLETTE, *Biochem. Pharmacol.*, **16**, 1877 (1967).

E. J. HEWITT, Nutrition of Legumes (E. G. Hallsworth, ed.), p. 15. Butterworths Scientific Publications, London (1958).

G. E. HOCH, H. N. LITTLE and R. H. BURRIS, *Nature*, **179**, 430 (1957).

G. E. HOCH, K. C. SCHNEIDER and R. H. BURRIS, *Biochim. Biophys. Acta.*, **37**, 273 (1960).

O. HOLM-HANSEN, G. C. GERLOFF and F. SKOOG, *Physiol. Plant.*, **7**, 665 (1954).

E. K. JACKSON and H. J. EVANS, *Plant Physiol.*, **41**, 1673 (1966).

E. K. JACKSON, G. W. PARSHALL and R. W. F. HARDY, *J. Biol. Chem.*, **243**, 4952 (1968).

D. C. JORDAN and E. H. GARRARD, *Can. J. Bot.*, **29**, 360 (1951).

N. P. KEFFORD, J. BROCKWELL and J. A. ZWAR, *Aust. J. Biol. Sci.*, **13**, 456 (1960).

D. KEILEN and Y. L. WANG, *Nature*, **155**, 227 (1945).

I. R. KENNEDY, Ph.D. dissertation. Primary Products of Symbiotic Nitrogen Fixation: A Study of the Rate of [15]N Distribution, and of Some Transformation Mechanisms. University of Western Australia, Perth (1965).

I. R. KENNEDY, C. A. PARKER and D. K. KIDBY, *Biochim. Biophys. Acta.*, **130**, 517 (1966).

P. A. KETCHUM, H. Y. CAMBIER, W. A. FRAZIER III, C. MADANSKY and A. NASON, *Proc. Natl. Acad. Sci. U.S.A.* (In press) (1970).

M. KLIEWER and H. J. EVANS, *Plant Physiol.*, **38**, 55 (1963a).

M. KLIEWER and H. J. EVANS, *Plant Physiol.*, **38**, 99 (1963b).

M. KLIEWER and H. J. EVANS, *Arch. Biochem. Biophys.*, **97**, 428 (1962).

R. V. KLUCAS and R. H. BURRIS, *Biochim. Biophys. Acta.*, **136**, 399 (1967).

R. V. KLUCAS and H. J. EVANS, *Plant Physiol.*, **43**, 1458 (1968).

R. V. KLUCAS, B. KOCH, S. A. RUSSELL and H. J. EVANS, *Plant Physiol.*, **43**, 1906 (1968).

B. L. KOCH and H. J. EVANS, *Plant Physiol.*, **41**, 1748 (1966).

B. KOCH, H. J. EVANS and S. RUSSELL, *Proc. Natl. Acad. Sci. U.S.A.*, **58**, 1343 (1967a).

B. KOCH, H. J. EVANS and S. RUSSELL, *Plant Physiol.*, **42**, 466 (1967b).

B. KOCH, P. WONG, S. A. RUSSELL, R. HOWARD and H. J. EVANS, *Biochem. J.* (In press) (1970).

H. KUBO, *Acta Phytochim (Japan)*, **11**, 195 (1939).

A. P. LEVIN, H. B. FUNK and M. D. TENDLER, *Science*, **120**, 784 (1954).

T. T. LILLICH and G. H. ELKAN, *Can. J. Microbiol.*, **14**, 617 (1968).

H. N. LITTLE and R. H. BURRIS, *J. Amer. Chem. Soc.*, **69**, 838 (1947).

W. D. LOOMIS and J. BATTAILE, *Phytochemistry*, **5**, 423 (1966).

R. H. LOWE and H. J. EVANS, *J. Bacteriol.*, **83**, 210 (1962).

R. H. LOWE and H. J. EVANS, *Soil Sci.*, **94**, 351 (1962).

R. H. LOWE, H. J. EVANS and S. AHMED, *Biochem. Biophys. Res. Commun.*, **3**, 675 (1960).

W. L. LOWTHER and J. F. LONERAGEN, *Plant Physiol.*, **43**, 1362 (1968).

R. MALKIN and J. C. RABINOWITZ, *Biochem. Biophys. Res. Comm.*, **23**, 822 (1966).

R. MALKIN and J. C. RABINOWITZ, *Annu. Rev. Biochem.*, **36**, 113 (1967).

M. MALPIGHI, Anatome plantarum. London (1675).

H. S. MASON, *Ann. Rev. Biochem.*, **34**, 595 (1965).

E. G. MULDER, *Plant Soil*, **1**, 94 (1948).

E. G. MULDER and W. L. van VEEN, *Plant Soil*, **13**, 265 (1960).

D. N. MUNNS, *Plant Soil*, **28**, 129 (1968).

C. W. NAGEL and M. M. ANDERSON, *Arch. Biochem. Biophys.*, **112**, 322 (1965).

A. NASON, A. D. ANTOINE, P. A. KETCHUM, W. A. FRAZIER III and D. K. LEE, *Proc. Natl. Acad. Sci. U.S.A.*, **65**, 137 (1970).

P. S. NUTMAN, *Biol. Rev. Cambridge Phil. Soc.*, **31**, 109 (1956).

P. S. NUTMAN, *Symp. Soc. Gen. Microbiol.* (P. S. Nutman and B. Mosse, eds.), **13**, p. 51. Cambridge University Press (1963).

P. S. NUTMAN, *Proc. Roy. Soc.*, **172**, 417 (1969).

J. S. PATE, B. E. S. GUNNING and L. G. BRIARTY, *Planta*, **85**, 11 (1969).

J. E. RICHMOND and K. SOLOMON, *Biochim. Biophys. Acta.*, **17**, 48 (1955).

R. SCHAEDE, *Planta*, **31**, 1 (1940).

E. A. SCHWINGHAMER, H. J. EVANS and M. D. DAWSON, *Plant Soil.* (In press) (1970).

D. SHEMIN and C. S. RUSSELL, *J. Amer. Chem. Soc.*, **75**, 4873 (1953).

A. L. SHUG, P. B. HAMILTON and P. W. WILSON. Inorganic Nitrogen Metabolism (W. D. McEroy and B. Glass, eds.), p. 344. Johns Hopkins Press, Baltimore (1956).

J. D. SMITH, *Biochem. J.*, **44**, 585 (1949a).

J. D. SMITH, *Biochem. J.*, **44**, 591 (1949b).

J. I. SPRENT, *Planta (Berl.)*, **88**, 372 (1969).

M. P. STARR and F. MORAN, *Science*, **135**, 920 (1962).

W. D. P. STEWART, Nitrogen Fixation in Plants. University of London, The Athone Press (1966).

P. R. STOUT and W. R. MEAGHER, *Science*, **108**, 471 (1948).

D. W, THORNE and R. H. BURRIS, *J. Bacteriol.*, **39**, 187 (1940).

H. G. THORNTON and H. NICOL, *J. Agri. Sci.*, **26**, 173 (1936).

G. L. TURNER and F. J. BERGERSEN, *Biochem. J.*, **115**, 529 (1969).

K. TUZIMURA and H. MEGURO, *J. Biochem.*, **47**, 391 (1960).

K. TUZIMURA and I. WATANABE, *Plant Cell Physiol.*, **5**, 157 (1964).

A. I. VIRTANEN, *Biol. Rev. Cambridge Phil. Soc.*, **22**, 239 (1947).

A. I. VIRTANEN, *Nature*, **155**, 747 (1945).

A. I. VIRTANEN and J. K. MIETTINEN, Plant Physiology. (F. C. Steward, ed.), III, p. 539. Academic Press, New York (1963).

A. I. VIRTANEN, J. ERKAMA and H. LINKOLA, *Acta. Chem. Scand.*, **1**, 861 (1947).

P. W. WILSON, The Biochemistry of Symbiotic Nitrogen Fixation. University of Wisconsin, Madison (1940).

S. B. WILSON and D. J. D. NICHOLAS' *Phytochemistry*, **6**, 1507 (1967).

P. W. WILSON, W. W. UMBREIT and S. B. LEE, *Arch. Mikrobiol.*, **8**, 440 (1937).

P. W. WILSON, W. W. UMBREIT and S. B. LEE, *Biochem. J.*, **32**, 2084 (1938).

P. P. WONG and H. J. EVANS, *Plant Physiol. Suppl.*, **44**, 35 (1969).

M. G. YATES and E. G. HALLSWORTH, *Plant Soil*, **19**, 284 (1963).

D. C. YOCH, J. R. BENEMANN, R. C. VALENTINE and D. I. ARNON, *Proc. Natl. Acad. Sci. U.S.A.*, **64**, 1404 (1969).

D. C. YOCH, J. R. BENEMANN, R. C. VALENTINE, D. I. ARNON and S. A. RUSSELL, *Biochem. Biophys. Res. Commun.*, **38**, 838 (1970).

C. S. YOCUM, *Science*, **146**, 432 (1964).

ADDENDUM

Recent investigations have provided evidence that NADPH may be more important as a physiological electron donor to nitrogenase than NADH. Benneman *et a..*, 1970 (personal communication from Dr. D. C. Yoch) have demonstrated that NADPH-dependent acetylene reduction by *Azotobacter* nitrogenase requires azotoflavin and *Azotobacter ferredoxin* (both from *Azotobacter vinelandii*) and ferredoxin-NADP reductase from spinach chloroplasts. Wong *et al.* (1970) have utilized an electron donor system for bacteroid nitrogenase that consisted of glucose-6-phosphate, glucose-6-phosphate dehydrogenase, NADP, purified azotoflavin (kindly supplied by Dr. D. C. Yoch) non-heme iron protein from soybean nodule bacteroids (Koch *et al.*, 1970), and a homogenous preparation of ferredoxin-NADP reductase from spinach chloroplasts. The addition of these components to a partially purified bacteroid nitrogenase, and an ATP-generating system resulted in consistent and reproducible nitrogenase-dependent acetylene reduction. Koch *et al.* (1970) have identified a flavoprotein in bacteroid extracts that appears to be analogous to azotoflavin from *Azotobacter* (Benneman *et al.*, 1969) and Mr. Danny Israel from our laboratory recently has discovered an active NADPH dehydrogenase in bacteroid extracts that very likely may function in the coupled nitrogenase system in place of ferredoxin-NADP reductase from spinach. It would appear, therefore (Fig. 6.14), that the most likely physiologically important electron donor system for bacteroid nitrogenase involves a sequence of electron flow from NADP to either bacteroid non-heme iron protein or to a flavoprotein analogous to azotoflavin (both proteins are essential but the sequence of electron acceptance is not known). Recent experiments by Wong *et al.* (1970) suggest that electron transport from NAD to bacteroid nitrogenase *via* flavin nucleotides or dyes such as benzyl viologen has little or no physiological significance. Recent experiments also provide no support for an important physiological role of β-hydroxybutyrate as an energy source for maintenance of nitrogenase activity.

REFERENCES

F. J. BERGERSEN and G. L. TURNER, *Biochim. Biophys. Acta*, **141**, 507 (1967).

J. R. BENNEMAN, D. C. YOCH, R. C. VALENTINE and D. I. ARNON, *Biochim. Biophys. Acta*, **226**, 205 (1971).

B. KOCH, P. WONG, S. A. RUSSELL, R. HOWARD and H. J. EVANS, *Biochem. J.*, **118**, 773 (1970).

P. WONG, H. J. EVANS, R. KLUCAS and S. RUSSELL, *Plant and Soil.* (In press) (1970).

CHAPTER 7

Physiological Chemistry of Non-Leguminous Symbioses

W. S. SILVER

Department of Botany and Bacteriology
University of South Florida
Tampa, Florida 33620 U.S.A.

7.1. INTRODUCTION

A general lack of familiarity of biochemists and microbiologists with
non-cultivated plants coupled with interest in the leguminous
symbioses, which were of obvious agronomic importance, has
delayed study of the non-leguminous plants from the viewpoint of
the physiology and biochemistry of nitrogen fixation, the physiological basis of the symbioses, and their ecological importance.
Although the non-leguminous symbioses were discussed in the
monograph of Fred *et al.*, (1932), it was not until the pioneering
work of Bond and his associates (much of this recently reviewed by
Bond 1967a, 1968), that the attention of the scientific community
was drawn to these diverse and important organisms. Other reviews
have appeared (for example, Lange, 1966; Allen and Allen, 1957)
but the emphasis has been on the biological aspects of the symbioses.
In this review the emphasis will be related to the biochemistry of
nitrogen fixation; however, due to a scarcity of data of a funda-
mentally biochemical nature, much biological information must be
included to present a unified picture of the current state of our
knowledge.

7.2 BIOLOGICAL ASPECTS OF NON-LEGUMINOUS SYMBIOSES

It is obvious that, since bacteria and other microorganisms are
ubiquitous in our biosphere, they would exist in close association
with a variety of plants. These associations may either involve little
specificity between host and parasite, termed here 'loose' (as in the
rhizosphere, phyllosphere and ectotrophic mycorrhiza), or involve
considerable specificity, termed here 'tight' (as in the root nodule
and foliar symbioses and endotrophic mycorrhiza), depending upon
the habitat and the length of time the microbe and plant have been

subjected to the forces of evolution. In this discussion it is primarily the latter type which will be considered. A useful, albeit somewhat arbitrary, subdivision of the latter distinguishes four types: (1) 'true' root nodule symbiosis, (2) superficial root-nodule symbiosis, (3) foliar or leaf-nodule symbiosis and (4) a catch-all for those noted in non-seed bearing plants: lower plant symbioses. The leguminous symbiosis is treated in Chapter 6 and will not be considered here.

7.2.1 'True' root-nodule symbiosis in non-leguminous plants

The genera of Angiosperms shown to possess nodulated roots, which presumably function in nitrogen fixation, have been tabulated by Bond (1967a). An adaptation of this is presented in Table 7.1. All those listed are woody dicots and that approximately one-third of these widely distributed plants are nodulated. It is not unlikely that a more thorough search for additional nodulated species will extend this list, and that the percentage of nodulated species within a genus is more a measure of how extensive the search has been.

Are there any common biological properties associated with these symbioses which permit generalizations? Only *Myrica* and *Alnus* have been studied in any detail so it is difficult to make generalizations which apply to all species. However, these two may be assumed to be typical of mutualism in the root nodules of non-legumes.

The nature of the endophyte can be inferred from indirect evidence, for authenticated isolations of pure cultures have not been

TABLE 7.1. Nodulated genera of non-leguminous angiosperms forming root nodules. Incidence refers to the ratio of species bearing nodules to the total number of known species.

GENUS	INCIDENCE
Alnus	25/35
Arctostaphylos	1/40
Casuarina	14/45
Ceanothus	30/55
Cercocarpus	1/20
Coriaria	12/15
Discaria	1/10
Dryas	3/4
Elaeagnus	9/45
Hippophae	1/1
Myrica	12/35
Purshia	2/2
Shepherdia	2/3

made. Both light microscopy and electron micrographs of infected roots of *Myrica gale* (Fletcher, 1957), *Myrica cerifera* (Silver, 1964), and *Alnus glutinosa* (Becking, *et al.*, 1964) reveal filamentous structures characteristic of 'actinomycetes'. Essentially, only unconfirmed sporadic reports of successful isolations have appeared for the alder endophyte (see Bond, 1967*a*, for references and for further discussion). The best evidence that it is an actinomycete is the morphology *in situ* (Becking *et al.*, 1964; Becking, 1970*a*). Becking has noted three morphological forms of the endophyte: 0.5 μm hyphae, larger (3-4 μm) vesicles and 0.5-1.0 μm bacteria-like cells which packed the host cell (Becking, 1970*a*). Since the hyphae contained mesosomes and lacked a nuclear membrane, the endophyte cannot be fungal (i.e. eucaryotic). Becking was unable to isolate the alder symbiont although he did successfully propagate it in root nodule callus tissue in a non-infective form. (This experiment is analogous to that of Holsten *et al.*, (1970) who were able to establish symbiosis *in vitro* for the *Rhizobium*-soybean system.) The endophyte in nodules of *Myrica* also has a hyphal appearance (Fig. 7.1) but no obvious vesicles (Silver, 1964), while the endophyte of *Hippophae* appears similar to that of the alder (Gatner and Gardner, 1970). Becking has noted fine structure similar to that of *Alnus* for *Casuarina equisetifolia* and *Coriaria myrtifolia* and, primarily on the basis of fine structure, has bravely ventured to classify all the non-legume root nodules endophytes as species of a new genus called *Frankia* (Becking, 1970*b*).

An actinomycete capable of inducing nodulation of seedlings of *Casuarina equisetifolia* has been isolated recently by K. L. Tyson (personal communication) from nodules of trees growing in southern Florida. It induced nodulation on sterile seedlings, which then grew vigorously in an N-free medium compared with uninoculated controls. Unfortunately, determinations of nitrogen content were not made. This must be independently verified in order to be accepted, as suggested by Bond (1967*a*), and for the present one must conclude that positive identifications of endophytes have yet to be made. Until pure cultures are available, the possibility that the non-leguminous endophytes may fix N_2 in the non-symbiotic condition cannot be ascertained and classification schemes must remain tenuous.

There is some degree of specificity between symbiont and host plant. A summary of these reports (Table 7.2) reveals that although cross inoculation may induce nodulation in *Myrica*, for example, such nodules are rather inactive in N_2 fixation, indicating host

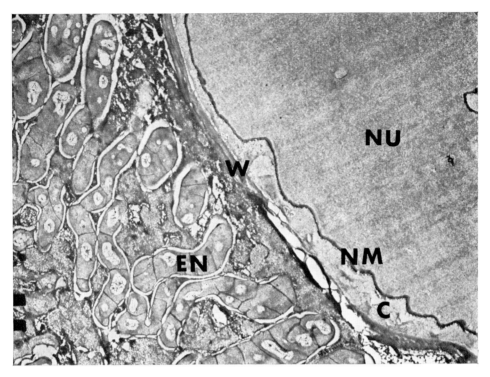

Figure 7.1 Section through root nodule of *Myrica cerifera*.
Experimental conditions: fixation, 2% permanganate; embedded with methacrylate; 0.1 μm thick section cut with a Porter–Blum Microtome; Phillips Model EM 100 electron microscope. The marker represents 1 μm.

NU = nucleus, NM = nuclear membrane, W = plant cell wall, EN = endophyted, C = cytoplasm.

(*To face page 248*)

TABLE 7.2. Effect of cross-inoculation on nodulation and nitrogen fixation in various genera of nodulating angiosperms.

Plant genus	Species of cross inoculation test	Compatability nodulation	N_2 fixation	Reference	
Alnus	*A. glutinosa* and *A. rubra*	poor	poor	Becking	1966*b*
	A. glutinosa and *A. jorullensis*	poor	poor	Rodriquez-Barrueco	1966
Casuarina	All combinations tested	complete	complete	Mowry Bond Hewitt and Bond	1933 1957 1961
Ceanothus	*A. velutinus* and *A. velutinus*	poor unless habitat soil is employed		Bond	1967
Cercocarpus	Not tested, only species, *C. betuloides*, reported	- - - - - - - - - - - - - - - -		Vlamis *et al.*,	1958
Coriaria	*C. myrtifolia* and *C. japonica*	None—i.e. host adaptation marked, habitat		Bond	1962
	Seven species	soil required		Allen *et al.*,	1966
Discaria	Not tested, only one species, *D. toumatou*, reported	- - - - - - - - - - - - - - - -		Morrison	1961
Dryas	Not tested, only one species, *D. drummondii*, investigated; other species are known	- - - - - - - - - - - - - - - -		Lawrence *et al.*,	1967
Elaeagnus	*E. pungens* and *E. angustifolia*	none	- - -	Gardner	1958
	Hippophae and *E. angustifolia*	good	good		
Hippophae	*H. rhamnoides* and *Shepherdia*; *H. canadensis*	complete	complete	Gardner and Bond	1957
Myrica	*M. gale* and *M. cerifera*	incomplete	incomplete	Bond	1967*b*
	M. gale and *M. cordifolia*	incomplete	incomplete	Gardner and Bond	1966
Purshia	Two species known, cross inoculation tests not reported	- - - - - - - - - - - - - - - -	- - - - - - - - - - - - - -		

specificity reminiscent of that noted for legumes. However, in families such as the *Eleagnaceae* there appears to be little specificity of host and endophyte even across generic lines, for nodules of *Hippophae* induced nodulation of *Eleagnus* and *Shepherdia*. Thus, varying degrees of species specificity exist among the wide range of families in which nodulated roots are found.

7.2.2   Superficial root-nodule symbiosis in non-leguminous plants

The roots of most plants growing under natural conditions enter into an association with fungi to form an extensive net of fungal and root tissue termed mycorrhizae. Several types of mycorrhizae have been distinguished: (a) ectotrophic, (b) ectendotrophic (not considered here) and (c) endotrophic, the latter containing extensive hyphal growth within the cortex of the root (Gerdemann, 1968). As discussed below (7.4.2), among mycorrhizal associations there is little evidence for N_2 fixation as a general phenomenon among fungi not associated with the rhizosphere.

Podocarpus, a gymnosperm, bears numerous nodule-like structures on its roots. Evidence obtained, particularly with the electron microscope, indicates that the endophyte is a non-septate filamentous organism, presumably a phycomycete (Nobbe and Hiltner, 1899; Shibata, 1902; Schaede, 1943; Bergersen and Costin, 1964; Becking, 1965; Becking, 1966) and not a bacterium as has sometimes been reported (Spratt, 1912; McLuckie, 1923; Phillips, 1932). Becking noted that the endophyte, which could not be isolated, appeared similar to the vessicular-arbuscular fungi described by Mosse (1963). Two brief reports claimed that actinomycetes, isolated from nodulated plants, caused nodulation of sterile seedlings of *P. macrophylla* and that N_2 fixation might have occurred (Uemura, 1967; Uemura et al., 1970). However, mycorrhiza in *Podocarpus totara* were successfully synthesized with endogone spores obtained from *Coprosma* (Bayliss, 1969).

Although it appears that the *Podocarpus* symbiont is a fungus, there is no reason why a plant species cannot contain two types of symbiotic associations. Indeed, *Casuarina* has been reported to contain mychorrhiza as well as the more familiar actinomycetous root nodules (Huguenin, 1969), and Morrison and English (1967) have described two microsymbionts in *Agathis australis,* the Kauri pine. For convenience, a tabulation of genera of gymnosperms known to bear root nodules and hence possibly involved in nitrogen fixation is presented in Table 7.3. More details of the botanical

TABLE 7.3. Nodulated gymnosperms[*] in which nitrogen fixation may occur.

Family	Genus	Species	Reference	
Cycadaceae	Cycas	revoluta[a]	Wattanabe and Kiyohara	1963
	Cycas	circinalis[1]	Douin	1953
	Bowenia	serrulata[2]	Bowyer and Skerman	1968
	Ceratozamia	mexicana[a]	Bond	1967a
	Encephalartos	villosus[a]	Bond	1967a
	Macrozamia	communis[a,2]	McLuckie	1922
			Bergersen et al.,	1965
	Zamia	floridana[1]	Chaudhuri and Akhtar	1931
Podocarpaceae	Podocarpus[a]		Phillips	1932
			Bond	1958
			Bergersen and Costin	1964
			Becking	1965
	Saxegothaea		Scott	1969
	Microcachrys		Scott	1969
	Dacrydium		Scott	1969
	Phyllocladus		Spratt	1912
Araucariaceae	Agathis	australis	Morrison and English	1967
	Araucaria	sp.	Bevege and Richards	personal communication

[a] N_2 fixation detected with ^{15}N (see sec. 7.2.2).

[1] Anabaena cycadeae is the endophyte.

[2] A species of Nostoc was isolated.

[*] According to Rendle, A. B. The Classification of Flowering Plants, Vol. 1, Cambridge University Press, 1956.

aspects of these symbioses may be found in the review of Allen and Allen (1964).

7.2.3   Foliar associations

The leaves of certain genera of the family *Rubiaceae: Pavetta, Psychotria*, and *Coprosma;* and the family *Myrisinaceae: Ardisia, Amblyanthus* and *Amblyanthopsis;* and the family *Dioscoreaceae: Dioscorea*, bear externally visible nodules on the leaf surface (Bond, 1967a; Lange, 1966; Schaede, 1962). These associations are of interest from several viewpoints: (a) the relative specificity between symbiont and host plant, (b) the possible involvement of N_2 fixation, and (c) their ecological importance in their natural tropical habitat. Points (b) and (c) are of more than passing interest in light of their wide distribution in tropical and subtropical areas (Bremenkamp, 1934) and high incidence of nodulation (Bremenkamp, 1933; Petit, 1966; Humm, 1944) among the described species.

A list of representative species from three families containing nodulated members, as well as the possible identity of the endophyte is given in Table 7.4. With the exception of the isolates from *P. bacteriophila* (Centifanto and Silver, 1964; Centifanto, 1964) and *P.*

TABLE 7.4. Bacteria isolated from leaf nodule plants.

Plant species	Identity of isolant	N_2 fixation in pure culture	Reference	
Ardisia crispa A. Dc	*Bacillus foliicola*	—	Miehe	1914
	Bacillus foliicola	—	de Jongh	1938
	Bacterium foliicola	+	Glaubitz	1957
	Ac−1	—	Gordon	1963
	Ac−2	+	Gordon	1963
	Ac−3	+	Gordon	1963
A. crispa and *A. puntata*	*Bacterium foliicola*	—	Yamada	1960
A. hortorum Maxim	*Xanthomonas hortoricola*	+	Hanada	1954
A. humilis Vahl	'L forms'	?	Bose	1955
Pavetta grandiflora	Pg 1	+	Gordon	1963
	Pg 2	—	Gordon	1963
Psychotria bacteriophila Vatke (punctata)	*Klebsiella rubiacearum*	+	Centifanto and Silver	1964
Psychotria bacteriophila Vatke	*Mycobacterium rubiacearum*	+	v. Faber	1914
(punctata) and other species	*Mycoplana rubra*	—	de Vries and Derx	1950
P. calva Hiern	*Flavobacterium* sp.	?	Adjanohoun	1957
P. emetica	Pe 1	+	Gordon	1963
	Pe 2	—	Gordon	1963
	Pe 3	+	Gordon	1963
P. nairobiensis	Pn 1	—	Gordon	1963
	Pn 2	—	Gordon	1963
	Pn 3	—	Gordon	1963
	Pn 4	+	Gordon	1963
	Chromobacterium lividum	+	Berrelheim, Gordon, Taylor	1968

nairobiensis (Bettelheim, Gordon and Taylor, 1968), and von Faber's description of his culture from *Pavetta zimmermanniana* (von Faber, 1914) the isolates are poorly described and there is little evidence that the organisms described were indeed the specific endophytes. It is of interest that *Klebsiella rubicearum* isolated from *P. bacteriophila* and *Chromobacterum lividum* from *P. nairobiensis* were capable of fixing nitrogen in pure culture. There is no doubt that *Klebsiella* originally isolated from *P. bacteriophila* was the endophyte of the plant variety from which it was isolated (Cetifanto and Silver, 1964; Centifanto, 1964). However, Becking reported that with cultures and plants available to him, abrasions of the leaves of 'cripples' did not induce nodulation (Becking, 1970c). Further clarification of this disagreement awaits reisolation of endophyte from the host plant. The universality of this trait as well as the fixation of nitrogen in the plant (Silver *et al.*, 1963; Centifanto, 1964; Becking, 1970c) is yet to be established throughout the many hundreds of known nodulated species.

An interesting biological aspect of the symbiosis which may not be closely allied to nitrogen fixation is that the symbiosis is self-perpetuating or cyclic: seeds contain the endophyte which previously had reached floral primordia during the formation of fruits (von Faber, 1914). It appears that both in *Ardisia* (Miehe, 1916; de Jongh, 1938) and *Psychotria* (Humm, 1944; Silver *et al.*, 1963) a plant growth regulator is involved. In the latter, plants devoid of leaf nodules exhibit a dwarf growth habit which invariably reverts to normal after periods as long as several years. Concomitant with this, nodulation appears (Centifanto, 1964). Since the application of gibberellin reversed to some extent the dwarf characteristics, it appears that the endophyte may either produce a plant growth regulator which the plant cannot synthesize or it reverses the effect of some endogenous inhibitor.

Lersten and Horner (1967) studying the structure of and development of nodulation in *P. bacteriophila* with electron microscopy, have presented evidence that the Gram-negative nodule bacteria are intimately associated with a plant reticulum, and that bacteria of various ages are present in each nodule. That is, a functional population is present for relatively long periods of time. Whether this population is involved in nitrogen fixation in the host plant will be discussed below (7.5.1).

7.2.4   Associations in lower plants
For purpose of this chapter the algal symbiosis noted in cycads, which are gymnosperms, will be considered together with those found in

non-seed bearing plants. Table 7.3 lists examples where the identity of the endophyte has been established. In all of these symbioses the major endophyte is a species of blue-green algae, usually of the well known nitrogen-fixing genera, *Nostoc* and *Anabaena*, and on this basis, Schaede (1962) has classified these associations as blue-green algal symbioses.

In *Azolla,* the aquatic fern, the endophyte has been identified as *Anabaena azollae* (Venkataraman, 1962). The endophyte occupies the extracellular space of the cavity at the base of each upper floating leaf. There is good biochemical as well as ecological evidence for nitrogen fixation in this fern (see below 7.4.3).

In liverworts, *Nostoc sphaericum* has been noted to be the endophyte (Watanabe and Kiyohara, 1963; Pankow and Martens, 1964), and in a variety of marine and non-marine lichens various nitrogen fixing blue-green algae have been implicated (see Table 7.5). In the lichens, nitrogen fixation occurs at a high rate in the cephalodia, and the nitrogen is secreted at an equal rate to the lichen thallus (Millbank and Kershaw, 1969). However, only 3% of the excreted nitrogen was accumulated by the algal cells (Kershaw and Millbank, 1970).

7.3 PHYSIOLOGICAL AND BIOCHEMICAL ASPECTS OF NITROGEN FIXATION IN TRUE ROOT-NODULE SYMBIOSIS

Experimental evidence accumulated since the turn of the century has confirmed the supposition that nitrogen fixation can explain the observation that nodulated non-legumes thrive in a low N habitat, are often pioneers in the natural reclamation of marginal environments, and that the presence of root-nodules is correlated with the conversion of dinitrogen into a nutritionally useful form. Initially, experimentation was restricted to classical pot experiments; later the use of $^{15}N_2$ permitted a more critical test of the supposition; most recently the use of the acetylene reduction assay has proven invaluable.

7.3.1 Studies with whole plants

7.3.1.1 *Nitrogen.* Bond and his colleagues have published extensively on the effects of nodulation on plant vigor and nitrogen content for a variety of non-legumes. Since much of this has been reviewed extensively by Bond (1963, 1967a, 1968), one example will be cited here. Uninoculated plants quickly showed deficiency signs typical of nitrogen deprivation when grown in an N-free nutrient

TABLE 7.5. Lower plants in which nitrogen fixation may occur*

Type	Plant Genus	Endophyte	N$_2$ fixation determined by	Reference
Lichen	*Collema*	Nostoc	Growth of phycobiont, micro-Kjeldahl ^{15}N$_2$	Henriksson (1951) Bond and Scott (1955)
	Leptogium	Nostoc	^{15}N$_2$	Bond and Scott (1955) Scott (1956)
	Lichina	Calothrix	^{15}N$_2$	
	Peltigera	Nostoc	^{15}N$_2$	Scott (1956) Millbank and Kershaw (1969)
	Stereocalum	Nostoc	^{15}N$_2$	Fogg and Stewart (1968)
Liver Wort	*Anthoceros*	Nostoc	Inference	Leitgeb (1878)
	Blasia	Nostoc	Inference	Molisch (1925)
			^{15}N$_2$	Bond and Scott (1955)
			^{15}N$_2$	Watanabe and Kiyohara (1963)
	Cavicularia	Nostoc	Inference	Molisch (1925)
			^{15}N$_2$	Watanabe and Kiyohara (1963)
Fern	*Azolla*	Anabaena	Growth	Oes (1913) Nickell (1958)

* Extensive descriptions of the botanical aspects of these symbioses may be found in Schaede (1962).

TABLE 7.6. Effect of nodulation on growth and N-content of *Shepherdia* grown in N-free nutrient solution for four months post inoculation (From Gardner and Bond, 1957).

Type plant (number)	Mean shoot height (cm)	Mean dry weight (mg)	Mean N (mg)
Nodulated (5)*	7.6	248	7.32
Non-nodulated (5)	3.6	64	1.00

* The source of inoculum was *Hippophae*.

solution and the N content of uninoculated plants was extremely low. A typical example, *Shepherdia*, is documented in Table 7.6. In this case nodulation increased the dry weight almost four fold and the N content more than seven fold. In a variety of nodulating plants the N content of nodulated plants is invariably much greater than that of uninoculated controls.

Several lines of evidence indicate that the root-nodules are the sites of N_2 fixation: the high N content of nodules compared to roots and shoots, the inability of non-nodulated tissue to fix N_2, the distribution of ^{15}N following exposure of the root system to an $^{15}N_2$ enriched atmosphere (Bond, 1958). Examples of the latter are given in Table 7.7. In all species studied the ^{15}N enrichment was greater in root-nodules than in roots or shoots. Alder nodular tissue was more than 8 times enriched as was root tissue. That the nodules are indeed the site of fixation is attested by the lack of activity of denodulated plants and the considerable activity of excised nodules (Bond, 1958).

Since combined N would often be available to a plant in the field, the effect of combined N on N_2 fixation has been assessed. Stewart

TABLE 7.7. Distribution of ^{15}N enrichment in nodulated plants exposed to $^{15}N_2$ enriched gas for 3-4 days. (From Bond, 1958).

Plant series	Atom % excess ^{15}N in		
	Nodules	Roots	Shoots
Alnus	1.086	0.125	0.157
Myrica	0.492	0.217	0.133
Casaurina	1.464	0.286	0.687
Shepherdia	0.216	0.083	0.045

and Bond (1961) exposed potted *Alnus* and *Myrica* to various levels of ^{15}N-labeled ammonium sulfate for ten to twelve weeks, and determined the dry weight of plants and nodules, total N, ^{15}N enrichment and the extent of nodulation. ammonium N did not eliminate N_2 fixation although it was lower in *Myrica* per unit weight of nodular tissue; in *Alnus*, N_2 fixation was enhanced because nodule development was greater. Pizelle (1966) noted that nitrate at > 0.5 me/l markedly reduced the mean number of nodules on roots of *A. glutinosa* seedlings potted by a 'split-root' technique, which permitted internal controls on each plant, which are very desirable because considerable variations occur between plants. It would be of interest to know whether the availability of combined N in the field would result in more numerous but smaller nodules, and whether the occurrence and extent of nodulation can be used as a marker for N_2 fixation irrespective of the N status of the habitat as was noted in these pot experiments.

Stewart (1962) reported a quantitative study of the transfer of N from root-nodules to other parts of the alder plant during the first season of growth. Plant dry weight, nodule weight, and nitrogen increased progressively during the growing season until September, young nodules being the most efficient. About 90% of the nitrogen fixed was transfered without delay and reached even higher proportions during the later stages of the experiment.

In addition to the seasonal effect on N_2 fixation, a diurnal fluctuation has been noted in both alder and *M. gale* (Wheeler, 1969). In these experiments the acetylene reduction assay revealed that there was a peak of activity at about 13.00 hours for both plant species. The nitrogen content of bleeding sap as well as the ethanol-soluble nitrogen in nodules had similar maxima. Although this could be interpreted as a reflection of peak photosynthesis the author was unable to detect the expected accumulation of soluble or reserve carbohydrate. However, it is reasonable to assume that ATP and reducing power would not be limiting at about midday.

7.3.1.2 pH relationships. Many of the nodulated non-legumes are found in acidic soils, so the effect of hydrogen ion concentration on both nodulation and plant growth has been studied. Bond, Fletcher and Ferguson (1954) noted that about half of inoculated *Myrica gale* L, and *Alnus glutinosa* (L) Gaertn nodulated at pH 4.2 whereas *Hippophae rhamnoides* L. and red clover did not; however, growth at this pH was satisfactory for all species tested and was adequate for *Myrica* and *Alnus* even at pH 3.3. The effect of pH on nodulation is therefore primarily upon the endophyte, a situation not unlike that noted in certain legumes. Once nodulation had occurred, the effects of pH on subsequent plant growth showed the following optima during the first seasons growth in N-free nutrient: *M. gale*, 5.4 (Bond, 1951); *A. glutinosa*, 4.2-5.4 (Bond, Fletcher and Ferguson, 1954); *H. rhamnoides*, pH 6.3 (Bond, Fletcher and Ferguson, 1954). These pH optima correlate generally with the acidity of typical habitats. Undoubtedly the pH effect is very complex for in addition to the effects of acidity on nodulation, the availability of N in its various forms and of many micro- and macronutrients, may also be pH dependent.

7.3.1.3 Minerals (which relate to N_2 fixation). The mineral nutrition of non-legumes has been the subject of numerous investigations. Only those minerals having a function related to N_2 fixation will be considered here.

Molybdenum Although Mo is now known to function in a variety of other biochemical transformations (Anderson, 1956), it is a required element for N_2 fixation in presumably all N_2 fixing systems, including the symbiotic ones. A Mo requirement for growth on dinitrogen has been shown for *Alnus* (Becking, 1961; Hewitt and Bond, 1961); *Casuarina* (Hewitt and Bond, 1961); and *Myrica* (Bond and Hewitt, 1961). Deficiency effects included reduced growth and N content, chlorosis, leaf scorching, and, in soil-grown but not liquid-cultivated plants, an increase in number and reduction in size of root-nodules. Nodular tissue had an increased affinity for Mo over other tissue; the effect was much less pronounced in soil-grown plants.

Cobalt. The importance of cobalt (Co) for the growth of nodulated legumes (Ahmed and Evans, 1961; Delwiche *et al.*, 1961) and the usual lack of effect in non-nodulated plants grown on fixed N was attributed to a Co requirement by the root-nodule symbiont, *Rhizobium* (Lowe and Evans, 1962). In non-legumes a similar Co requirement for synthesis of the B_{12} coenzyme by the yet to be isolated endophyte would be expected. Evans and co-workers (Evans and Kliewer, 1964) have assayed the cobamide coenzyme content of a variety of nitrogen-fixing symbionts and microorganisms including non-legumes. *Alnus oregona* and *Ceanothus velutinus* had levels of the cobamide coenzyme similar to that of legumes and rhizobia. Since this substance was restricted to the nodular tissue of legumes it may be assumed that its presence reflects the metabolism of the symbiont rather than the host plant, because higher plants are known to contain little or no vitamin B_{12} compounds. Whether the B_{12}-coenzyme may be involved in nucleotide reductase and methylmalonyl mutase as appears to be the case in soybean bacteroids (Chapter 6), can only remain inferrential until the organisms are available in pure culture. A list of species with a requirement for Mo or Co is summarized in Table 7.8.

TABLE 7.8. Some mineral requirements of nodulated non-legumes growing in the absence of fixed nitrogen.

Plant	Requirement	
	Mo	Co
Alnus glutinosa	+	+
Alnus oregona	. . .	i
Alnus rubra	. . .	i
Casuarina cunninghamiana	+	+
Ceanothus velutinus	. . .	i
Coriaria myrtifolia	. . .	n.m.
Hippophae rhamnoides	. . .	i
Myrica gale	+	+

+ = required
. . . = not known
i = inferred since B_{12} coenzyme detected
n.m. = B_{12} content of nodular tissue not markedly greater than that of roots.

Iron. The effects of iron deficiency is readily demonstrated in most plants irrespective of the source of nitrogen because iron is essential in both respiration and photosynthesis as a constituent of the cytochromes. In addition, iron, in a non-heme form, is found in ferredoxin. In so far as the author is aware, no reports comparing the iron requirements of nodulated and non-nodulated non-legumes have appeared although one would expect a greater requirement for nodulated plants.

7.3.2 Nitrogen fixation by excised tissue

The use of $^{15}N_2$ with detached nodules greatly simplifies the study of symbiotic N_2 fixation. Aprison and Burris (1952) noted that, although there was considerable variation among samples, rapid handling of field grown soybean plants and removal of root-nodules just prior to exposure to $^{15}N_2$ permitted the detection of N_2 fixation when analyses were performed on the acid soluble fraction. However, even under these conditions, the rate of fixation declined rapidly after the first hour. Aprison, Magee and Burris (1954) later noted a marked deleterious effect of slicing or homogenization upon N_2

fixation. A noteworthy characteristic of excised non-leguminous root-nodules was that removal of nodules from the plants had a much lesser adverse effect on N_2 fixation (Bond, 1959) than with legume nodules. This prompted their subsequent use in attempts to achieve active cell-free homogenates (Bond, 1964; Sloger and Silver, 1965).

A summary of nitrogen fixation by excised nodules of a variety of plant genera is presented in Table 7.9. Although a variety of experimental conditions have been used, it is apparent that there was marked ^{15}N enrichment in most species tested within 3-6 hours of incubation. In those species where exposure was 24 hours or longer, fixation by free-living bacteria may be of great significance, especially in soil (as in *Comptomia peregrina* and *Podocarpus rospiglosii*). If such were the case a kinetic study would reveal increasing enrichment with time, not the usual decrease after about 8 hours as in *M. cerifera* (Sloger and Silver, 1965; Silver and McHue, 1970). Sprent (1969) however, has noted that the alleged advantage of the use of excised non-legume nodules over soybeans is more an experimental artifact than an inherent property of the tissue since detached soybean nodules reduced acetylene at a constant rate for at least 8 hours providing they were not allowed to dry out and adequate O_2 was available to the tissue.

7.3.2.1 Distribution of fixed nitrogen. One early result of the use of excised nodules was an investigation of the distribution of ^{15}N in nodules of *Myrica gale* L. (Leaf, Gardner and Bond, 1958). The exposure of freshly excised nodules to a gas mixture of 10% N_2 (36 atm. % [15]N excess), 20% O_2 and 70% Ar resulted in high enrichment into the dicarboxylic acids, their amides, alanine and tryptophan (Table 7.9), all except the latter being biosynthetically close to NH_3. On this basis the authors concluded that asparagine and glutamine are repositories of excess N, and NH_3 was an obligate intermediate of N_2 fixation, as in the legumes. For comparison, data for N_2 fixation by excised soybean nodules is included in Table 7.10.

A recent report, pertinent to a discussion of the amino acid pool of non-leguminous root-nodules, is a survey of Wheeler and Bond (1970) of the free amino acids in eleven species of nitrogen-fixing, non-legumes. Citrulline was present only in alder, in confirmation of an earlier report by Miettinen and Virtanen (1952); in all other species except *Coriaria* asparagine was the predominant form of N. Asparagine and glutamine accounted for 92% of the total amino acid N in *Casuarina.* and a high proportion in most of the other species studied. Considering the diverse plant families analyzed, the

TABLE 7.9. $^{15}N_2$ incorporation by excised nodules of non-legumes.

Plant species	Amount of tissue	Enrichment of ^{15}N gas, atom %	Exposure conditions	Atom % excess ^{15}N	Reference
Alnus glutinosa	0.8 g (f.w.)	36	3 hrs 24°C	0.090	Bond 1957
Casuarina cunninghamiana	1.5 g (f.w.)	36	19 hrs 24°C	0.314	Bond 1957
Ceanothus velutinus	6.1 mg N	34.3	5 hrs 24°C	0.091	Bond 1967b
Coriaria myrtifolia	4.4 mg N	99.3	5 hrs 25°C	0.328	Bond 1967b
	7.8 mg N	34.3	5 hrs 24°C	0.307	Bond 1967b
Comptonia peregrina	1.9% (of dry wt.)	16.2	6 days	0.151 (in water) 2.224 (in soil)	Ziegler and Hüser 1963
Hippophae rhamnoides	0.5 g (f.w.) [= 2.9 mg N]	16.2	6 hrs 24°	0.310	Bond 1957
Myrica cerifera	4 g (f.w.)	30	4 hrs 28°C	0.644	Sloger 1968
Myrica gale	0.5 g	18	24 hrs 21°C	0.463	Bond and MacConnell 1955
Podocarpus lawrencei	5 g f.w. roots [= 1.9 g nodules]	90.6	4 hrs 23°C	0.076	Bergersen and Costin 1964
Podocarpus rospigliosii	0.6 mg N (Nodules)	35	24 hrs	0.45	Becking 1965

TABLE 7.10. Distribution of ^{15}N in nodules of soybean and alder.

Substance	Atom % ^{15}N excess		
	Soybean*	Myrica†	Alnus‡
Ammonia	0.127	0.736	—
γ-aminobutyric acid	—	—	0.504
Aspartic acid	0.049	0.427	0.910
Glutamic acid	1.172	0.602	1.078
Asparagine (amide N)	0.110	0.506	—
Asparagine (amino N)	0.058	0.149	—
Citrulline	—	—	0.732
Glutamine (amide N)	—	1.450	—
Glutamine (amino N)	—	0.460	—
Glycine	0.109	0.054	—
Alanine	0.089	0.810	—
Tryptophan	—	0.359	—
Arginine	—	0.002	—
Serine-threonine	0.235	—	—
Leucine	0.095	—	—
Lysine	0.069	—	—
Isoleucine	0.095	—	—
Histidine	0.050	—	—

* from Aprison, Magee and Burris (1954). Excised nodules exposed to 10% ^{15}N$_2$ (31 atm % ^{15}N excess), 20% O$_2$, 70% He for 1 hr. Samples were 1-3 g/flask.

† from Leaf, Gardner and Bond (1958a). 1.3 g nodules in 1 ml culture solution exposed to 10% N$_2$ (36 atm % ^{15}N excess), 20% O$_2$ and 70% Ar.

‡ from Leaf, Gardner and Bond (1958b). 1 g nodules exposed to 20% O$_2$, 10% N$_2$ (36 atm % ^{15}N excess), 70% Ar for 90 min at 25°C.

dominant position of asparagine and glutamine is striking and should be investigated further.

7.3.2.2 The effect of the gaseous environment on N$_2$ fixation by excised nodules. Since the endophytes of the non-legumes have not been isolated, it is not known what their O$_2$ relationships are. The effect of altering the gaseous environment on N$_2$ fixation by excised tissue was measured in order to determine whether the process in excised tissue was ogygen dependent. As may be noted from the tabulation in Table 7.11 the general pattern is not unlike that found in a typical legume, there being a broad maximum over the range 10-30% O$_2$, although *Alnus* and *Hippophae* seemed more sensitive and *Casuarina* less sensitive to a higher pO$_2$. It appears then that the system is well adapted to the pO$_2$ of the normal environment. Since respiration was not directly affected at high oxygen pressure (MacConnell, 1956), the effect is direct and is probably explainable

TABLE 7.11. Effect of oxygen pressure on N_2 fixation by excised nodules of non-legumes.

Percent O_2	Myrica[a] gale	Myrica[b] cerifera	Casuarina[a] cunninghamiana	Alnus[c] glutinosa	Hippophae[a] rhamnoides
0	2	0	—	0	—
5	30	—	32	43	79
10	58	77	62	92	100
20	100	—	100	71	83
30	82	100	100	7	38
40	24	9	75	3	21

[a] Bond, 1961.
[b] Sloger, 1968.
[c] Bond, 1959*b*.

by the now known extreme sensitivity of the Fe-protein of nitrogenase to oxidation (Mortensen, Morris and Jeng, 1967). In whole plant experiments the incidence of nodulation was dependent upon the pO_2 (MacConnell, 1959).

The behavior of excised nodules of non-legumes is similar to that of legumes with respect to other gases as well. Both H_2 and CO inhibited N_2 fixation in *Alnus* and *Myrica*, the inhibition being quite similar to that of pea and soybean controls (Bond, 1961). However, about six times more H_2 was needed than CO on a volume basis to achieve similar inhibition.

7.3.3 *Nitrogen fixation by subcellular fragments*

The adverse effect of tissue fragmentation upon nitrogen fixation by soybean root-nodules first reported by Aprison, Magee and Burris (1954) was also noted by Koch, Evans and Russell (1967) whose preparations in the early phases of their work, though active if used within several hours after preparation, were inactive following overnight storage under Ar in the cold or at room temperature. The presence of large amounts of greenish and brownish pigments revealed by chromatography of crude extracts on a DEAE column prompted these authors to attempt to remove phenolic substances which are often associated with fragmented plant tissue. This led to the very important observation that treatment of ascorbate-buffered homogenates with polyvinyl pyrrolidone (PVP) permitted the preparation of active extracts on a routine basis. This landmark report allowed the subsequent purification of the enzyme complex into two component proteins and established that the bacteroids were the site of nitrogen fixation in legumes (see Chapter 6).

Prior to these observations with legumes, several reports appeared describing fragmented nodules of non-legumes capable of nitrogen fixation (and subsequently acetylene reduction) (Bond, 1964; Sloger and Silver, 1965, 1966; Sloger 1968). Generally the activity of breis was much lower than that of intact nodules. The presence of a reducing agent and the absence of O_2 was essential during homogenization; however, O_2 was required, presumably for the production of ATP, during exposure of the homogenate to $^{15}N_2$. The phenomenon was independently noted by Bergersen for soybean breis (Bergersen, 1966). Although it is reasonable to assume that the use of PVP to sequester phenolics should increase the activity of homogenized root-nodules of non-legumes, Sloger (1968) was unable to demonstrate this for *Myrica*; since the technic was ineffective for soybean extracts as well in his hands, it may be assumed that the details of preparation handling rather than the technic *per se* was at fault. Further reports of fragmented systems from *Myrica* or other non-legumes have not appeared. Now that easily worked tissue like that of soybean can yield stable cell free preparations, the use of the woody tissue of non-legumes seems much less attractive for studies of the enzymes involved unless comparative aspects are the prime considerations.

7.3.4   *Is haemoglobin involved in nitrogen fixation in the root-nodules of non-legumes?*

The relative importance of the plant and the endophyte in the synthesis of leghaemoglobin in legumes has been the subject of many investigations and opinions concerning its function in the nodule and possible involvement in nitrogen fixation are varied (see Chapter 6).

In contrast, knowledge about the role of, or even presence of haemoglobin in the root-nodules of non-legumes is very scanty. There is spectroscopic evidence for the presence of haemoglobin (observed as 'pyridine haemochromogen') in *Alnus glutinosa, Hippophae rhamnoides* and *Myrica gale*, but not in a podocarp and a cycad (*Encephalartos*) (Egle and Munding, 1951). Davenport (1960) noted, in addition to the three species mentioned above, that *Casuarina cunninghamiana* contained haem proteins at levels comparable to that of the pea. Although one report failed to confirm these observations (Moore, 1964) it is likely that haemoglobin is at least present in the root nodules of several of the non-legumes. A more intensive study is needed.

7.3.5 Miscellaneous physiological effects associated with nodulation

The interaction of a parasitic organism with a host plant invariably leads to marked alterations in the metabolism of the host. In several of the non-legumes there is an interesting reversal of the orientation of the root (in this case nodule-roots, the rootlets growing from the apices of root-nodules), so that they grow in the upward direction (Bond, 1959b). In the root-nodules of legumes and plant galls, infection induces a marked elevation in the concentration of indole acetic acid (IAA) (Pate, 1958), as well as general hypertrophy. In contrast, infection of the root in *Myrica cerifera* and *Casuarina cunninghamiana* leads to a great increase in IAA oxidase and no detectable IAA (Bendana, 1963; Silver *et al.*, 1966). As a consequence, the nodule-root, being physiologically like a stem, grows upward. Unlike these plants and similar to the legumes, the nodules of *Alnus serrulata* were relatively rich in IAA. Although this species does not form nodule-roots, it is of interest that subsequently Rodriquez-Barrueco (1966) found downward-growing nodule-roots in another species, *A. jorullensis*, as would be expected from the report of Silver *et al.*, (1966).

The endophyte does not penetrate into the nodule-root (Bond, 1951). However, removal of the latter halves the amount of nitrogen fixed by excised nodules (Bond, 1961). Although the significance of this observation is not clear, it may be that upward orientation, which in nature places the tissue close to or at ground surface, increases respiration sufficiently so that more ATP is available for nitrogen fixation.

Dullaart investigated the auxin content of nodules and roots of *A. glutinosa* using a sensitive spectrophotometric method as well as the *Avena* bioassay (Dullaart, 1970). He reported six times more IAA (the major auxin found) in nodules than in roots although the IAA levels were considerably lower than that noted in a different species by others (Silver *et al.*, 1966).

Further similarity between the symbiosis in alder and legumes was attested by the inhibitory effect of nitrate on IAA stimulated nodulation (Pizelle, 1970). In a very much different plant, *Cycas circinalis*, there was no difference in the IAA content of nodules and roots (only about 20-50 μg/kg fresh weight for both tissues: Dullaart, 1968).

It is evident that a detailed study of the effects of infection on the metabolism of root-nodule tissue is warranted whether the point of interest be nitrogen fixation, geotropism, or the general host-parasite relationship.

7.4 PHYSIOLOGICAL AND BIOCHEMICAL ASPECTS OF NITROGEN FIXATION IN SUPERFICIAL ROOT-NODULE SYMBIOSIS

Since relatively little is known about the biochemical and physiological aspects of the symbioses in this group, only three will be discussed: podocarps, mycorrhizae and cycads.

7.4.1 Podocarps

Ever since the studies of Nobbe and Hiltner (1899) demonstrated the inability of non-nodulated *Podocarpus* to maintain vigorous growth in nitrogen-free sand culture, nitrogen fixation has been inferred in this symbiosis. It was not until recently, however, that a direct demonstration using $^{15}N_2$ was made (Bergersen & Costin, 1964). They noted that *P. lawrencei* nodules, exposed to a N_2-O_2-Ar mixture for four hours at 23°C, accumulated 0.076 atm. % excess ^{15}N in the acid soluble fraction (= 0.6 μg N fixed/1.98 g fresh wt nodules). The fixation rate was low, for under similar conditions soybean nodules would fix about twenty times that amount. In another species, *P. rospigliosii*, similar low rates of fixation were obtained by Becking (1965). As would be expected, growth experiments revealed markedly inferior growth in an N-free nutrient as compared with combined nitrogen. Sloger (1968) failed to detect actylene reduction by nodulated roots of *P. macrophylla*, yet Richards and Bevege obtained $^{15}N_2$ fixation by nodulated roots of *P. spinulosus* (personal communication). It would appear that the phenomenon is real although the rates are quite low. It is of interest that the podocarp systems are the only known examples where eucaryotic microorganisms seem to be involved in nitrogen fixation.

7.4.2. Mycorrhizae

Melin first proposed in 1925 that mycorrhizae might fix nitrogen in the plant-soil environment (cited in Melin, 1959). Bond and Scott (1955) failed to find evidence for this in *Pinus sylvestis* using $^{15}N_2$, and Sloger and Silver (1967) also found no evidence for fixation using the very sensitive acetylene reduction method. In view of the well known effect of pine mycorrhizae on phosphate uptake (Harley and Lewis, 1969) and the fact that supplementation with exogenous glucose was essential for $^{15}N_2$ fixation in nitrogen-depleted *Araucaria* seedlings (B. N. Richards and D. I. Bevege, personal communication), one must conclude that positive reports for nitrogen fixation probably reflect the activity of rhizosphere bacteria and that ectotrophic mycorrhizae are not involved.

7.4.3   Cycads

Although considerable work has been done on the biology of the blue-green algal containing coralloid roots of cycads (Schaede, 1962) little has been reported on nitrogen fixation by the symbiotic system. Watanabe and Kiyohara (1963) detected $^{15}N_2$ enrichment by *Cycas revoluta* as well as by isolated symbiont, *Nostoc cycadae*. The rates were very low in the symbiotic system. In other cycads (*Encephalartos* and *Macrozamia*) only the isolated symbiont was active. Similar studies with the liverwort systems in *Cavicularia* and *Blasia* also revealed low enrichments in the symbiotic system.

Much more substantial evidence for nitrogen fixation was obtained by Bergersen, Kennedy and Wittman (1965), who obtained rates comparable to those detached soybean nodules in short term experiments with a different species, *Macrozamia communis*. It is significant that the coralloid root tissue evolved hydrogen, a trait characteristic of nitrogen fixing tissues, under those conditions. ^{15}N was distributed to various parts of an intact plant exposed to the isotope for 48 hours (Table 7.12) showing that the nitrogen sequestered was available to the plant and that the amounts involved were significant.

TABLE 7.12. Distribution of ^{15}N in *Macrozamia* (from Bergersen, Kennedy and Wittman, 1965).

Plant Part	Total N (mg)	^{15}N Atoms % Excess	^{15}N (mg Excess)
Leaves	60.2	0	0
Petioles	15.2	0.065	9.88
Swollen leaf bases and stems	101.9	0.057	66.24
Capralloid roots	15.3	0.716	109.55
Hypocotyl and swollen tap root	21.7	0.064	13.89
Roots	13.4	0.113	15.14
Totals	227.7	—	214.70

The intact plant was exposed to 12% N_2 (44 atom % excess) 25% O_2 and 63% Ar.

7.5   PHYSIOLOGICAL AND BIOCHEMICAL ASPECTS OF FOLIAR NITROGEN FIXATION

7.5.1   The isolated endophyte

Von Faber (1912) was the first to approach experimentally the possible role of the endophyte in the folliar symbiosis in the *Rubiaceae*. The organism which he isolated from *Pavetta zimmermanniana* grew in an N-free broth as well as a complex plant decoction medium. The isolate, which he named *Mycobacterium rubiacearum*, grew slowly on N-free solid or liquid medium when incubated aerobically. Kjeldahl analysis revealed a small increase of N after 20 days incubation (complex medium, 13.32 mgN; mineral N-free medium, 6.2 mgN; mineral N-free + 0.1% asparagine, 9.2 mgN). It was presumed that the increase in N was due to N_2 fixation. The fact that the complex medium permitted double the increase of that obtained in the N-free broth casts doubt upon the claim of the occurrence of N_2 fixation. A comparison of the description of the organism with the properties of *Klebsiella rubiacearum* isolated from *Psychotria bacteriophila* however, leaves little doubt that they are closely related if not identical, (Centifanto and Silver, 1964; Centifanto, 1964). The rather poor growth and increase in N encountered by von Faber was undoubtedly due to the use of aerobic incubation conditions. Static, cotton-plugged cultures of *K. rubiacearum* grew very slowly until there was an accumulation of H_2 and CO_2 in the gas phase over the culture. As N_2 fixation occurred, the H_2 concentration fell (Centifanto and Silver, unpublished observations). It is fortuitous that von Faber selected asparagine to supplement the N-free, mineral broth, for in the case of *Klebsiella* at least its deamination product, aspartic acid, does not repress N_2 fixation as does ammonia (Yoch and Pengra, 1966; Neelands, 1967).

A study of the physiology of the isolate obtained from *P. bacteriophila* revealed no marked differences from typical laboratory strains of *K. pneumoniae* (Centifanto and Silver, 1964; Centifanto, 1964). It is of interest that the vigorous evolution of gas (H_2 and CO_2) by *K. rubiacearum* which is typical in N-free sucrose mineral broth is apparently not evident when the culture is run through the usual enteric differential tests for several reference laboratories classified the culture as an anaerogenic strain.

That *K. rubiacearum* is a free-living N_2-fixing bacterium does not prove that it fixes N_2 in the plant. In an attempt to test this directly and to discover whether N_2 fixation was necessary for the reversal of the dwarf syndrome in 'bacteria-free' plants, Silver and Neelands obtained strains by UV and X irradiation which were unable to fix N_2 (Neelands, 1967). In these cultures inability to fix N_2 appeared to be associated with the loss of part or all of the nitrogenase complex, for pyruvate metabolism was normal yet asparagine-grown, nitrogen-sparged cultures failed to fix $^{15}N_2$ or reduce acetylene to ethylene (Neelands, 1967). The status of these cultures in purposely infected host plants could not be followed up due to the reversion of available dwarf plants to the normal, nodulated condition.

Characterization of the properties of the cell-free nitrogenase from the parent N_2-fixing strain revealed little difference from the enzyme in other free-living bacteria (Heeb, 1968), or the enzyme from *K. pneumoniae* (Detroy *et al.*, 1968; Kelly, 1969).

Some of the cultures isolated by Gordon (1963) from species of *Psychotria*, *Pavetta* and *Ardisia* were capable of N_2 fixation in pure culture (see Table 7.5). Although a detailed description of one isolate, *C. lividum* has appeared (Bettleheim, Gordon and Taylor, 1968), the relationship of this organism to *K. rubiacearum* and others reported previously (Gordon, 1963), as well as the universality of N_2 fixing bacteria as the endophytes in the foliar symbioses, is not known.

7.5.2 Nitrogen fixation in detached and fragmented leaves

Attempts to detect nitrogen fixation in detached leaves have not usually been successful. Silver, Centifanto and Nicholas (1963) reported evidence for N_2 fixation by leaf homogenates using the isotopic method, but subsequently this was not routinely obtained (Silver, unpublished data). Furthermore, detached leaves were inactive as determined by the more sensitive acetylene reduction assay even after 18 hours exposure (Sloger, 1968; Silver, 1969). Also, Bond (1959a) mentioned in an earlier report there was no evidence for $^{15}N_2$ enrichment when a detached leaf of *P 'bacteriophila* was exposed to an appropriate gas mixture. An uncontrollable critical factor, in so far as N_2 fixation is concerned, is the size of the effective population in the plant tissue being tested.

7.5.3 Nitrogen fixation in the intact plant

Indirect yet valid arguments may be made for the occurrence of nitrogen fixation in the plant from the readily observable vigorous growth of nodulated plants over long periods of time in N-free nutrient solution (Centifanto and Silver, unpublished observations). In addition, the observation that nodulated *Psychotria* appears deep green in the field amidst other obviously N-deficient plants in low N soil of Papua (Purseglove, personal communication), cannot be dismissed. However, in another leaf nodulated species, *Psychotria calva*, the nitrogen content of leaves was not higher than that of other trees growing in the area (Löhr, 1968). The rate of fixation must be low. A technical problem which has thus far prevented critical testing with whole plants is the difficulty of exposing attached leaves (not the entire plant) to an ^{15}N enriched gas phase for prolonged periods. The author's experience has been that attached leaves enclosed in a glass gassing vessel for more than several days do not remain vigorous.

There is good reason to believe that, whether or not N_2 fixation occurs in the plant, the physiological basis for the mutualistic relationship between bacterium and plant is related to plant growth regulators (Humm, 1944; Silver, 1969). The dwarf growth habit may be noted in plants growing on fixed nitrogen, and its remission is invariably correlated by reversion to the nodulated state irrespective of the nutrition (Centifanto, 1964). This most interesting problem as well as the possible contribution of nitrogen fixation to the plant is approachable experimentally using dwarf plants inoculated with nitrogenless, mutants marked with appropriate non-selective genetic markers.

There has been renewed interest in the symbiosis in *Gunnera* in which symbiotic blue-green algae are located in glands at the base of leaves (Schaede, 1962). Nitrogen-free sand cultures of *Gunnera dentata* increased in nitrogen from 0.78 to 6.84 mg/plant in 10 weeks, a highly significant increase; and nitrogen fixation was confirmed with $^{15}N_2$ and acetylene reduction (Silvester and Smith, 1969). Now that methods are available for the rapid isolation of the blue-green symbionts (of for example, *Gunnera, Macrozamia, Encephalartos* and *Bowenia*) in axenic culture (Bowyer and Skerman, 1968), these symbioses will undoubtedly receive more attention in the future.

7.6   PHYSIOLOGICAL ECOLOGY OF N_2 FIXATION BY NON-LEGUMES

7.6.1   *Introduction*

The possible involvement of nitrogen fixation in the growth of certain plants in natural areas can be obvious even to the botanical novice for, in habitats where soil N is limiting, plants which use dinitrogen appear dark green amidst a generally pale green and yellow background. The difference is often as obvious as is a comparison between plantings of nodulated and non-nodulated legumes in N-deficient soils. This is, in itself, presumptive evidence that a plant species is capable of N_2 fixation. Furthermore, such plants are commonly surrounded by grasses healthier appearing than more distant areas. The author noted a striking example of this along a roadside planting of *Casuarina* adjacent to Waiku paddock on the island of Hawaii where the pasture was vigorous only along the roadside exposed to the leachings of the massive planted windbreak. The designation of many species of nodulated non-legumes as pioneer plants and their use in this capacity acknowledges that the occurrence of N_2 fixation is accepted even in the absence of supporting quantative data.

7.6.2   Temperate habitats

Probably more work has been done on alder than any other of the non-legume symbiosis (Trappe *et al.*, 1968). Becking (1970*a*) cited the experiments of Dinger who demonstrated that leaves of denodulated plants had less N than those of nodulated specimens. It is somewhat remarkable that this was demonstrated only seven years after the early experiments of Hellriegel and Wilfarth on legumes.

In New Zealand, where there are few native legumes, non-legumes like *Coriaria* have been implicated as the main source of some 82 lb N/acre/year over periods of 5 to 14 years. After 55 years a mixed forest had developed whereas at 14 years approximately 75% of the cover was *Coriaria* (Silvester, 1969). In this habitat *Coriaria* plays a role as an important pioneer plant similar to that of *Alnus* in other locations.

Even though one might suspect the favorable nutritive quality of leaf litter under alder, the complex interaction in nature makes interpretation difficult. Crocker and Major (1955) and Lawrence *et al.*, (1968) have taken advantage of the accurate geochronology available in zones of deglaciation at Glacier Bay, Alaska and have reconstructed the succession during revegetation. In this area the rate of recession was remarkably rapid having traveled approximately 96 km in about 150 years even though the receding glacier was perhaps 800 m thick! The effect of *Alnus*, as well as *Dryas* and *Shepherdia* to a lesser extent, on soil reaction, soil N and vegetation was very marked. A striking feature was the remarkable acidifying effect of alder which lowered the pH of the soil horizon from 8.0 to 5.0 within 30-35 years. The abundant leaf fall was especially well retained due to the growth habit and sociability of the plant. It was estimated that the mean net annual rate of accumulation was about 55 lb N/acre (Crocker and Major, 1955).

Under more controlled environmental conditions, Mikola (1966) documented that the N sequestered by alder was available to pine, for both the N and chlorophyll content as well as the needle length were improved when plantings were supplemented with alder litter. In green house experiments, alder litter proved even superior to $(NH_4)_2SO_4$ supplementation when the litter was added at a rate of 200 kg N/m^2 which corresponds to 500 kg/ha (Mikola, 1966). High rates of fixation were reported by Zavitkoviki and Newton (1968) who estimated that fixation of 100 kg/ha likely during the first year in the field. All these observations verify that the classic pot experiments showing a beneficial effect of alder on spruce (Virtanen, 1957) may be applied to natural systems as well.

In the *Myricenaceae* the distribution of the species in nature is also indicative of a common capacity for nitrogen fixation. Schramm (1966) has pointed out that *M. aspenifolia* (*Comptonia peregrina*) is an excellent pioneer in the revegetation of anthracite waste spoils in Pennsylvania. This plant is common in the Indiana sand dunes, along road cuts in Wisconsin, and *M. cerifera* has been tested for beach erosion control in the Carolinas. An attempt to assess the contribution of *M. cerifera* to the N of a Florida soil based on the acetylene reduction assay of 'all' of the nodular tissue of a single plant in the field, yielded a minimal figure of 3 lbs N/acre annum (= 0.55 kg/ha/yr) (Silver and Mague, 1970).

Prior to a direct demonstration of $^{15}N_2$ fixation by *Ceanothus* (Delwiche, Zinke and Johnson, 1965), there were reports of their ecological importance in mountainous areas of western North America. For example, *Ceanothus* exerted a favorable effect upon yellow pine (Wahlenberg, 1930); tomato plants grown in soil previously supporting *Ceanothus* had twice the N content of plants grown in control soil (Hellmers and Kelleher, 1959); nodulated *C. interriums* grew as well as N-supplemented plants in N-deficient soils (Vlamis, Schultz and Biswell, 1958). The rates of fixation obtained by Delwiche, Zinke and Johnson (1965) range from 10 to 212 mmoles/hr/g fresh weight for twelve species of *Ceanothus*. All were ecologically significant rates. Webster (1968) estimated that 21.6 mg N/g dry wt/day were fixed by *Ceanothus* nodules as determined by the ^{15}N method.

7.6.3   Tropical and sub-tropical habitats

The great rapidity with which nitrogen is turned over and leached from tropical soils is a problem of global importance, and means have been sought to increase the nitrogen content of such soils (see for example Bartholomew and Clark, 1965). An as yet largely unassessed source of tropical soil nitrogen may be that sequestered by root- and leaf-nodulated plants, which are widely distributed throughout the southern parts of Asia, Africa and America, and much of Oceania.

7.6.3.1 *Leaf symbionts and epiphytes.* In warm areas of heavy rainfall the leaf rather than the root system may be an important source of nitrogen returned to the soil. Though carefully collected data documenting this are scarce, agriculturists in Ceylon reputedly value leaves of the nodulated species, *Pavetta indica*, as green manure (Rao, 1928). In *Gunnera*, which contains blue-green algal symbionts in glands at the base of the leaf, as much as 64 lb N/acre/year may be added to the ecosystem (Silvester and Smith, 1969). This would explain its occurrence as a numerous pioneer plant in wet areas of New Zealand. Although the same may be true for the leaf nodule symbionts of *Psychotria,* there are yet no direct data to substantiate this (see 7.5.3).

In addition to the microhabitat within the plant where the rate of nitrogen fixation must indeed be at best very low in some species, the external leaf surface (phyllosphere) represents an area of intense microbial activity (Ruinen, 1961). One of the most frequently noted colonizers of the phyllosphere was *Beijerinckia*, as well as other nitrogen fixing or oligonitrophilic microbes (Ruinen, 1956; Bond, 1959). Experiments *in vitro* in which leaves were floated on water and analyzed after extensive time intervals indicated that the autochonous microflora as well as natural populations supplemented with known nitrogen fixing bacteria caused a gain of 20-105% N within two weeks (Ruinen, 1965). The moist, lush leaf surface in the tropics where moisture does not limit bacterial populations (Leben and Daft, 1969) may account for considerable gains of nitrogen due to non-symbiotic bacteria, a situation analogous to that in the rhizophere (Evans, Hill and Campbell, 1971). Preliminary studies of the nitrogen budget of a tropical rain forest in Puerto Rico have shown a net gain of 161 lb N/acre/year, the greatest part of which was due to nitrogen fixation by the epiphylleae (Edmisten, 1970). In addition to the sporadic reports of a variety of nitrogen-fixing bacteria on the leaf surface (Ruinen, 1956; Bhurat and Sen, 1968; Shende *et al.*, 1968; Meikeljohn, 1962), stipules (Stevenson, 1953) and stems (Balasundaram and Sen, 1968) of tropical plants, the phenomenon may also be of importance in temperate climates. In the latter under field conditions at temperatures $< 11°C$ as much as 1.79 μg N/day was fixed on shoots of Douglas fir (Jones, 1970).

7.6.3.2 Root symbionts and epiphytes. Casuarina, the
dominant species of nodulated non-legume, is distributed widely in
the tropics and subtropics, its extremely light seeds being readily
dispersed by the wind. This, coupled with the capacity for nitrogen
fixation, excellent seed germination, and for some species, resistance
to high salinity, has aided its fortuitous as well as intentional spread
as a pioneer plant. Indeed, *Casuarina* has been noted as the first
higher plant to populate newly formed coral atolls in the Pacific
Ocean (F. R. Fosberg, personal communication). Quantitative
estimates of the contribution of *Casuarina* to soil nitrogen have been
made: 58 Kg N/ha/annum for *C. equisetifolia* in Senegal
(Dommergues, 1963); 58.3 Kg N/ha/annum for *C. deplancheana* in
New Caledonia (Huguenin, 1969).

In addition to the effect of nodulated non-legumes on soil
nitrogen levels, free-living nitrogen-fixing bacteria of the rhizosphere
and rhizoplane may also contribute to the nitrogen economy of the
habitat. Although the rhizosphere microflora are by no means
restricted to nitrogen fixers, large populations of *Azotobacter paspali*
were associated with the root system of *Paspalum notatum* in the
usual habitat (Dobereiner, 1966), and were re-established as a highly
specific population after one year in the rhizosphere of plants
transplanted into soil initially free of the microbe (Dobereiner and
Campelo, 1971). The association of free-living nitrogen-fixing bacteria such as *Azotobacter, Klebsiella* and *Derxia* in the soil and on or
near the surfaces of indigenous plants in diverse ecosystems is outside
the scope of this discussion and will not be considered further.

7.7 SPECULATIONS ON THE EVOLUTION OF SYMBIOTIC NITROGEN FIXATION IN ROOT-NODULE AND FOLIAR SYMBIOSES OF NON-LEGUMES

A question which seems paramount in a discussion of the evolution
of nitrogen-fixing symbiosis is whether the ability of an endophyte
to fix nitrogen is intimately related with and a *sine qua non* for the
occurrence of the symbiosis of whether it is a likely outcome of the
physiology of the microbes which are predominant in that type of
ecological niche. Perhaps the easiest to visualize are the foliar systems
which will be discussed first.

7.7.1 Evolution of foliar symbioses
Leben (1965) pointed out that our view of the flora of the plant
surface, i.e. epiphytes, has been distorted by a preoccupation with

the pathological state. The plant surface supports populations of two types of microbial epiphytes: 'residents,' which multiply on the surface of healthy plants without causing noticeable damage, and 'casuals,' whose presence is fortuitous since they subsist saprophytically on debris largely foreign to the plant. In temperate zones there are few visible aggregations of resident epiphytes for moisture is limiting (Leben and Daft, 1967). However, in the tropics a rich and somewhat specific microflora is found (see 7.6.3.1). That nitrogen fixers and oligonitrophiles predominate is not surprising for carbon is available in the form of sucrose and polymeric sugars whereas nitrogen is limited. In addition to the sparcity of fixed nitrogen exerting a strong selective pressure for the above types, a high C/N ratio leads to an enhanced production of capsular slime by the epiphyte. The importance of the capsule in the adherence of the epiphyte to the host plant in the rain forest is obvious.

It is not unreasonable to assume that in many species of tropical plants relatively specific ecological niches on the surface of and within the plant were occupied by diverse microbial species. Ziegler (1958) has shown that leaf hydathodes and bud apices of *Coffea arabica* contain several types of bacteria. Coffee is a member of the *Rubiaceae* and is therefore phyllogenetically closely related to the leaf-nodule bearing species, *Psychotria* and *Pavetta*. In the former the symbiont may be observed penetrating stomatal regions in leaf primordia and masses of bacterial cells surround the growing point (Centifanto, 1964). In other species of related plants the interaction is similar (Faber, 1914; Gordon, 1963). An evolutionary progression from loose associations found in the epiphytes to the foliar symbioses is not difficult to envisage.

However, the absolute requirement for the endophyte for normal growth and survival of *Psychotria* is more difficult to explain. The effect of gibberellins on non-nodulated plants coupled with the known ability of bacteria and fungi to synthesize plant hormones suggests that the hormonal effect rather than nitrogen fixation is of paramount importance in this symbiosis (Silver, 1969). How this became an obligate property during its evolution is more difficult to visualize.

That nitrogen fixation by the endophyte is not required per se in all the foliar symbioses is indicated by the inability of the endophyte of *Ardisia* to fix nitrogen (de Jongh, 1938; Yamada, 1960). De Jongh (1938) has suggested that peroxidase levels are important, and Yamada (1960) has presented evidence that a plant growth regulator, probably IAA, may be involved. Yamada has assumed that the host plant lost the ability to synthesize IAA and that the synthesis can occur only when specific bacteria are present in the meristem. In view of the complex interaction between plant growth regulators and the significance of peroxidase levels (Galston and Davis, 1969) this must certainly be a great oversimplification of the complex interaction between host and endophyte in *Ardisia*.

7.7.2 Evolution of root-nodule symbioses

Since the symbionts of the non-legumes have not been isolated a discussion of their evolution as symbionts must be largely based on conjecture and analogies with the legumes. Whether or not the endophyte can fix nitrogen in the free living form cannot be answered until the organisms are isolated.

One might assume that the actinomycetes infecting the root of a non-legume penetrate the root by the combined action of IAA production and some pectinolytic ability. A streptomycete isolated from root-nodules of *Ceanothus* induced root swelling but not infection (Wollum, Youngberg and Gilmour, 1966). The factors involved were not mentioned. The production of plant growth regulators by soil microorganisms is not an unusual occurrence (Katznelson and Cole, 1965) so the endophytes could easily produce them. The presence of and importance of hemoglobin in non-legumes is not clear (7.2.4). A gradual progression early in the evolution of plants from 'loose' to 'tight' associations seems plausible. Parker (1957) envisages the major stages in the process as follows:

Casual associations of N_2-fixing bacteria and plants →
Loose associations on plant surfaces →
Symbiosis within the plant in cortical tissues →
Symbiosis in organized, highly adapted tissues.

With such tenuous information it does not appear profitable to speculate further at this time on these symbioses in the way that has been done for legumes (Dilworth and Parker, 1969). It is obvious that insufficient of the microbiology and biochemistry of these diverse systems is known.

REFERENCES

E. ADJANOHOUN, *Compte. Rend.*, **245**, 576 (1957).
S. AHMED and H. J. EVANS, *Proc. Natl. Acad. Sci.*, *U.S.A.*, **47**, 24 (1961).
E. K. ALLEN and O. N. ALLEN, Encyclopedia of Plant Physiology. (W. Ruhland, ed.). Springer Verlag, Berlin, **8**, 48 (1958).

E. K. ALLEN and O. N. ALLEN, Microbiology and Soil Fertility, p. 77. Oregon State University Press, Corvallis, Oregon, (1964).

J. D. ALLEN, W. B. SILVESTER and M. KALIN, *N.Z. J. Bot.*, **4**, 57 (1966).

A. J. ANDERSON, Inorganic Nitrogen Metabolism. (W. D. McElroy and B. Glass, eds.), p. 3. John Hopkins Press, Baltimore, (1956).

M. H. APRISON and R. H. BURRIS, *Science*, **115**, 264 (1952).

M. H. APRISON, W. E. MAGEE and R. H. BURRIS, *J. biol. Chem.*, **208**, 29 (1954).

V. R. BALASUNDARAM and A. SEN, *Indian J. Microbiol.*, **8**, 33 (1968).

W. V. BARTHOLOMEW and F. E. CLARK, (eds.) Soil Nitrogen, Am. Soc. Agron. Madison, Wisconsin. (1965).

G. T. S. BAYLIS, *Nature, Lond.*, **221**, 1267 (1969).

J. H. BECKING, *Pl. and Soil*, **15**, 217 (1961).

J. H. BECKING, *Pl. and Soil*, **23**, 213 (1965).

J. H. BECKING, *Annls. Inst. Pasteur, Paris*, **111**, 295 (1966a).

J. H. BECKING, *Annls. Inst. Pasteur, Paris*, **111**, 211 (1966b).

J. H. BECKING, *Pl. and Soil*, **32**, 611 (1970a).

J. H. BECKING, *Int. J. Sys. Bacteriol.*, **20**, 201 (1970b).

J. H. BECKING, Biological Nitrogen Fixation in Natural and Agricultural Habitats. p. e29. IBP-PP Technical Meeting Abstracts, Wageningen. (1970c).

J. H. BECKING, W. E. DE BOER and A. L. HOUWINK, *Antonie van Leeuwenhoek*, **30**, 343 (1964).

F. E. BENDAÑA, Effect of nodules on geotropisms of roots. 98 pp. Dissertation, Univ. of Florida, Gainesville. (1963).

F. J. BERGERSEN, *Biochim. Biophys. Acta*, **130**, 304 (1966).

F. J. BERGERSEN and A. B. COSTIN, *Aust. J. biol. Sci.*, **17**, 44 (1964).

F. J. BERGERSEN, G. S. KENNEDY and W. WITTMAN, *Aust. J. biol. Sci.*, **18**, 1135 (1965).

K. A. BETTELHEIM, J. F. GORDON and J. TAYLOR, *J. gen. Microbiol.*, **54**, 177 (1968).

M. C. BHURAT and A. SEN, *Indian J. agric. Sci.*, **38**, 319 (1968).

G. BOND, *Ann. Bot. Lond. N.S.*, **15**, 447 (1951)

G. BOND, *Ann. Bot. N.S.* **21**, 513 (1957).

G. BOND, Nutrition of the Legumes. (E. G. Hallsworth, ed.), p. 216. Academic Press, New York, (1958).

G. BOND, *Advmt. Sci.*, 382 (1959a).

G. BOND, Utilization of Nitrogen and its Compounds by Plants. *Symp. Soc. Exp. Biol.*, **13**, 59 (1959b).

G. BOND, *Z. allg. Mikrobiol.*, **1**, 93 (1961).

G. BOND, *Nature, Lond.*, **193**, 1103 (1962).

G. BOND, Symbiotic Associations. *Symp. Soc. Gen. Microbiol.*, (P. S. Nutman and Barbara Mosse, eds.), **XIII**, 72 (1963).

G. BOND, *Nature, Lond.*, **204**, 600 (1964).

G. BOND, *Ann. Rev. Pl. Physiol.*, **18**, 107 (1967a).

G. BOND, *Phyton B. Aires*, **24**, 57 (1967b).

G. BOND, Recent Aspects of Nitrogen Metabolism in Plants. (E. J. Hewitt and C. V. Cutting, eds.), p. 15. Academic Press, London and New York, (1968).

G. BOND, W. W. FLETCHER and T. P. FERGUSON, *Pl. Soil*, **5**, 309 (1954).

G. BOND and E. J. HEWITT, *Nature Lond.*, **190**, 1033 (1961).

G. BOND and J. T. MAC CONNEL, *Nature, Lond.*, **176**, 606 (1955).

G. BOND and G. D. SCOTT, *Ann. Bot. Lond. N.S.*, **19**, 67 (1955).

S. R. BOSE, *Nature, Lond.*, **175**, 395 (1955).

J. W. BOWYER and V. B. D. SKERMAN, *J. gen. Microbiol.*, **54**, 299 (1968).

C. E. B. BREMENKAMP, *J. Bot. Lond.*, 271 (1933).

C. E. B. BREMENKAMP, *Repert. Spec. nov. regn. vegetabilis*, **37**, 1 (1934).

Y. M. CENTIFANTO, Ph.D. Dissertation, University of Florida, Gainesville. 133 pp. (1964).

Y. M. CENTIFANTO and W. S. SILVER, *J. Bact.*, **88**, 776 (1964).

H. CHAUDHURI and A. R. AKHTAR, *J. Indian Bot. Soc.*, **10**, 43 (1931).

R. L. CROCKER and J. MAJOR, *J. Ecol.*, **43**, 427 (1955).

H. E. DAVENPORT, *Nature, Lond.*, **186**, 653 (1960).

PH. DE JONGH, Verhandelingen der koninklijke Nederlandsche Akademic Van Wetenschappen, Afdeeling Natuurkunde, Tweede Sectie. Deel XXXVII, No. 6. (1938).

C. C. DELWICHE, C. M. JOHNSON and H. M. REISENHAUR, *Pl. Physiol.*, **36**, 73 (1961).

C. C. DELWICHE, P. J. ZINKE and C. M. JOHNSON, *Pl. Physiol.*, **40**, 1045 (1965).

R. W. DETROY, D. F. WITZ, R. A. PAREJKO and P. W. WILSON, *Proc. Natl. Acad. Sci. U.S.A.*, **61**, 537 (1968).

J. T. DE VRIES and H. G. DERX, *Annales Bogorienses*, **1**, 53 (1950).

M. J. DILWORTH and C. A. PARKER, *J. Theoret. Biol.*, **25**, 208, (1969).

J. DOBEREINER, *Pesq. agropec. bras.*, **1**, 357 (1966).

J. DOBEREINER and A. B. CAMPELO, In Biological nitrogen fixation in natural and agricultural habitats. (T. A. Lie and E. G. Mulder, eds.), *Plant and Soil*, special IBP issue.

Y. DOMMERGUES, *Agrochimica*, **7**, 335 (1963).

J. DULLAART, *Acta. Bot. Neerl.*, **17**, 496 (1968).

J. DULLAART, *J. exp. Bot.*, **21**, 975 (1970).

J. EDMISTEN, A Tropical Rain Forest: a study of irradiation and ecology at El Verde, P.R. (H. T. Odum, ed.), p. H-211. Division Technical Information, U.S. Atomic Energy Comm., (1970).

K. EGLE and HERTA MUNDING, *Naturwiss.*, **38**, 548 (1951).

H. J. EVANS, S. HILL and N. E. R. CAMPBELL, *Can. J. Microbiol.* In Press.

H. J. EVANS and M. KLIEWER, *Ann. N.Y. acad. Sci.*, **112**, 735 (1964).

F. C. VON FABER, *Jb. wiss. Bot.*, **51**, 285 (1912).

F. C. VON FABER, *Jb. wiss. Bot.*, **54**, 243 (1914).

W. W. FLETCHER, *Ann. Bot., n.s.*, **19**, 501 (1955).

G. E. FOGG and W. D. P. STEWART, *Brit. Antarctic Surv. Bull.*, **15**, 39 (1968).

E. B. FRED, I. L. BALDWIN and E. McCOY, *Root Nodule Bacteria and Leguminous Plants*, Madison, Wisc. (1932).

A. W. GALSTON and P. J. DAVIS, *Science*, **163**, 1288 (1969).

I. C. GARDNER, *Nature, Lond.*, **181**, 717 (1958).

I. C. GARDNER and G. BOND, *Can. J. Botany*, **35**, 305 (1957).

I. C. GARDNER and G. BOND, *Naturwiss.*, **53**, 61 (1966).

E. M. S. GATNER and I. C. GARDNER, *Arch. Mikrobiol.*, **70**, 183 (1970).

W. GLAUBITZ, Diplom-Arbeit, Hamburg, 82 pp. (1957).

J. F. GORDON, Ph.D. Thesis, Imperial College, London., 370 pp. (1963).

K. HANADA, *Jap. J. Bot.*, **14**, 235 (1954).

J. L. HARLEY and D. H. LEWIS, *Adv. Microbiol. Physiol.*, (A. H. Rose and J. F. Wilkinson, eds.), Academic Press, New York, **3**, 53 (1969).

MARY JO. HEEB, Master's Thesis, University of Florida, Gainesville 50 pp. (1968).

H. HELLMERS and J. M. KELLEHER, *Forest Science*, **5**, 3 (1959).

E. HENRIKSSON, *Physiologia, Pl.*, **4**, 542 (1951).

E. J. HEWITT and G. BOND, *Pl. Soil*, **14**, 159 (1961).

R. D. HÓLSTEN, R. C. BURNS, R. W. F. HARDY and R. R. HERBERT, In press (1970).

B. HUGUENIN, These, Universite-de Rouen. 56 pp. (1969).

H. J. HUMM, *J. N.Y. Bot. Gdn.*, **45**, 193 (1944).

K. JONES, *Ann. Bot. n.s.*, **34**, 239 (1970).

H. KATZNELSON and SHIRLEY E. COLE, *Can. J. Microbiol.*, **11**, 733 (1965).

M. KELLY, *Biochim. Biophys. Acta*, **191**, 527 (1969).

K. A. KERSHAW and J. W. MILLBANK, *New Phytol.*, **69**, 75 (1970).

B. KOCH, H. J. EVANS and S. RUSSELL, *Proc. natn. Acad. Sci. U.S.A.*, **58**, 1343 (1967).

R. T. LANGE, *Symbiosis.*, (S. M. Henry, ed.), **1**, 99 (1966).

D. B. LAWRENCE, R. E. SCHOENIKE, A. QUISPEL and G. BOND, *J. Ecol.*, **55**, 793 (1967).

G. LEAF, I. C. GARDNER and G. BOND, *Biochem. J.*, **72**, 662 (1958a).

G. LEAF, I. C. GARDNER and G. BOND, *J. exp. Bot.*, **9**, 320 (1958b).

C. LEBEN, *A. Rev. Phytopath.*, **3**, 209 (1965).

C. LEBEN and G. C. DAFT, *Can. J. Microbiol.*, **13**, 1151 (1967).

H. LEITGEB, *Akad. d. Wissensch. Wien, I Abt.*, **77**, (1878).

N. R. LERSTEN and H. T. HORNER, Jr., *J. Bact.*, **94**, 2027 (1967).

E. LOHR, *Physiol. Plant.*, **21**, 1156 (1968).

R. H. LOWE and H. J. EVANS, *J. Bact.*, **83**, 210 (1964).

J. T. MAC CONNELL, *Ann. Bot., n.s.*, **23**, 261 (1959).

J. MC LUCKIE, *Proc. Linn. Soc. N.S.W.*, **47**, 319 (1922).

J. MC LUCKIE, *Proc. Linn. Soc. N.S.W.*, **48**, 82 (1923).

J. MEIKELJOHN, *Emp. J. exp. Agric.*, **30**, 117 (1962).

E. MELIN, Encyclopedia of Plant Physiology, (W. Ruhland, ed.), Springer Verlag, Berlin, **11**, 605 (1959).

H. MIEHE, *Jahrb. Wiss. Bot.*, **53**, 1 (1914).

J. K. MIETTINEN and A. I. VIRTANEN, *Physiol. Plant.*, **5**, 540 (1952).

P. MIKOLA, Report Project E8-F5-46, Grant E6-Fi-131. U.S. Public Law 60, Helsinki (1966).

J. W. MILLBANK and K. A. KERSHAW, *New Phytol.*, **68**, 72 (1969).

H. MOLISCH, *Sci. Rep. Tohoku Univ.*, **1**, 169 (1925).

A. W. MOORE, *Can. J. Bot.*, **42**, 952 (1964).

T. M. MORRISON, *Nature, Lond.*, **189**, 945 (1961).

T. M. MORRISON and D. A. ENGLISH, *New Phytol.*, **66**, 245 (1967).

L. E. MORTENSON, J. A. MORRIS and D. Y. JENG, *Biochim. Biophys. Acta*, **141**, 516 (1967).

B. MOSSE, Symbiotic Associations. (P. S. Notman and B. Mosse, eds.), p. 146. University Press, Cambridge, (1963).

H. MOWRY, *Soil Sci.*, **36**, 409 (1933).

M. NEELANDS, Master's Thesis, Univ. of Florida, Gainesville, (1967).

L. G. NICKELL, *Amer. Fern J.*, **48**, 103 (1958).

F. NOBBE and L. HILTNER, *Landw. Vers. Sta.*, **51**, 241 (1899).

A. OES, *Z. Bot.*, **5**, 145 (1913).

H. PANKOW, and B. MARTENS, *Arch. Microbiol.*, **48**, 203 (1964).

C. A. PARKER, *Nature, Lond.*, **179**, 593 (1957).

J. S. PATE, *Aust. J. biol. Sci.*, **11**, 516 (1958).

E. PETIT, *Bull. State Botanic Gdn.*, Brussels, **36**, 65 (1966).

J. PHILLIPS, *Ecology*, **13**, 189 (1932).

G. PIZELLE, *Annls. Inst. Pasteur. Paris*, **111**, 259 (1966).

G. PIZELLE, *Bull. Acad. Soc. Lorraines Sci.*, **9**, 174 (1970).

A. RAO, *Agri. J. India*, **18**, 132 (1928).

K. C. RODRIGUEZ-BARRUECO, *Phyton B. Aires*, **23**, 103 (1966).

J. RUINEN, *Nature, Lond.*, **177**, 220 (1956).

J. RUINEN, *Pl. Soil*, **15**, 81 (1961).

J. RUINEN, *Pl. Soil*, **22**, 375 (1965).

R. SCHAEDE, *Planta*, **33**, 703 (1943).

R. SCHAEDE, Die pflanzlichen Symbiosen, 2nd ed., Gustav Fischer Verlag, Stuttgart, (1962).

J. R. SCHRAMM, *Trans. Am. Phil. Soc.*, **56**, 1 (1966).

G. D. SCOTT, *New Phytol.*, **55**, 111 (1956).

G. D. SCOTT, Plant Symbiosis. p. 17. St. Martin's Press, New York, (1969).

S. T. SHENDE, V. R. BALASUNDARAM, M. C. BHURAT and A. SEN, *Indian J., Agri. Sci.*, **38**, 298 (1968).

K. SHIBATA, *Jb. wiss. Bot.*, **37**, 643 (1902).

W. S. SILVER, *J. Bact.*, **87**, 416 (1964).

W. S. SILVER, *Proc. Roy. Soc. B.*, **172**, 389 (1969).

W. S. SILVER, F. E. BENDAÑA and R. D. POWELL, *Physiologia Pl.*, **19**, 207 (1966).

W. S. SILVER, Y. M. CENTIFANTO and D. J. D. NICHOLAS, *Nature, Lond.*, **199**, 396 (1963).

W. S. SILVER and T. MC HUE, *Nature, Lond.*, **227**, 378 (1970).

W. B. SILVESTER, *N.Z. Soil News*, **9**, 142 (1969).

W. B. SILVESTER and D. R. SMITH, *Nature, Lond.*, **224**, 1231 (1969).

C. SLOGER, Dissertation, Univ. of Florida, Gainesville, 95 (1968).

C. SLOGER and W. S. SILVER, Non-Heme Iron Proteins: Role in Energy Conversion. (A. San Pietro and L. P. Vernon, eds.), p. 299. Antioch Press, Yellow Springs, Ohio, (1965).

C. SLOGER and W. S. SILVER, *Bact. Proc.*, 112 (1967).

C. SLOGER and W. S. SILVER, IX Internat. Cong. of Microbiol., Abstracts, Moscow (1966).

E. R. SPRATT, *Ann. Bot.*, **25**, 801 (1912).

J. I. SPRENT, *Planta*, **88**, 372 (1969).

G. B. STEVENSON, *Ann. Bot. n.s.*, **17**, 343 (1953).

W. D. P. STEWART, *Jour. Exptl. Botany*, **13**, 250 (1962).

W. D. P. STEWART and G. BOND, *Pl. Soil*, **14**, 347 (1961).

J. M. TRAPPE, J. F. FRANKLIN, R. F. TARRANT and G. M. HANSEN, (eds.) *Biology of Alder*. USDA Forest Service, Portland, Oregon, (1968).

S. UEMURA, *Proc. First Symp. on Nitrogen Fixation and Nitrogen Cycle* (JIBP-PP). (H. Takahashi, ed.), p. 37. Sendai, (1967).

S. UEMURA, *Proc. Second Symp. on Nitrogen Fixation and Nitrogen Cycle* (JIBP-PP). (H. Takahasi, ed.), p. 1. Sendai, (1970).

G. S. VENKATARAMAN, *Indian J. agric. Sci.*, **32**, 22 (1962).

A. I. VIRTANEN, *Physiologia Pl.*, **10**, 164 (1957).

J. VLAMIS, A. M. SCHULTZ and H. H. BISWELL, *Calif. Agr., Jan.*, **11**, (1958).

W. G. WAHLENBERG, *J. Agric. Res.*, **41**, 601 (1930).

A. WATANABE and T. KIYOHARA, *Microalgae and Photosynthetic Bacteria* (Suppl. to *Pl. Cell Physiol.*). 189 (1963).

S. R. WEBSTER, MS Thesis, Oregon State University, 81 pp (1968).

C. T. WHEELER, *New Phytol.*, **68**, 675 (1969).

C. T. WHEELER and G. BOND, *Phytochem.*, **9**, 705 (1970).

A. G. WOLLUM, II, C. T. YOUNGBERG and C. M. GILMOUR, *Soil Sci. Soc. Am., Proc.*, **30**, 463 (1966).

T. YAMADA, *Bull. Fac. Ed., Chiba University*, **9**, 1 (1960).

D. C. YOCH and R. M. PENGRA, *J. Bact.*, **92**, 618 (1966).

J. ZAVITKOVSKI and M. NEWTON, Biology of Alder, (J. M. Trappe, *et al.* eds.), p. 209 (1967).

H. ZIEGLER, *Z. Naturforsch.*, **199**, 508 (1958).

H. ZIEGLER and R. HUSER, *Nature, Lond.*, **199**, 508 (1963).

APPENDIX 1

A Review of Research by Soviet Scientists on the Biochemistry of Nitrogen Fixation

M. G. YATES

University of Sussex,
Falmer, Brighton BN1 9QJ,
Sussex, England.

A1.1   INTRODUCTION

The biochemistry of nitrogen fixation has advanced impressively during the last decade since the publication by Carnahan, Mortenson,

Mower and Castle (1960) of a method to obtain cell free nitrogen fixation in extracts from *Clostridium pasteurianum*. There seems to have been very little contact between East and West in this field although Soviet scientists have actively studied the problem and some of their claims and suggestions have been far reaching: for instance, Turchin, Berzenova and Zhidkikh (1963) claim that plants which do not fix nitrogen contain a dormant nitrogenase. It seems relevant, therefore, to review Soviet research in the subject. Soviet laboratories have in many instances confirmed results published earlier from Western laboratories but some results contradict Western observations and, in other cases, the Soviet scientists have used techniques that have not been used extensively in the West to study nitrogen fixation. For obvious reasons Soviet work that falls into these latter two categories is discussed here in more detail than that which merely confirms Western observations. The majority of Soviet work on the biochemistry of nitrogen fixation has been published in the past few years; reviews by Soviet authors that appeared in the mid nineteen sixties cited largely Western work (Ivanov, Il'ina and Sitonite, 1964; Karmaukov, 1965; Lyubimov, 1965); therefore, this review will be mainly confined to major publications emerging from the U.S.S.R. since 1964. As far as the writer is aware the coverage is reasonably comprehensive; nevertheless some important work may have been missed. In the author's opinion some claims appear with insufficient experimental data to assess their validity. Such papers will be accepted at their face value unless there exist other published data which partly or wholly contradict them, in which case the conflict will be mentioned and discussed.

Nitrogen fixation has been studied more intensively in Russia than in most countries during this century. Much of their effort has been directed towards the agricultural aspect: the contribution by nitrogen fixation to the nitrogen balance in soils and the application of nitrogen fixers as fertilisers have been two major lines of enquiry. On the academic side Soviet scientists have made significant contributions to nitrogen fixation: for example: the discovery of an organism classified as *Mycobacterium flavum* (Federov and Kalininskaya, 1961a) which is the only established aerobic nitrogen-fixing bacterium outside the Azotobacteriaceae. Other nitrogen-fixing mycobacteria isolated are *M. roseo-album* (Federov and Kalininskaya 1961b) and *M. azot absorptium* (L'vov and Lyubimov, 1965). The late Academician Ierusalimskii with Khmel and other colleagues (Ierusalimiskii, Zaitseva and Khmel, 1961; Khmel, Gabinskaya and Ierusalimskii, 1965; Khmel and Gabinskaya, 1965; Khmel and

Ierusalimskii, 1967; Khmel and Andreeva, 1967) have studied continuous culture of *Azotobacter vinelandii* and also the effect of aeration upon growth when fixing nitrogen. Such data as the nutritional state of the population (carbon-limited, oxygen-limited etc.) which are of critical importance in assessing continuous culture studies, are often not clear from these publications. These topics have been expanded and discussed by Dalton and Postgate (1969*a* and *b*) and will not be mentioned again except in connection with more biochemical studies. Kalininskaya has investigated a topic that may be of considerable importance in the ecology of nitrogen fixation: symbiosis during growth between nitrogen-fixing and other bacteria (symbiotrophic nitrogen fixers: Kalininskaya, 1967*a, b, c*; 1969). Such studies have also been made by Japanese investigators (Okuda, Yamagachi and Kobayashi, 1960, 1961*a, b*). This work so far may be broadly defined as biological. On the chemical side, the work of Vol'pin, Shur and their colleagues upon the reduction of nitrogen in the presence of titanium complexes (see Vol'pin and Shur, 1966) has excited considerable interest in the West, both as a chemical process and for its relevance as a model for the biological process (Brintzinger, 1966; Van Tamelen, Boche and Greeley, 1968; Van Tamelen and Seeley, 1969). These matters are discussed in Chapter 2; the purpose of this article is to review the work that has been reported by Soviet scientists directed to the understanding of the biochemistry of nitrogen fixation.

A1.1.1   The choice of organism
In nearly all studies the organism used was *Azotobacter vinelandii*. From the biochemical viewpoint this choice presents difficulties since Azotobacter nitrogenase, although less oxygen-sensitive in crude cell-free extracts than are nitrogenases from anaerobic nitro-gen-fixing bacteria, is particulate in crude extracts and physically associated with many other enzymes, including dehydrogenases and the respiratory chain. Moreover, cell-free extracts that fix nitrogen are extremely difficult to obtain reproducibly unless sodium dithio-nite is used as a reductant; only in the last two years have Soviet scientists reported upon nitrogen fixation using this chemical. Exceptions to this general pattern of work on Azotobacter include reports of nitrogen fixation in *Clostridium pasteurianum* (Ivanov and Bondar, 1968), in a photosynthesising sulfur bacterium, *Chromatium minutissimum* (Ivanov and Demina, 1966*a, b,* Ivanov, 1967; Ivanov, Demina and Gogoleva, 1967; Demina, Ivanov and Medvedev, 1969) and in extracts from lupin nodules (Manorik, Krikunets and

Starchenkov, 1965, Manorik and Starchenkov, 1969). In a few instances, extracts from *C. pasteuranium* have been examined, but only to provide supplementary evidence in publications that deal mainly with extracts from *A. vinelandii* (Moshkovskii. Ivanov, Stukan, Matkhanov, Mardanyan, Belov and Goldanskii, 1967; Ivanov, Moshkovskii, Stukan, Matkhanov, Mardanyan and Belov, 1968).

THE NITROGENASE REACTION

It is generally accepted in the West that nitrogenase contains non-haem iron and molybdenum; Soviet scientists do not generally disagree with this. In addition, an electron-donating system, usually sodium dithionite, an energy source (ATP) and an anaerobic environment are all necessary for nitrogen fixation in cell-free extracts from Azotobacter. These conditions have been used in some Soviet investigations but other authors claim to have observed nitrogen fixation in cell-free extracts without one, two or all three of these requirements; indeed in one instance, nitrogen fixation was apparently observed in the absence of a recognisable nitrogenase fraction (Ivanov, Matkhanov, Belov and Gogoleva, 1967). On the other hand, Ivanov, Gogoleva and Demina (1969) reported that ADP was necessary for nitrogen fixation with washed *A. vinelandii*. These observations are discussed under separate headings below.

A1.2.1   The nature of the reductant
   A1.2.1.1   With aerobic bacteria. Syrtsova, Yakovlev and Rachek (1968); Lyubimov, L'vov, Kirshteine (1968); Alfimova, Gvozdev and Linde (1968); Ivanov, Rakhleeva, Demina, Belov and Gogoleva (1968); Slepko, Uzenskaya, Linde, Alfimova, Rakhman and Likhtenshtein (1969); Novikov, Syrtsova, Likhtenshtein, Trukhtanov, Rachek and Goldanskii (1969); Syrtsova, Likhtenshtein, Pisarskaya, Ganelin, Frolov and Rachek (1969); Ganelin, L'vov, Kirshteine, Lyubimov and Kretovich (1969); Ivleva, Medzhidov, Likhtenshtein, Sadkov and Yakovlev (1969) and Ivleva, Likhtenshtein and Sadkov (1969) used sodium dithionite as a reductant for nitrogen fixation in cell-free extracts from *A. vine-landii*. On the other hand, Il'ina (1968a, 1969) used sodium pyruvate, Ivanov *et al.* (1967) used sodium succinate, Yakovlev, Syrtsova and Vorob'ev (1966) used hydrogen gas, Yakovlev, Vorob'ev, Levchenko, Linde, Slepko and Syrtsova (1965) used sucrose together with Burk's medium, as did Nicholas and Fisher (1960); all claimed to have observed nitrogen fixation. Other Western

investigators were not able to substantiate the claims of Nicholas and Fisher (1960) with extracts from either *A. vinelandii* (Bulen, Burns and Lecomte, 1964) or *A. chroococcum* (Kelly and Postgate: unpublished work in the ARC Unit of Nitrogen Fixation). Therefore, these Soviet reports conflict with what is known or would be expected from Western work and so will be discussed further. Il'ina grew *A. vinelandii* in a sucrose medium (resembling Burk's but containing small amounts of sodium citrate and sodium acetate) under either static conditions for 15 days or with vigorous aeration for 19 hours. Cells were suspended in culture fluid containing 25 mM phosphate buffer, pH 8.0 and disrupted in a press after freezing. The crude cell-free supernatant, after spinning at 14,000 x g fixed nitrogen (according to the ammonia produced) in the presence of oxygen and dinitrogen (1:9, v:v), pyruvate (50 μMoles) and NAD (1 mg) in 1 to 3 ml of assay mixture. Maximum specific fixation was 0.04 μMoles N_2/mg protein/hr: approximately one tenth of that usually obtained in Western laboratories with dithionite and crude extracts from *Azotobacter*. ATP or ADP enhanced fixation. The author emphasised the absence of undamaged bacteria in the extracts and claimed that oxidative phosphorylation provided energy for the nitrogen fixation observed. This is a feasible explanation and such conditions should be obtained in principle. Yates and Daniel (1970) obtained preparations from *A. chroococcum* that reduced acetylene to ethylene in the presence of any one of several physiological electron donors including sodium pyruvate, provided ATP or ADP was present. In their preparations, however, acetylene reduction was inhibited by a pO_2 of 0.05 atmospheres.

Ivanov *et al.* (1967) tested horse-heart succinoxidase with preparations of *A. vinelandii* cytochromes c_4 and c_5 (isolated according to Tissieres and Burris, 1956) and claimed to observe nitrogen fixation under anaerobic conditions over periods from four to twelve hours. Their analytical data were impressive: for example, 1 mMole of NH_3 was produced or 1.6 atom % $^{15}N_2$ enrichment was obtained (no protein concentrations were quoted so specific activities cannot be assessed). Such a system should be free from conventional nitrogenase and the authors claimed that the results with this 'model system' indicated that cytochromes were involved in electron transport to nitrogenase, an hypothesis that is supported, according to the authors, by the observation that 10^{-3} M KCN inhibited 80% of nitrogen fixation. The authors assumed that cytochrome oxidase was the site of action of KCN; however, CN^- is an alternative substrate for nitrogenase (Hardy and Knight, 1967) and competes with nitrogen for electrons (Burris, 1969). It is surprising that such a system, as described above, fixes nitrogen, in view of what is known about the nitrogenase reaction and its need for ATP. It could not be confirmed in the reviewer's laboratory using the succinoxidase or NADH oxidase from beef heart mitochondria and a cytochrome fraction from *A. chroococcum*.

Yakovlev *et al.* (1965) obtained cell-free extracts from *A. vinelandii* by lysozyme treatment or mechanical disintegration. Crude supernatants, after centrifuging at 8000 x g for 15 minutes fixed nitrogen over a 2-hour period under oxygen/nitrogen mixtures (1:4) containing 8% $^{15}N_2$ with sucrose and Burk's medium; the highest enrichment achieved was 0.635 atm. % in the ammonia produced. Yakovlev *et al.* (1966) stirred a similar crude supernatant fraction from *A. vinelandii* with small amounts of DEAE cellulose. This treatment removed a component essential for nitrogen fixation, but addition of 'ferridin', a non-haem iron protein isolated from *A. vinelandii* by the same authors, restored nitrogen fixing ability provided hydrogen (in a gas ratio $N_2:H_2:Ar,3:1:6$ v/v) was present. When oxygen (0.2 atmospheres) replaced hydrogen no nitrogen fixation was observed, although the authors claimed that the crude extract before DEAE treatment was active when oxygen was present (under conditions similar to those described above which were used by Yakovlev *et al.* (1965) 0.93 atm. per cent enrichment of the ammonia was obtained). Bulen, Burns and Lecomte (1964) failed to show nitrogen fixation in crude extracts from *A. vinelandii* with hydrogen as the electron donor unless hydrogenase and ferredoxin from *C. pasteurianum* were used. Kelly and Postgate, in unpublished work in the ARC Unit, attempted, without success, to use hydrogen as an electron donor for nitrogen fixation with either whole cells or cell-free extracts from *A. chroococcum*. Yates and Daniel (1970) could not include hydrogen amongst the substrates that provided electrons for acetylene reduction in preparations from *A. chroococcum*.

In a further article Syrtsova, Yakovlev and Rachek (1968) (see section A.5) claimed that 'ferridin' was identical with fraction 2 of *A. vinelandii* nitrogenase (Bulen and Lecomte, 1966; Kelly, Klucas and Burris, 1967). Although the elution characteristics of ferridin and fraction II from DEAE cellulose are similar, this contention is unlikely since the nitrogenase function of fraction 2 is irreversibly destroyed by a short (2 to 3 minutes) exposure to oxygen and strict anaerobic precautions are necessary during the purification. Yakovlev *et al.* (1966) do not mention any such precautions during a purification procedure involving two ammonium sulfate fractionations and DEAE-cellulose chromatography and yet they claim that 'pure' ferridin is active in nitrogen fixation.

Sodium succinate or ethanol were used successfully as electron donors for nitrogen fixation with washed $A.$ $vinelandii$ cells provided ADP and oxygen were present (Ivanov, Gogoleva and Demina, 1969). The most active preparations gave 0.09 atom % $^{15}N_2$ enrichment. This system appears to have similar requirements to the cell-free system reported by Il'ina above and can be rationalised in a similar way: ADP and oxygen were necessary for oxidative phosphorylation which, in turn, supplied the energy for nitrogen fixation. Presumably the bacteria were sufficiently damaged during washing to respond to externally added ADP. Comparable preparations which reduced acetylene have occasionally been obtained in the writer's laboratory from $A.$ $chroococcum$ but anaerobic conditions were necessary for maximum acetylene reduction.

Levchenko, Ivleva and Yakovlev (1969) claimed that small cell organelles which they called 'mitochondria' (first mentioned by Yakovlev and Levchenko, 1964) contained the site of NADH, glutamine, malate and pyruvate dehydrogenases. They used iodo-nitrotetrazolium indicator and they determined the sites of formazan formation by electron microscropy. They also claimed that substituting nitrogen for argon in gas mixture caused a decrease in the rate of formazan formation in the 'mitochondria' with the four substrates, NADH, malate, pyruvate or glutamate and suggested that the 'mitochondria' contained nitrogenase and that the dehydrogenases transferred electrons to nitrogen. On the other hand, glucose-6-phosphate dehydrogenase was sited in the cytoplasm and did not contribute electrons to nitrogenase. In contrast, Yates and Daniel (1970) found that glucose-6-phosphate was one of the best electron donors for acetylene reduction by preparations from $A.$ $chroococcum.$

$A1.2.1.2$ $Reductant$ in $anaerobic$ $bacteria.$ In a study of washed $C.$ $pasteurianum$ $cells,$ Ivanov and Bondar (1968) measured the rates of reduction of triphenyl tetrazolium bromide to formazan with various carbon substrates and compared these rates under argon or a 1:1 (v:v) mixture of argon and dinitrogen. Nitrogen gas markedly reduced the rate of formazan formation suggesting a competition for electrons between triphenyl tetrazolium bromide and nitrogenase. NH_4Cl at 50 mM inhibited the reduction of triphenyl tetrazolium bromide, hydrogenase activity and nitrogen fixation. Therefore, the authors suggested that carbon substrates could donate electrons to hydrogenase and thus to nitrogenase, or tetrazolium salt when the latter was present. By contrast, in cell-free extracts from *C. pasteurianum,* sodium pyruvate and, to a lesser extent, sodium α-ketobutyrate were the only carbon sources that supported nitrogen fixation. (Carnahan, Castle, Mortenson and Mower, 1960).

A1.2.1.3   Reductant in photosynthetic bacteria. Arnon, Losada, Nozaki and Tagawa (1961) showed that NADH or hydrogen supported low levels of nitrogen fixation (0.048 and 0.043 atm. % enrichment) in cell-free extracts of *Chromatium.* Ivanov and Demina (1966) studied the effects of argon, helium, dinitrogen or hydrogen atmospheres on the reduction of triphenyl tetrazolium chloride to formazan in washed suspensions of *Chromatium minutissimum.* They claimed that the reduction of tetrazolium salt to formazan was depressed under dinitrogen provided that any one of the following potential electron donors was present: hydrogen, pyruvate, oxalo-acetate, α-ketoglutarate, succinate, malate, citrate or ethanol. This depression, they suggested, was due to competition for electrons between dinitrogen and tetrazolium salt and reflected nitrogenase activity; it was independent of light, therefore the authors postulated that nitrogen fixation was a dark process, related to photosynthesis which acted as a supplier of electrons and protons. Ivanov (1967) in his review of biological nitrogen fixation as an oxidation-reduction process extended the range of carbon substrates which act as donors in the electron transport system to nitrogenase in *C. minutissimum* to include glucose, xylose, galactose, sucrose, glycerol and hippuric acid. He suggested that these substrates donated electrons to nitrogenase via hydrogenase, a mechanism similar to that postulated by Ivanov and Bondar (1968) in *C. pasteurianum (section A1.2.1.2).*

A1.2.1.4   Reductant in symbiotic systems: root nodule bacteroids. Several Soviet investigators have indicated that endogenous or specific dehydrogenases act as electron carriers to nitrogenase in nodule bacteroids: Manorik, Krikunets and Starchenkov (1965) and Manorik and Starchenkov (1969) measured nitrogen fixation using $^{15}N_2$ with endogenous substrates in preparations from soybean and lupin nodules and concluded that the bacteroid fraction was most active and that bacteroids were the probable sites of nitrogen fixation. Peive, Zhiznevskaya, Yagodin and Dubrovo (1964) estimated dehydrogenase activity by the rate of reduction of triphenyl-tetrazolium bromide in nodules at different stages of growth of broad beans and found that malate dehydrogenase was the most active dehydrogenase and that it was most active at the budding stage. Unfortunately they did not measure directly the rate of nitrogen fixation, but suggested that increased dehydrogenase activity increased nitrogen fixation. Kretovitch, Evstigneeva, Aseeva, Zargaryan and Martynova (1969) investigated dehydrogenase activities in extracts from effective and ineffective root nodules of two strains of lupin. Malate dehydrogenase was again the most active but its specific activity was 16-fold greater in extracts from effective than in ineffective nodules. Other enzymes were also more active in effective nodules: NADP-glutamate dehydrogenase (10-fold); glutamine synthetase (7-fold); succinic dehydrogenase (3-fold); aspartate amino-transferase and alanine amino-transferase (2-fold). It appears from these results that the metabolic rate of effective nodules is greater than that of ineffective nodules and no specific importance need by attached to malate dehydrogenase in nitrogen fixation. For instance, Klucas and Evans (1968) found that soybean nodule extracts contained high levels of β-hydroxybutyrate dehydro-genase and that β-hydroxybutyrate could support acetylene re-duction by these extracts under anaerobic conditions provided ATP and catalytic amounts of benzyl viologen were added. Manorik and Lisova (1968) found that vitamin B_{12} or Co added to soil increased endogenous dehydrogenase levels and nitrogen fixation in lupin nodules and suggested that the effect of these nutrients upon nitro-gen fixation was through their effect upon dehydrogenase activity. Yagodin and Ovcharenko (1969) found a similar effect of cobalt on dehydrogenase activity of the root nodules of beans, soybeans and lupins. Dorosinski, Zagore and Buziashuili (1966) claimed to have observed a correlation between the endogenous dehydrogenase activi-ties of rhizobia and their effectiveness towards nodule formation; they proposed that this should be a test for selecting effective strains.

A1.2.2   The energy requirement

In all the cell-free systems that have been studied by Western scientists, nitrogenase activity requires energy (ATP); this is true whether the electron source is a carbon substrate such as sodium pyruvate, a natural component of the electron transfer chain such as NADH or an artificial donor such as sodium dithionite or reduced benzyl viologen. Furthermore, there is considerable evidence that this energy expenditure is required in the growing cell. Most of the Soviet investigations have confirmed this requirement for ATP (see references in the first paragraph in section A1.2.1.1) and the successful use of ADP, in some instances, rather than ATP can be rationalised in terms of its ready conversion to ATP by the preparation used. Even the low level of activity without added ATP or ADP observed by Yakovlev *et al.* (1965, 1966) might be accounted for by sufficient ATP produced from endogenous ADP in the presence of oxygen (section A1.2.1.1). There are two notable exceptions: the first is the work of Ivanov *et al.* (1967) with the 'model systems' (section A1.2.1.1) comprising horse heart succin-oxidase and *A. vinelandii* cytochromes, when substantial nitrogen fixation was claimed under anaerobic conditions without an added energy source. The second is by Ivanov, Rakhleeva and Demina *et al.* (1968) who showed that an ADP-dependent hydrogen evolution occurred under argon or an ADP-dependent nitrogen fixation occurred under dinitrogen (98% $^{15}N_2$: argon mixture; 3:1) when sodium dithionite was added to extracts from *A. vinelandii*. This took place without an ATP-generating system, commonly used in such experiments. 10 μM ADP was the optimum level; activity fell away sharply on either side. The reported activity rates varied considerably from 70 to 18 μl. H_2 evolved in 12 min, or 50 to 15 μg of ammonia in two hours or a maximum of 0.035 atm. % excess $^{15}N_2$ fixed. This is the first report that low concentrations of ADP stimulate nitrogen fixation, but it is surprising that measurable rates of nitrogen fixation occurred as no source of ATP was added. Western scientists have observed inhibition by higher concentrations (1 to 5 mM) of ADP: Moustafa and Mortenson (1967) with extracts from *C. pasteurianum* showed that ADP was a specific inhibitor of ATP-dependent hydrogen evolution; Biggins and Postgate (1969) confirmed this with extracts from *M. flavum* and Kelly (1969) showed a similar effect with extracts from *A. chroococcum*.

A1.2.3   The anaerobic nature of nitrogen fixation
There is considerable evidence that nitrogen fixation is an anaerobic process even in aerobic organisms (see Postgate, 1969). Soviet investigations have been consistent with this view in that they generally used anaerobic condition in their experiments. Three exceptions exist: both Il'ina (1968a) and Manorik and Starchenkov (1969) used a low pO_2 (0.1 atmospheres) in their studies of nitrogen fixation in extracts from *A. vinelandii* and in bacteroids from lupin or soybean nodules respectively. These findings are not imcompatible with Western work: Bergersen in his work with soybean nodule bacteroids (Bergersen, 1966*a*, *b*; Bergersen and Turner, 1967, 1968) showed a low-level oxygen requirement for nitrogen fixation and attributed this to a need for energy to be supplied by oxidative phosphorylation; Sloger and Silver (1966) showed a similar oxygen requirement with ground nodular tissue from *Myrica cerifera*. The

extracts produced by Il'ina or by Yakovlev *et al.* (1965, 1966, sections A1.2.1.1) might be compared with bacteroid preparations.

A1.2.4 Metal ions

Since Bortels (1930) first showed a molybdenum requirement in nitrogen-fixing *Azotobacter* it has been confirmed many times for both *Azotobacter* and other micro-organisms. Esposito and Wilson (1956) also showed that nitrogen-fixing *Azotobacter* require more iron than *Azotobacter* grown upon fixed nitrogen. It is now generally accepted that these two elements are specifically involved in nitrogenase activity. Soviet investigators (Federov, 1952, 1963; Gradola-Krylova, 1967; Kurdina, 1968), confirmed the molybdenum requirements for nitrogen-fixing *Azotobacter;* Ivanov, Il'ina and Sitonite (1964) claimed a similar requirement for the anaerobe *C. pasteurianum:* both hydrogenase and nitrogenase activity were inhibited by trichloromethylsulfenyl benzoate but molybdate ions restored both enzyme activities. The inhibitor did not affect *C. pasteurianum* when grown with fixed nitrogen. There was insufficient evidence to evaluate this result critically. In several publications Il'ina (1966a, b, c; 1967a, b; 1968b) has established that *M. flavum* and *M. roseo-album* require molybdenum for nitrogen fixation, but at levels (5-25 $\mu g/1$) several times less than other nitrogen-fixing micro-organisms. To this she attributes their ability to grow in acid podzolic soils where molybdenum is scarce.

Ivanov *et al.* (1964) speculated that molybdenum might be an integral part of enzymes involved in electron transport to nitrogenase. Il'ina (1966c) claimed that, when *M. roseo-album* was grown on an N_2 free medium, endogenous dehydrogenase activity was higher in bacteria grown with adequate molybdenum. However, in a more detailed investigation, Il'ina (1968b) found that this effect of molybdenum on endogenous dehydrogenase varied with the age of the culture; she showed that, in general, hydrogenase and the dehydrogenases of lactate, oxaloacetate, fumarate, succinate, pyruvate and malate were all more active (between 10 and 300%) in molybdenum-deficient than in molybdenum-sufficient cultures. Adding Mo to the assays had no effect upon the individual dehydrogenases. On the other hand, molybdenum stimulated nitrogen fixation 10-fold with most substrates. The author concluded that molybdenum was involved only in nitrogenase activity.

The effect of cobalt upon nitrogen fixation during the growth of micro-organisms has received considerable attention; it stimulates nitrogen fixation in *Azotobacter* (Falcone, Shug and Nicholas, 1962;

Saubert and Strijdom, 1968; Kurdina, 1969), in *C. pasteurianum* (Nicholas, Fisher, Redmond and Osborne, 1964; Evans, Russell and Johnson, 1965), in *Mycobacterium* (Il'ina, 1967*a*), in *Beijerinckia* and *Calothrix antarctica* (Saubert and Strijdom, 1968) and in root nodules (Ahmed and Evans, 1959; Hallsworth, Wilson and Greenwood, 1960; Manorik and Lisova, 1968 and 1969; Yagodin and Ovcharenko, 1969). Il'ina (1967*a*) claimed that cobalt had a specific supplemental effect on nitrogen fixation during growth in Mycobacteria since ammonia-grown cultures did not respond to the element. On the other hand, Manorik and Lisova (1968), and Yagodin and Ovcharenko (1969) claimed that cobalt supplied to the soil during growth stimulated endogenous and succinic dehydrogenase activities in lupin nodules and thereby indirectly stimulated nitrogen fixation. Evans *et al.* (1965) suggested that ammonium ions may complex cobalt from glass walls of bacterial fermentors and thus reduce the apparent requirements when nitrogen-fixing bacteria are grown on ammonium salts. Hallsworth, Wilson and Adams (1965) showed that non-nodulated subterranean clover responded to cobalt, although not to the same extent as the nodulated plant.

Soviet investigators have tested other trace elements for their effect on nitrogen fixation (e.g.: Cu, Zn, Mn, B); Il'ina (1966*c* and 1967*b*) claimed that copper (13 to 26 μg/l) stimulated nitrogen fixation in growing *M. flavum,* particularly if Mo, Zn, Mn and B were added to the medium, but not in *M. roseo-album.* Copper did not have a sparing effect on Mo; high levels of copper (1320 μg/l) competitively antagonised molybdenum and inhibited nitrogen fixation.

A1.3   ENZYMES ASSOCIATED WITH NITROGEN FIXATION: ELECTRON TRANSPORT

A1.3.1   *Hydrogenase*

Nitrogen-fixing organisms show two kinds of hydrogenase activity: an ATP-activated system, specifically associated with nitrogenase, which evolves hydrogen (Bulen, Burns and Lecomte, 1965; Kelly, 1966; Mortenson, 1966) and a 'conventional' hydrogenase that can transfer electrons to nitrogenase through ferredoxin (Mortenson, 1964) or a similar electron carrier, such as flavodoxin (Knight and Hardy, 1966). The work of Ivanov and Bondar (1968) with *C. pasteurianum* (section A.2.1.2) supports this.

In *Azotobacter* the conventional hydrogenase catalyses the absorption but not the evolution of hydrogen. This enzyme is slightly more active in nitrogen-fixing than in ammonia-grown *Azotobacter* (Lee, Wilson and Wilson; 1942; Bulen *et al.* 1965). It is not associated with dithionite-activated nitrogenase preparations from Azotobacter (Bulen *et al.* 1965) but there is some evidence that it has a role in association with nitrogen fixation *in vivo*. Yakovlev and Levchenko (1964, 1966) and Yakovlev *et al.* (1965) studied the intracellular location of enzymes that reduced tetrazolium nitro-blue or neotetrazolium chloride in N_2-grown *A. vinelandii;* they found that hydrogenase was located in the cytoplasm and in organelles which they called 'mitochondria'. N_2 inhibited 'mitochondrial'-located hydrogenase, as judged by the fact that N_2 inhibited the reduction of tetrazolium nitro blue by hydrogen in the mitochondria, but it did not inhibit the cytoplasmic hydrogenase. The authors suggested that nitrogenase was in 'mitochondria' and that hydrogenase donated electrons to nitrogenase in washed bacteria. They claimed that hydrogen also promoted the reduction of $^{15}N_2$. Yakovlev *et al.* (1966), as mentioned and discussed earlier in section A1.2.1.1, isolated a non-haem iron protein, 'ferridin', from acetone-dried *Azotobacter vinelandii* and showed that it stimulated nitrogenase in the presence of hydrogen in DEAE-treated extracts from this organism.

Ivanov and Demina (1966) found that the reduction of triphenyl tetrazolium chloride by hydrogen with washed cells of *Chromatium minutissium* was depressed in the presence of dinitrogen and deduced that hydrogen denoted electrons to dinitrogen through hydrogenase and nitrogenase.

A1.3.2   Cytochromes

Several investigators have shown interest in the relationship between cytochromes and nitrogenase in *Azotobacter*. In the West, Shug, Hamilton and Wilson (1956) and Wilson (1958) reported that dinitrogen could oxidise the cytochromes of *A. vinelandii*, and, as a parallel case, Bergerson and Wilson (1959) claimed that leghaemoglobin was an electron carrier to nitrogenase in bacteroids. In the light of recent studies (see Chapter 5) there appear two other ways in which cytochromes influence nitrogenase activity:

    (a) the respiratory chain, through oxygen uptake, acts to protect nitrogenase from inhibition by oxygen in cell suspensions (Dalton and Postgate, 1969a; Drozd and Postgate, 1970; Yates 1970).

    (b) purified nitrogenase from *Azotobacter* is easily and irreversibly damaged by oxygen (Bulen and Lecomte, 1966); in crude extracts a conformational protection exists which may, in living cells, be controlled such that they can cause the nitrogenase to 'switch off' and so prevent damage from exposure to oxygen (Dalton and Postgate, 1969a). Kelly, Klucas and Burris (1967) found that either a cytochrome fraction from *A. vinelandii* or mammalian cytochrome *c*, could protect purified nitrogenase fractions against loss of activity during storage: this might be connected with the conformational protection mentioned above since cytochromes are components of the conformationally-protected particles in crude Azotobacter extracts.

Speculation and claims that cytochromes are involved as electron carriers in nitrogen fixation have been made in several publications by Ivanov and his colleagues (Ivanov *et al.* 1964; Ivanov, Sitonite and Belov, 1965; Ivanov *et al.* 1967; Ivanov, Moshkovskii, Stukan, Matkhanov, Mardanyan and Belov, 1968; Ivanov, Rakhleeva, Demina *et al.* 1968; Ivanov, Gogoleva and Demina, 1969; Ivanov and Demina, 1966; Ivanov, Demina and Gogoleva, 1967; Demina, Ivanov and Medvedev, 1969; Gogoleva and Ivanov, 1969). These authors suggest that dinitrogen and oxygen compete for electrons from the respiratory chain which are released by cytochrome oxidase and, therefore, that electron transport systems to nitrogenase and to oxygen have a common pathway containing the cytochromes. Their evidence is based largely upon observations obtained using known inhibitors of mitochondrial electron transport or oxidative phosphorylation. Ivanov *et al.* (1965) used the reduction of triphenyl tetrazolium bromide to formazan as an estimate of NADH dehydrogenase or 'diaphorase' (Straub, 1939) activity in *A. vinelandii*. They found that hydroxylamine inhibited diaphorase activity and also nitrogen fixation. They assumed that tetrazolium salt was reduced by enzymes of the respiratory chain and, since *A. vinelandii* possessed an hydroxylamine reductase, that hydroxylamine inhibited diaphorase activity by competing for electrons. In addition, they showed that diaphorase activity was decreased by 10% when uv-irradiated dinitrogen replaced non-irradiated nitrogen in experiments. The authors argue that this evidence indicates that nitrogen and oxygen compete for respiratory chain electrons and that hydroxylamine reductase is a third competitor, as a result of which activity nitrogenase is inhibited. In a second experiment, alcohol dehydrogenase activity in *A. vinelandii* was measured by the rate of reduction of triphenyl tetrazolium bromide to formazan. The activity varied under atmospheres containing different ratios of oxygen to dinitrogen. When the percentage of dinitrogen increased the rate of reduction of tetrazolium salt also increased and the time lag before this reduction started, decreased. This again suggested competition for electrons between oxygen and dinitrogen and, since dinitrogen was the weaker electron acceptor, when the dinitrogen to oxygen ratio was high the rate of tetrazolium reduction, as the third competitive process, also increased. If this argument is valid then replacing dinitrogen by an inert gas should increase further the rate of tetrazolium reduction. However, when helium replaced dinitrogen, the rate of reduction of tetrazolium was unaffected but the time lag before the reduction started, increased rather than decreased. On the other hand, argon, in the absence of oxygen or dinitrogen, did decrease the time lag and increase the rate of tetrazolium reduction when compared with an argon/dinitrogen mixture (4:1 v/v). The authors interpreted the results as meaning that dinitrogen was not an 'inert' gas in these experiments.

The evidence in this paper (Ivanov *et al.*, 1965) gives no indication at which site electron transport to nitrogen and to oxygen diverge: the three systems, nitrogen reduction, oxygen uptake and hydroxylamine reductase could be three separate electron transport systems involving separate NADH dehydrogenase activities, or, if a single NADH dehydrogenase catalyses all three reactions this need not involve cytochromes. Secondly, the high level of oxygen (0.3 atmospheres) used in some experiments could cause inhibition of nitrogenase by the 'switch off' mechanism (Chapter 5) rather than by competition for electrons. This is even more likely to happen in the presence of an alternative electron acceptor such as triphenyltetrazolium bromide since this would decrease the rate of oxygen uptake and increase the intracellular oxygen concentration.

Ivanov *et al.* (1968) studied the effect of known inhibitors of mitochondrial electron transport on suspensions of washed *A. vinelandii*, previously grown in N_2-free medium containing ^{57}Fe. After experiments specimens were freeze-dried and gamma resonance (Mössbauer) spectroscopy of the freeze-dried material showed that the percentage of ferrous ions in the bacteria decreased in the following order of experimental treatment: washed bacteria under argon, bacteria under N_2 + KCN, bacteria under N_2 + antimycin A, bacteria under $N_2 \pm$ sodium amytal. The authors claimed that this sequence suggested that nitrogenase can accept electrons from the respiratory chain, thereby maintaining a steady state level of reduced iron. Interception of this flow by inhibitors which blocked at specific sites along the respiratory chain would increase the amount of reduced iron. Sodium amytal had no effect since its site of action was by-passed by using sodium succinate as substrate. Ivanov *et al.* (1969) adopted a similar approach but measured $^{15}N_2$ reduction by washed suspensions of *A. vinelandii* with sodium succinate or ethanol in the presence of the electron transfer inhibitors amytal, antimycin A, 2-heptyl-4-hydroxyquinoline-N-oxide, phenazine methosulfate, KCN or sodium azide. All these inhibitors caused marked inhibition of nitrogen fixation, leading the authors to reassert their view that the respiratory chain was responsible for electron transfer to nitrogenase. Ivanov and Demina (1966) and Ivanov, Demina and Gogoleva (1967) investigated the reduction of triphenyltetrazolium bromide under argon or dinitrogen with washed cell suspensions of *Chromatium minutissimum.* They claimed that KCN (as low as $10^{-4}M$), NaN_3 ($10^{-3}M$) and antimycin A (1 to 2.6 µg/ml) all increased the rate of formazan production under nitrogen thus indicating that electron transfer to nitrogen in the absence of inhibitors occurred through the cytochromes.

Gogoleva and Ivanov (1969) claimed that uncouplers of oxidative phosphorylation, including neotetrazolium, triphenyltetrazolium chloride, 2.4 dinitrophenol, thyroxin and quinacrin-HCl at low concentrations increased the rate of nitrogen fixation by washed suspensions of *A. vinelandii.* At high concentrations these compounds were inhibitory. A similar claim was made by Demina, Ivanov and Medvedev, (1969): that quinacrin-HCl (10^{-5} M) stimulated hydrogenase activity and electron transfer to nitrogenase in *C. minutissimum;* higher concentrations of quinacrin-HCl ($10^{-3}M$) inhibited hydrogenase activity and nitrogen fixation. The authors claim that, by analogy with electron transport and oxidative phosphorylation in mitochondria, uncouplers increase electron flow along the electron transport chain to nitrogenase. Gogoleva and Ivanov also observed that ADP, in the absence of uncouplers, increased nitrogen fixation in *A. vinelandii,* presumably by increasing the rate of electron transport and by supplying ATP. In their experiments respiratory chain inhibitors differed in their effect depending on whether the electron source was ethanol or sodium succinate; this led the authors to postulate two distinct pathways of electron transfer in nitrogenase in *Azotobacter.*

The interpretations of the results described above can be criticised on the grounds that inhibitors of electron transport are likely to inhibit nitrogenase by causing it to 'switch off' in the presence of excess oxygen rather than by impeding electron flow; since the inhibitors prevent oxygen uptake they would cause excess of oxygen to accumulate and induce the 'switch off'. Yates (1970) demonstrated such an effect when ATP caused a simultaneous inhibition of oxygen uptake and acetylene reduction by $A.$ $chroococcum$ treated with lysozyme and ethylene-diamine-tetracetic acid. Two other objections can be raised against the experiments with $A.$ $vinelandii$ described in this series. A serious objection is that whole cells of $Azotobacter$ will not usually fix nitrogen anaerobically because $Azotobacter$ possess no means of producing adequate energy under anaerobic conditions. Secondly, the concentrations of electron transport inhibitors used in some of these experiments were not mentioned (see Ivanov et $al.,$ 1969). Daniel and Yates in unpublished work in the ARC Unit found that amytal, antimycin, A, 2-heptyl-4-hydroxyquinolin-N-oxide and quinacrin-HCl all inhibited acetylene reduction under argon and oxygen uptake under air in preparations from $A.$ $chroococcum.$ However, the concentrations used to obtain inhibition were 100-fold greater than those necessary to inhibit mitochondrial respiration; it was felt that at such high concentrations non-specific inhibition might occur.

Additional evidence for involving cytochromes in nitrogen fixation was obtained by Lisenkova and Khmel (1967) who found that continuous cultures of N_2-grown $A.$ $vinelandii$ contained a higher concentration of cytochromes, particularly of cytochrome a_2, than did ammonia-grown $A.$ $vinelandii$; (70 p Moles a_2/mg of protein in ammonia-grown cells compared with 260 p Moles a_2/mg of protein in N_2-grown bacteria). The authors interpreted this as evidence for participation of cytochromes in electron transport to nitrogenase. Knowles and Redfearn (1968) reported an even greater concentration difference for cytochrome a_2 when $A.$ $vinelandii$ was grown with N_2 compared with urea; however, they suggested that the Cyt b_1 — Cyt a_1 pathway (Jones and Redfearn, 1967) might be necessary for the production of additional high energy intermediates necessary for nitrogen fixation (see also p. 184).

Gvozdev, Sadkov, Yakovlev and Alfimova (1968) presented the opposing view to the theories of Ivanov and his collaborators after attempting similar experiments to those of Wilson (1958): they exposed extracts of $A.$ $vinelandii$ containing nitrogenase to H_2 or N_2 and found that N_2 did not affect the cytochrome spectra. They concluded that cytochromes were 'very unlikely' to be involved in electron transfer to nitrogenase. These observations agree with carefully controlled tests in our laboratory with whole cells and cell-free extracts of $A.$ $chroococcum$ obtained from continuous cultures. In a further study along similar lines, Ivleva, Sadkov and Yakovlev (1969) claimed that electrons were transferred to nitrogenase in cell-free extracts of *A. vinelandii* through flavoproteins in the respiratory chain but not through the cytochromes.

A1.4 MECHANISMS OF THE NITROGENASE REACTION AND THE NATURE OF THE ACTIVE SITE

A1.4.1 *Kinetic studies*
Dr R. C. Burns at Yellow Springs, Ohio (Burns, 1969) measured the activation energy for nitrogenase activity in extracts from *A. vinelandii* at different temperatures by either N_2 reduction, ATP hydrolysis or ATP-dependent H_2 evolution. In each case biphasic Arrhenius plots were obtained yielding maximum activation energies of 14.6 K cal/mole above 21° and 39 K cal/mole below 21°. In a similar study with cell-free extracts from *A. vinelandii*, Alfimova *et al.* (1968) showed that hydrogen evolution occurred under vacuum with additions of sodium dithionite and ATP. The hydrogen evolved was identified and measured by mass spectrometry; using this as a measure of nitrogenase activity, the authors obtained a single phase Arrhenious plot and calculated the activation energy to be 21.4 K cal/mole.

Slepko *et al.* (1969), using cell-free extracts from *A. vinelandii* measured the effect of increasing amounts of ADP upon the rate of ATP-dependent hydrogen evolution in the absence of an ATP-generating system. They also measured the change in the rate of dithionite-dependent ATP hydrolysis and the ATP/H_2 ratio (the moles of ATP hydrolysed/mole of H_2 evolved) with time. The rate of phosphate release rapidly decreased with time and the rate of hydrogen evolution also decreased with increasing additions of ADP. The authors felt that these two observations were related and suggested that the rate of ATP-hydrolysis diminished with time because of product (ADP) inhibition. This agrees with the observations of Moustafa and Mortenson (1967) who showed that ADP was a specific inhibitor of nitrogenase in extracts from *C. pasteurianum*. In the experiments of Slepko *et al.*, however, this product inhibition by ADP apparently inhibited the ATPase site more than the hydrogen-evolving site since the ATP/H_2 ratio fell dramatically in time from as high as 30 to as low as 1. This decrease was biphasic, very rapid in the first five minutes and then gradually for the next twenty minutes. The results suggested that a measurable level of ADP is necessary for maximum efficiency of nitrogenase. In the ARC Unit a different approach has been used to find whether a low, as opposed to an infinite, ratio of ATP to ADP concentration is necessary for maximum nitrogenase efficiency during the reaction: nitrogenase activity in extracts from *A. chroococcum* was measured in the presence of ATP-generating systems with different amounts of creatine kinase. High concentrations of creatine kinase would ensure that ATP was re-formed more rapidly than it was hydrolysed by nitrogenase activity, low levels of creatine kinase would ensure the opposite and a measurable concentration of ADP would then be present throughout the reaction. High concentrations of creatine kinase (50 to 300 μg protein/ml) inhibited the nitrogenase activity in crude extracts but did not do so with purified nitrogenase fractions. Slepko *et al.,* used relatively crude *Azotobacter* extracts in their experiments.

A1.4.2   Gamma resonance (Mossbauer) and electron spin resonance (ESR) studies

These physical techniques have been used in several Soviet investigations (Ivanov *et al.,* 1968; Ivanov, Rakhleeva, Demina *et al.,* 1968; Novikov, Syrtsova, Likhtenshtein, Trukhtanov, Rachek and Goldanskii, 1969: Syrtsova, Likhtenshtein, Pisarskaya, Ganelin, Frolov and Rachek, 1969; Ivleva, Medzhidov, Likhtenshtein, Sadkov and Yakovlev, 1969; Ivleva, Likhtenshtein and Sadkov, 1969; Matkhanov, Ivanov, Vanin and Belov, 1969). The articles by Ivanov and his collaborators have already been discussed in connection with cytochromes (section 3.2). Novikov *et al.* obtained a fraction which contained nitrogenase from *A. vinelandii* by extracting an acetone powder of the bacteria with phosphate buffer, followed by an ammonium sulfate fractionation and molecular sieve chromotography on Sephadex G25. Subsequent chromatography on Biogel P200 yielded an homogenous fraction which fixed 120 nMoles N_2/mg protein in 15 min. The gamma resonance spectrum of this fraction showed temperature-independent quadrupole splitting and isomeric shift of the resonance lines characteristic of the high-spin ferric state in asymetric ligand surroundings. When sodium dithionite was added an intense doublet appeared indicating that the iron was converted to a high-spin ferrous state. The authors argued that the position of the centre of the doublet suggested that the iron was not in the purely ionic ferrous state and that electrons may have shifted to the 3^d orbital, resulting in the iron acquiring a valency less than 2. In a further article Syrtsova *et al.* (1969) determined the distribution of molybdenum and nitrogenase activity in a similar preparation from *A. vinelandii* grown in the presence of ^{99}Mo, ^{57}Fe and ^{35}S. They also determined the ratio of iron to labile sulfur in their nitrogenase fraction to be 2.8 to 3.6. However, approximately one-third of this iron reacted with a concentration of *o*-phenanthroline which did not inhibit nitrogenase activity and, from this datum, the authors deduced that the ratio of iron to labile sulfur in nitrogenase was 2:1. The authors proposed a scheme for the active site of nitrogenase in which Fe[1] is co-ordinated in a planar network to four organically bound sulfur atoms. Ivleva, Likhtenshtein and Sadkov (1969) studied the nitrogenase activity in a supernatant fraction from *A. vinelandii* which was obtained after sonicating cells suspended in a 0.025 M phosphate buffer pH 7.0 for 4 to 5 min and centrifuging at 65,000 x g for 2 hours. They found that nitrogenase activity (as determined by ^{15}N$_2$ uptake in the presence of dithionite and ATP) was inhibited by diazobenzene sulfonic acid (10^{-3}M), nitric oxide (10^{-7}M), *o*-phenanthroline (5×10^{-4}M) or *p*-hydroxymercuribenzoate (10^{-4}M). They also showed that the cell-free supernatant yielded ESR signals at g = 2.00 and g = 1.94 under argon and that diazobenzene sulfonic acid (10^{-3}M) reduced the g = 2.00 signal. NO (8×10^{-7}M), on the other hand, abolished the g = 2.00 and g = 1.94 signal and created a new signal at g = 2.03, which the authors claim is characteristic of an iron-nitric oxide complex associated with cysteine groups. Further addition of *o*-phenanthroline or *p*-hydroxymercuribenzoate did not alter the g = 2.03 signal appreciably. The authors speculated that the iron-sulfur planar complex mentioned above was not the active nitrogenase site but an electron-transfer site through which electrons pass to molybdenum and finally to the nitrogen-reducing site. Ivleva, Likhtenshtein, Sadkov and Yakovlev (1969) observed how the ESR spectrum of a crude supernatant from N$_2$-grown *A. vinelandii* varied in nitrogen-fixing conditions upon addition of CO. In the presence of sodium dithionite alone under argon a g = 1.94 signal was observed, attributed to iron. When ATP was added another signal at g = 1.98 appeared. This signal decreased in the presence of CO or N$_2$ and was attributed to molybdenum. In the absence of dithionite the g = 1.94 signal disappeared when the *o*-phenanthroline and *p*-hydroxymercuribenzoate was added but CO had no effect, whereas when dithionite was present the reverse was true: *o*-phenanthroline and *p*-hydroxymercuribenzoate did not affect the signal but CO caused a split and a broadening at g = 1.94 and decreased a further signal at g = 2.01. N$_2$ caused a similar effect to CO provided all the components for nitrogen fixation were present. By applying the Van Vleck formula to data obtained in ESR studies, the authors calculated the distance between iron and the nearest molybdenum to be, at least, 6A°.

A1.5   SEPARATION OF NITROGENASE INTO ITS COMPONENT FRACTIONS

In a survey of the literature only one Soviet report was found of a serious investigation into the component parts of nitrogenase. This was by Syrtsova, Yakovlev and Rachek (1968), in which they claimed that 'ferridin' (isolated from *A. vinelandii* by Yakovlev *et al.,* 1966) was identical with fraction 2 of nitrogenase. Their purification procedure was as follows: an acetone powder of *A. vinelandii* cells was extracted with 0.05M phosphate buffer, pH 7.4, under argon at 4° for 12 hours. After centrifuging, the supernatant fixed 6 mg N_2/mg protein/hr with dithionite and ATP. This supernatant was made 50% saturated with ammonium sulfate under argon; the resulting precipitate was re-suspended in the same buffer and dialysed anaerobically, or passed through Sephadex G-25 in air at 10° to remove ammonium sulfate. Subsequent passage through Biogel P.200 under air at 10° removed components of molecular weight above 200,000, and also a cytochrome fraction. The ferridin containing fraction was then chromatographed on DEAE-cellulose equilibrated with 0.05M phosphate buffer, pH 7.4. A molybdenum-rich protein which the authors claimed to be fraction 1 of nitrogenase passed through the column without being retarded; 'ferridin' was eluted with 0.4 to 0.5M phosphate buffer. Alternatively, if the column was equilibrated in 0.025M Tris buffer, pH 7.2, the two fractions were separated by gradient elution with Tris buffer containing NaCl (0.5M). The uv and visible absorption spectrum of ferridin, with and without sodium dithionite, is similar to that of fraction 2 (see Bulen and Lecomte, 1966). The authors claim that the spectrum of fraction 1 is also similar to that of the molybdenum-rich protein. It is quite possible that the proteins isolated by these authors are the same as fraction 1 and fraction 2 but proof of such claims would be to obtain nitrogen fixation by re-combining the two fractions as did Bulen and Lecomte (1966); the Soviet authors have not reported this experiment. Ammonium sulfate treatment, such as described by these authors, has always caused a considerable loss of nitrogenase activity (60 to 90%) in the writer's laboratory. The authors describe the effects of oxidising agents and iron-chelating compounds on the absorption spectrum of ferridin and also claim that diluting ferridin with buffer to 0.8 mg/ml causes the protein to dimerise; suggesting that this may be the reason why nitrogenase activity is not proportional to protein concentration (Burns and Bulen, 1965). Ferridin had a molecular weight of 46,000 to 50,000 and contained 2 atoms of iron and two labile sulfide groups per molecule.

Ganelin, L'vov, Kirshteine and Kretovich (1969) claimed to have isolated the iron and molybdenum-containing component of nitrogenase from crude extracts from *A. vinelandii* by precipitating with ammonium sulfate (25 to 50%). Polyacrylamide gel disc electrophoresis yielded several iron-containing bands. The authors claimed that molybdenum was cleaved from the protein during electrophoresis at pH 9.0.

A1.6   CONTROL OF NITROGEN FIXATION *IN VIVO*

Lyubimov, L'vov, Kirshteine and Kretovich (1969) found adenosine and adenine deaminases and also an enzyme that converted ADP to AMP with the release of adenine, in crude cell-free extracts of *A. vinelandii* that were capable of fixing nitrogen. While assaying nitrogen fixation in these extracts with sodium dithionite and ATP under $^{15}N_2$ they found that some of the ammonia produced was not labelled with $^{15}N_2$. The authors suggested that deaminases were responsible for this since heating the crude extract of *A. vinelandii* at $60°$ under N_2 for 10 min destroyed the deaminase activity and increased the rate of fixation and total nitrogen fixed. They also suggested that nitrogen fixation is a self-regulating allosteric system, ammonia being an active inhibitor; this self-regulating mechanism is destroyed by heat.

A1.7   CONCLUSION

Soviet biochemists in nitrogen fixation have not yet caught up with their Western colleagues although many investigators have confirmed the requirement for sodium dithionite, ATP and anaerobic conditions for nitrogen fixation or ATP-dependent hydrogen evolution in cell-free extracts of *Azotobacter*. It is noteworthy that the acetylene-test for nitrogenase, which became widely used in the West within months of its discovery by Dilworth (1966) and Schollhorn and Burris (1966), has not been used in the U.S.S.R. despite the fact that its simplicity and, when combined with gas-liquid chromatography, its extreme sensitivity (Hardy and Knight, 1967) has led to important advances and revision of Western work. The use of tetrazolium salts is an oblique way of tackling the problem of nitrogen fixation and is subject to the criticisms already indicated, e.g. the presence of a high oxygen concentration complicates results by inhibiting nitrogen fixation. On the other hand, nitrogenase activity in *Azotobacter* extracts, in the absence of oxygen and artificial electron donors such as sodium dithionite or reduced benzyl viologen, is barely detectable, at least in Western laboratories, even with the extreme sensitivity of the acetylene method. Therefore, experimental results claiming to reflect nitrogen fixation in extracts from *A. vinelandii* in the absence of sodium dithionite or ATP must be regarded with scepticism. Soviet investigators have not yet reported upon the alternative substrates for nitrogenase and Syrtsova *et al.* (Section A1.5) are the only authors to report a constructive attempt to purify nitrogenase into its component proteins; such claims should be supported by positive nitrogen fixation assays with the isolated proteins since *Azotobacter* contain several non-haem iron proteins (Shethna, Wilson, Hansen and Beinert, 1964; Shethna, Wilson and Beinert, 1966; Shethna, DerVartanian and Beinert, 1968; Benemann, Yoch, Valentine and Arnon, 1969). Results using physical methods, such as Mossbauer and ESR spectrometry on unpurified proteins are difficult to interpret, but the emphasis placed by Soviet scientists upon these techniques in recent months may lead to significant results when applied to the purified nitrogenase proteins.

ACKNOWLEDGEMENT

I thank Dr A. J. Leggett for helping with translation and Miss S. Hill for reading the manuscript.

REFERENCES

S. AHMED and H. J. EVANS, *Biochem. Biophys. Res. Commun.*, 1, 271 (1959).

E. Ya. ALFIMOVA, R. I. GVOZDEV and V. R. LINDE, *Izvest Akad. Nauk. S.S.S.R.*, No. 6, 915 (1968).

D. I. ARNON, M. LOSADA, M. NOZAKI and K. TAGAWA, *Nature, (London)*, 190, 601 (1961).

J. R. BENEMANN, D. C. YOCH, R. C. VALENTINE and D. I. ARNON, *Proc. natl. Acad. Sci., U.S.*, 63, 1079 (1969).

F. J. BERGERSEN, *Biochim. biophys. Acta*, 115, 247 (1966*a*).

F. J. BERGERSEN, *Biochim. biophys. Acta*, 130, 304 (1966*b*).

F. J. BERGERSEN and G. L. TURNER, *Biochim. biophys. Acta*, 141, 507 (1967).

F. J. BERGERSEN and G. L. TURNER, *J. gen. Microbiol.*, **53**, 205 (1968).

F. J. BERGERSEN and P. W. WILSON, *Proc. natl. Acad. Sci., U.S.*, **45**, 1641 (1959).

D. R. BİGGINS and J. R. POSTGATE, *J. gen. Microbiol.*, **56**, 181 (1969).

H. BORTELS, *Arch. Mikrobiol.*, **1**, 333 (1930).

H. BRINTZINGER, *Biochemistry*, **5**, 3947 (1966).

W. A. BULEN, R. C. BURNS and J. R. LECOMTE, *Biochem. Biophys. Res. Commun.*, **17**, 265 (1964).

W. A. BULEN, R. C. BURNS and J. R. LECOMTE, *Proc. natl. Acad. Sci., U.S.*, **53**, 532 (1965).

W. A. BULEN and J. R. LECOMTE, *Proc. natl. Acad. Sci., U.S.*, **56**, 979 (1965).

R. C. BURNS, *Biochim. biophys. Acta*, **171**, 253 (1969).

R. H. BURRIS, Royal Society Discussion meeting on Nitrogen Fixation. *Proc. Roy. Soc. B.*, **172**, 339 (1969).

J. E. CARNAHAN, L. E. MORTENSON, H. F. MOWER and J. E. CASTLE, *Biochim. biophys. Acta*, **44**, 520 (1960).

H. DALTON and J. R. POSTGATE, *J. gen. Microbiol.*, **54**, 463 (1969a).

H. DALTON and J. R. POSTGATE, *J. gen. Microbiol.*, **56**, 307 (1969b).

N. C. DEMINA, I. D. IVANOV and G. A. MEDVEDEV, *Mikrobiologiya*, **38**, 428 (1969).

M. J. DILWORTH, *Biochim. biophys. Acta*, **127**, 285 (1966).

L. M. DOROSINSKII, I. V. ZAGORE and D. M. BUZIASHUILI, *Mikrobiologiya*, **35**, 319 (1966).

J. W. DROZD and J. R. POSTGATE, *J. gen. Microbiol.*, **60**, 427 (1970).

R. G. ESPOSITO and P. W. WILSON, *Biochim. biophys. Acta*, **22**, 186 (1956).

H. J. EVANS, S. A. RUSSELL and G. V. JOHNSON. Non-heme-iron proteins. (San Pietro ed.) p. 303. A. Antioch Press, Yellow Springs, Ohio, (1965).

A. B. FALCONE, A. L. SHUG and D. J. D. NICHOLAS, *Biochem. Biophys. Res. Commun.*, **9**, 126 (1962).

M. V. FEDEROV, *Biological Fixation of Atmospheric Nitrogen* Sel'hkozgiz, Moscow (1952).

M. V. FEDEROV and T. A. KALININSKAYA, *Mikrobiologiya*, **30**, 9 (1961a).

M. V. FEDEROV and T. A. KALININSKAYA, *Dokl. s. Kh. Akad. im K. A. Timiryazeva*, **70**, 145 (1961b).

V. L. GANELIN, N. P. L'VOV, B. E. KIRSHTEINE, V. I. LYUBIMOV and V. L. KRETOVICH, *Dokl. Akad. Nauk. S.S.S.R.*, **185**, 1169 (1969).

N. B. GRADOLA-KRYLOVA, *Biol. Azot. Rol. Zemled. Akad. Nauk. S.S.S.R. Inst. Mikrobiol.*, p. 52. (1967).

T. V. GOGOLEVA, and I. D. IVANOV, *Biol. Nauki*, **6**, 92 (1969).

R. I. GVOZDEV, A. P. SADKOV, V. A. YAKOVLEV and E. Ya. ALFIMOVA, *Izvest. Akad. Nauk. S.S.S.R.*, No. 6, 836 (1968).

E. G. HALLSWORTH, S. B. WILSON and W. A. ADAMS, *Nature, (London)*, **205**, 307 (1965).

E. G. HALLSWORTH, S. B. WILSON and E. A. N. GREENWOOD, *Nature (London)*, **187**, 79 (1960).

R. W. F. HARDY and E. Jr. KNIGHT, *Biochim. biophys. Acta*, **139**, 69 (1967).

N. D. IERUSALIMSKII, G. N. ZAITLEVA and I. A. KHMEL, *Mikrobiologiya*, **31**, 417 (1962).

T. K. IL'INA, *Mikrobiologiya*, **35**, 155 (1966a).

T. K. IL'INA, *Mikrobiologiya*, **35**, 422 (1966b).

T. K. IL'INA, *Mikrobiologiya*, **35**, 323 (1966*c*).
T. K. IL'INA, *Mikrobiologiya*, **36**, 626 (1967*a*).
T. K. IL'INA, *Mikrobiologiya*, **36**, 970 (1967*b*).
T. K. IL'INA, *Dokl. Akad. Nauk, S.S.S.R.*, **180**, 476 (1968*a*).
T. K. IL'INA, *Mikrobiologiya*, **37**, 217 (1968*b*).
T. K. IL'INA, *Agrokhem. Talajtan*, **18**, 39 (1969).
I. D. IVANOV, *Biol. Azot. Ego. Rol. Zemled. Akad. Nauk, S.S.S.R., Inst. Mikrobiol.* p. 35. (1967).
I. D. IVANOV and V. K. BONDAR, *Izvest. Akad. Nauk. S.S.S.R. Ser. Biol.*, No. 5, 752 (1968).
I. D. IVANOV and N. S. DEMINA, *Izvest. Akad. Nauk. S.S.S.R. Ser. Biol.*, No. 1, 115 (1966*a*).
I. D. IVANOV and N. S. DEMINA, *Mikrobiologiya*, **35**, 780 (1966*b*).
I. D. IVANOV, N. S. DEMINA and T. V. GOGOLEVA, *Mikrobiologiya*, **36**, 8 (1967).
I. D. IVANOV, T. V. GOGOLEVA and N. S. DEMINA, *Dokl. Akad. Nauk, S.S.S.R.*, **184**, 703 (1969).
I. D. IVANOV, T. K. IL'INA and Yu. P. SITONITE, *Mikrobiologiya*, **33**, 540 (1964).
I. D. IVANOV, G. I. MATKHANOV, Yu. M. BELOV and T. V. GOGOLEVA, *Mikrobiologiya*, **36**, 205 (1967).
I. D. IVANOV, Yu. Sh. MOSHKOVSKII, R. A. STUKAN, G. I. MATKHANOV, S. S. MARDANYAN and Yu. M. BELOV, *Mikrobiologiya*, **37**, 407 (1968).
I. D. IVANOV, E. E. RAKHLEEVA, N. S. DEMINA, Yu. M. BELOV and T. V. GOGOLEVA, *Izvest. Akad. Nauk. S.S.S.R., Ser. Biol.*, No. 3, 435 (1968).
I. D. IVANOV, Yu. P. SITONITE and Yu. M. BELOV, *Mikrobiologiya*, **34**, 193 (1965).
I. N. IVLEVA, G. I. LIKHTENSHTEIN and A. P. SADKOV, *Biofizika*, **14**, 779 (1969).
I. N. IVLEVA, A. A. MEDZHIDOV, G. I. LIKHTENSHTEIN, A. P. SADKOV and V. A. YAKOVLEV, *Biofizika*, **14**, 639 (1969).
I. N. IVLEVA, A. P. SADKOV and V. A. YAKOVLEV, *Izvest. Akad. Nauk. S.S.S.R.*, No. 5, 688 (1969).
C. W. JONES and E. R. REDFEARN, *Biochim. biophys. Acta*, **143**, 340 (1967).
Yu. I. KARMAUKOV, *Izvest. Akad. Nauk. S.S.S.R. Ser. Biol.*, **30**, 714 (1965).
T. A. KALININSKAYA, *Mikrobiologiya*, **36**, 345 (1967*a*).
T. A. KALININSKAYA, *Mikrobiologiya*, **36**, 526 (1967*b*).
T. A. KALININSKAYA, *Mikrobiologiya*, **36**, 621 (1967*c*).
T. A. KALININSKAYA, *Agrokhem. Talajtan*, **18**, 11 (1969).
M. KELLY, *9th Intern. Congr. Microbiol. Moscow*, p. 277. (1966).
M. KELLY, *Biochim. biophys. Acta*, **171**, 9 (1969).
M. KELLY, R. V. KLUCAS and R. H. BURRIS, *Biochem. J.*, **105**, 3c (1967).
I. A. KHMEL and N. B. ANDREEVA, *Mikrobiologiya*, **36**, 438 (1967).
I. A. KHMEL and K. N. GABINSKAYA, *Mikrobiologiya*, **34**, 763 (1965).
I. A. KHMEL, K. N. GABINSKAYA and N. D. IERUSALIMSKII, *Mikrobiologiya*, **34**, 689 (1965).
I. A. KHMEL and N. D. IERUSALIMSKII, *Mikrobiologiya*, **36**, 632 (1967).
R. V. KLUCAS and H. J. EVANS, *Plant Physiol.*, **43**, 1458 (1968).
E. Jr. KNIGHT and R. W. F. HARDY, *J. biol. Chem.*, **241**, 2752 (1966).
C. J. KNOWLES and E. R. REDFEARN, *Biochim. biophys. Acta*, **162**, 348 (1968).

V. L. KRETOVICH, Z. G. EVSTIGNEEVA, K. B. ASEEVA, O. N. ZARGARYAN and E. M. MARTYNOVA, *Izvest. Akad. Nauk. S.S.S.R. Ser. Biol.*, **No. 2**, 208 (1969).

N. V. KRYLOVA, Abstr. of Dissert. Moscow, 1964.

R. M. KURDINA, *Tr. Inst. Mikrobiol. Virusol. Akad. Nauk. Kaz. S.S.S.R.*, **11**, 80 (1968).

R. M. KURDINA, *Tr. Inst. Mikrobiol. Virusol. Akad. Nauk. Kaz. S.S.S.R.*, **11**, 84 (1969).

S. B. LEE, J. B. WILSON and P. W. WILSON, *J. biol. Chem.*, **144**, 273 (1942).

L. A. LEVCHENKO, I. N. IVLEVA and V. A. YAKOVLEV, *Izvest. Akad. Nauk. S.S.S.R. Ser. Biol.*, **No. 1**, 165 (1969).

L. L. LISENKOVA and I. A. KHMEL, *Mikrobiologiya*, **36**, 905 (1967).

N. P. L'VOV and V. I. LYUBIMOV, *Izvest. Akad. Nauk. S.S.S.R. Ser. Biol.*, **30**, 250 (1965).

V. I. LYUBIMOV, *Izvest. Akad. Nauk. S.S.S.R. Ser. Biol.*, **30**, 394 (1965).

V. I. LYUBIMOV, N. P. L'VOV, B. E. KIRSHTEINE and V. L. KRETOVICH, *Biokhimiya*, **33**, 364 (1968).

V. I. LYUBIMOV, N. P. L'VOV, B. E. KIRSHTEINE and V. L. KRETOVICH, *Izvest. Akad. Nauk. S.S.S.R. Ser. Biol.*, **No. 4**, 505 (1969).

A. V. MANORIK, V. M. KRIKUNETS and Yu. P. STARCHENKOV, *Dopovida Akad. Nauk. Ukr. R.S.R.*, **8**, 1096 (1965).

A. V. MANORIK and N. Yu. LISOVA, *Dopovida Akad. Nauk. Ukr. R.S.R. Ser. Biol.*, **30, No. 4**, 366 (1968).

A. V. MANORIK and N. Yu. LISOVA, *Dopovida. Akad. Nauk. Ukr. R.S.R.*, **31, No. 4**, 3 (1969).

A. V. MANORIK and E. P. STARCHENKOV, *Dokl. Akad. Nauk, S.S.S.R.*, **186**, 975 (1969).

G. I. MATKHANOV, I. D. IVANOV, A. F. VANIN and Yu. M. BELOV, *Biofizika*, **14**, 1124 (1969).

L. E. MORTENSON, *Biochim. biophys. Acta*, **81**, 473 (1964).

L. E. MORTENSON, *Biochim. biophys. Acta*, **127**, 18 (1966).

Yu. Sh. MOSHKOVSKII, I. D. IVANOV, R. A. STUKAN, G. I. MATKHANOV, S. S. MARDANYAN, Yu. M. BELOV and V. I. GOLDANSKII, *Dokl. Akad. Nauk., S.S.S.R.*, **174**, 215 (1967).

E. MOUSTAFA and L. E. MORTENSON, *Nature*, **216**, 1241 (1967).

D. J. D. NICHOLAS and D. J. FISHER, *J. Sci. Food Agric.*, **11**, 603 (1960).

D. J. D. NICHOLAS, D. J. FISHER, W. J. REDMOND and M. OSBORNE, *Nature (London)*, **201**, 793 (1964).

G. V. NOVIKOV, L. A. SYRTSOVA, G. I. LIKHTENSHTEIN, V. A. TRUKHTANOV, V. F. RACHEK and V. I. GOLDANSKII, *Dokl. Akad. Nauk, S.S.S.R.*, **181**, 1170 (1969).

A. OKUDA, M. YAMAGUCHI and M. KOBOYASHI, *Soil and Plant food*, **6**, 35 (1960).

A. OKUDA, M. YAMAGUCHI and M. KOBOYASHI, *Soil Sci. Plant Nutr.*, **7**, 115 (1961).

A. OKUDA, M. YAMAGUCHI, M. KOBOYASHI and J. KATAYAMA, *Soil Sci. Plant Nutr.*, **7**, 146 (1961).

Yu. V. PEIVE, G. Yu. ZHIZNEVSKAYA, B. A. YAGODIN and P. N. DUBROVO, *Agrokhimiya*, **12**, 27 (1964).

J. R. POSTGATE, Royal Society Discussion Meeting on nitrogen fixation: discussion. *Proc. Roy. Soc. B.*, **172**, 355 (1969).

S. SAUBERT and B. W. STRIJDOM, *S. Afr. J. Agric. Sci.*, **11**, 769 (1968).

R. SCHOLLHORN and R. H. BURRIS, *Federation Proc.*, **25**, 710 (1966).

Y. I. SHETHNA, P. W. WILSON, R. E. HANSEN and H. BEINERT, *Proc. natl. Acad. Sci., U.S.*, **52**, 1263 (1964).

Y. I. SHETHNA, P. W. WILSON and H. BEINERT, *Biochim. biophys. Acta*, **113**, 225 (1966).

Y. I. SHETHNA, D. V. DERVARTANYAN and H. BEINERT, *Biochem. Biophys. Res. Commun.*, **31**, 862 (1968).

A. L. SHUG, P. B. HAMILTON and P. W. WILSON. In a symposium on *Inorganic nitrogen metabolism*. (W. D. McElroy and B. Glass, eds.) p. 344. John Hopkins Press, Baltimore, (1956).

G. I. SLEPKO, A. M. UZENSKAYA, V. R. LINDE, E. Ya. ALFIMOVA, L. M. RAKHMAN and G. I. LIKHTENSHTEIN, *Dokl. Akad. Nauk. S.S.S.R.*, **144**, 473 (1969).

C. SLOGER and W. S. SILVER, *Proc. 9th Intern. Congr. Microbiol. Moscow*, 285 (1966).

F. B. STRAUB, *Biochem. J.*, **33**, 787 (1939).

L. A. SYRTSOVA, G. I. LIKHTENSHTEIN, T. N. PISARSKAYA, V. L. GANELIN, E. N. FROLOV and V. F. RACHEK, *Molekulyarnaya, Biologiya*, **3**, 651 (1969).

L. A. SYRTSOVA, V. A. YAKOVLEV and V. F. RACHEK, *Biokhimiya*, **33**, 753 (1968).

A. TISSIERES and R. H. BURRIS, *Biochim. biophys. Acta.*, **20**, 436 (1956).

F. V. TURCHIN, Z. N. BERZENOVA and C. G. ZHIDKIKH, *Dokl. Akad. Nauk. S.S.S.R.*, **149**, 741 (1963).

E. E. VAN TAMELEN, G. BOCHE and R. GREELEY, *J. Amer. Chem. Soc.*, **90**, 1677 (1968).

E. E. VAN TAMELEN and D. A. SEELEY, *J. Amer. Chem. Soc.*, **91**, 5194 (1969).

M. E. VOL'PIN and V. B. SHUR, *Nature, (London)*, **209**, 1236 (1966).

P. W. WILSON, Encyclopedia of Plant Physiology. (W. Ruhland, ed.), p. 9. Berlin, Germany, (1958).

B. A. YAGODIN and G. A. OVCHARENKO, *Izvest. Akad. Nauk. S.S.S.R. Ser. Biol.*, No. **1**, 113 (1969).

V. A. YAKOVLEV and L. A. LEVCHENKO, *Dokl. Akad. Nauk. S.S.S.R.*, **159**, 1173 (1964).

V. A. YAKOVLEV and L. A. LEVCHENKO, *Dokl. Akad. Nauk, S.S.S.R.*, **171**, 1224 (1966).

V. A. YAKOVLEV, L. A. SYRTSOVA and L. V. VOROB'EV, *Dokl. Akad. Nauk, S.S.S.R.*, **171**, 477 (1966).

V. A. YAKOVLEV, L. V. VOROB'EV, L. A. LEVCHENKO, V. R. LINDE, G. I. SLEPKO and L. A. SYRTSOVA, *Biokhimiya*, **30**, 1167 (1965).

M. G. YATES, *J. gen. Microbiol.*, **60**, 393 (1970).

M. G. YATES and R. M. DANIEL, *Biochim. biophys. Acta.*, **197**, 161 (1970).

APPENDIX 2

The Acetylene Test for Nitrogenase

JOHN POSTGATE

University of Sussex,
Falmer, Brighton, BN1 9QJ,
Sussex.

A2.1   INTRODUCTION

The 'acetylene test' has been cited by nearly every contributor to this book. Because of its ease and apparent specificity as a test for nitrogenase, it can be said to have revolutionized the study of biological nitrogen fixation. As Professor Burris described in Chapter 4, the test arose from the independent observations of Dilworth (1966) and Schollhorn and Burris (1966, 1967) that nitrogenase preparations reduced acetylene to ethylene; gas-liquid chromatography with a flame ionization detector was an obvious assay technique, having advantages of speed and sensitivity, as Hardy and Knight (1967) pointed out. The reaction was rapidly adopted in many laboratories, both as a field test and as an enzymological probe for nitrogenase, and references too numerous to repeat here, but cited by various contributors, testify to its value. This appendix provides a brief survey of the use of the technique; Hardy *et al.* (1968) described its application to numerous test systems and an account in detail of the methods used in the writer's laboratory has been given (Postgate, 1971).

A2.2   EQUIPMENT

Apart from the usual microbiological and enzymological equipment, a gas liquid chromatograph is required with a flame ionization detector and a column of suitable length and packing to separate methane, ethylene and acetylene within one or two minutes. Appropriate instruments, packing materials and operating conditions have been specified by the publications already cited. In addition, gas-tight syringes are needed—plastic, disposable ones have the advantage of avoiding any risk of cross-contamination of gas samples—and closures which will tolerate repeated injection and removal of gas samples without leakage. Since these are normally made of rubber or silicone, in which lower hydrocarbons are soluble, they must be disposable (Kavanagh and Postgate, 1970).

A2.3   GASES

Acetylene is not available as the pure, compressed gas because it is explosive. Most laboratory suppliers will provide acetylene diluted in argon or another inert gas at 1 to 4% (v/v). Some laboratories prepare the gas as needed from calcium carbide (CaC_2) and water, a technique which is readily adapted to preparing deuterated acetylene. Both commerical and 'home-made' preparations of acetylene contain traces of methane (usually below $1:10^4$ v/v), ethylene (usually below $1:10^6$ v/v) and phosphine (detectable by smell). The latter gives no signal on the flame-ionization detector and is not known to interfere with the test. Ethylene, in a state of purity adequate for calibration of the apparatus, is available commercially.

A2.4   TESTS OF SOIL, WATER AND TISSUE SAMPLES

The basic principle involved in handling such samples is to disturb them as little as possible, because aerobic N_2-fixing bacteria may 'switch off' the nitrogenase if subjected to excessive aeration (Drozd and Postgate, 1970a, b) and the partial pressure of oxygen in the normal atmosphere may be 'excessive' to organisms adapted to the lower pO_2 values which often occur in natural micro-environments. Hardy et al. (1968) have described procedures in detail for handling soil cores and other samples, including techniques for flushing them out with oxygen-argon mixtures.

A2.5   TESTS OF AEROBIC MICRO-ORGANISMS

Flask cultures, containing about 10 vol. gas space over the liquid layer, illuminated for photosynthetic microbes such as Cyanophyceas, are set up, if necessary with minimal amounts of fixed N to allow growth to visible turbidity. It is important to avoid dense populations at this stage because oxygen may be scavenged from solution and facultative anaerobes may then appear to fix aerobically. When grown, it is normally sufficient to replace the plug by a seal, inject acetylene to between 0.04 and 0.1 atm. and to continue incubation; a positive reaction, if present, will be manifest in a matter of hours, sometimes minutes. For slow-growing or very sparse populations, precautions must be taken to avoid contamination of the culture during injection of C_2H_2 and sampling; a cotton plug between seal and vessel is sufficient but time must then be allowed for equilibration of gases across the plug. For more precise estimates of the acetylene reduction rate, the atmosphere in the flask should first be replaced, aseptically if necessary, by a mixture of argon or helium + 0.2 atm. O_2.

The technique does not usually lend itself satisfactorily to the estimation of Most Probable Numbers of aerobes in natural samples because of the ambiguity introduced by possible interaction of oxygen-scavenging aerobes and facultatively anaerobic nitrogen fixers.

It is most important to include tests at relatively low pO_2 values because fixation by oxygen-sensitive aerobes might otherwise be missed. Fig. 5 of chapter 5 illustrates effects of pO_2 on activity with some aerobes; the curve for *Mycobacterium flavum* illustrates a situation in which acetylene reduction could be missed at normal pO_2 values.

A2.6   TESTS OF FACULTATIVE AND OBLIGATE ANAEROBES

The tubes designed by Pankhurst (1967) for the culture of exacting anaerobes have been adapted by Campbell and Evans (1969) for detection and estimation of anaerobic N_2-fixing bacteria. Essentially a side-arm to the main tube, containing alkaline pyrogallol, sustains anaerobic conditions during growth and permits injection of acetylene to about 0.05 atm. and removal of gas samples. A negative pressure in the tube, due to absorption of oxygen by the pyrogallol, may be avoided by gassing out as described by Campbell and Evans (1969) or by injecting an equivalent amount of N_2 (Postgate, 1971).

As with aerobes, for precise measurement of acetylene reduction rate the atmosphere should be replaced aseptically with argon or helium.

A2.7   TESTS OF ENZYME PREPARATIONS

In detail, the precise conditions for such tests differ from laboratory to laboratory. Conditions for use with extracts of free-living bacteria and of nodules are given by Professor Burris (section 4.2.6) and Professor Evans (Table 6.4). Vessels used for such tests are generally small, to conserve material, the atmosphere is of argon or helium, and the pC_2H_2 is well above the Michaelis constant for acetylene (about 0.5 mM, corresponding to a pC_2H_2 of about 0.01 atm.). For time-course experiments a sufficient excess pressure of gas is introduced to allow for removal of several gas samples. Nitrogenase preparations are usually sensitive to buffer concentration and the prescribed limits should not be exceeded; some preparations are sensitive to sodium dithionite (Smith and Evans, 1970) and this should be used with caution. ATP is often inhibitory above about 10 mM, which is why an ATP-generating system is widely used; it is not essential because adequate acetylene reduction may be obtained with sub-inhibitory amounts of ATP (Moustafa and Mortenson, 1967). Adaptation of the test for use with other electron donors such as NADH plus a viologen dye, or a chloroplast preparation, are largely self-evident; examples are given in Professor Evans's Figure 6.11 and Table 6.7.

A2.8   INTERPRETATION

In most systems so far studied the rate of acetylene reduction is three times the rate of nitrogen fixation, so a simple correction factor may normally be applied. With whole plants, large soil cores etc., the relatively greater solubility of C_2H_2 in water and lipid, compared with N_2, may enable it to diffuse more rapidly to the active sites. The correction factor of 3 will then not apply. With new potential nitrogen-fixing systems, the ethylene must be shown to arise from the acetylene and not be endogenous (as in certain fungal and plant systems: Ilag and Curtis, 1968; Yang, 1969). With microbes, suppression of the acetylene-reducing system by growth with ammonia is an important confirmatory test. In all systems, analyses for fixed N and $^{15}N_2$ incorporation, subject to the usual precautions, are desirable because an acetylene-reducing system which is independent of nitrogenase may yet be discovered.

REFERENCES

N. E. R. CAMPBELL and H. J. EVANS, *Canad. J. Microbiol.*, **15**, 1342 (1969).

M. J. DILWORTH, *Biochim. biophys. Acta*, **127**, 285 (1966).

J. W. DROZD and J. R. POSTGATE, *J. gen. Microbiol.*, **60**, 427 (1970a).

J. W. DROZD and J. R. POSTGATE, *J. gen. Microbiol.*, **63**, 63 (1970b).

R. W. F. HARDY and E. KNIGHT, *Biochim. biophys. Acta*, **139**, 69 (1967).

R. W. F. HARDY, R. D. HOLSTEN, E. K. JACKSON and R. C. BURNS, *Plant Physiol.*, **43**, 1185 (1968).

L. ILAG and R. W. CURTIS, *Science*, **159**, 1357 (1968).

E. P. KAVANAGH and J. R. POSTGATE, *Lab. Practice*, **19**, 159 (1970).

E. MOUSTAFA and L. E. MORTENSON, *Nature, (Lond.,)*, **216**, 1341 (1967).

E. S. PANKHURST, *Lab. Practice*, **16**, 58 (1967).

J. R. POSTGATE, *The acetylene reduction test for nitrogen fixation* in 'Methods in Microbiology' (J. R. NORRIS and D. RIBBONS, eds.) Vol. 6. London: Acad. Press. (1971).

R. SCHOLLHORN and R. H. BURRIS, *Fed. Proc.*, **25**, 710 (1966).

R. SCHOLLHORN and R. H. BURRIS, *Proc. nat. Acad. Sci., U.S.A.*, **57**, 213 (1967).

R. V. SMITH and M. C. W. EVANS, *Nature, (Lond.,)*, **225**, 1253 (1970).

S. F. YANG, *J. biol. Chem.*, **244**, 4360 (1969).

Index